AF565000

Nanotechnology

NIPA® GENX ELECTRONIC RESOURCES & SOLUTIONS P. LTD.
New Delhi-110 034

About the Editors

Dr. Namita Ashish Singh is currently working as Assistant Professor in Department of Microbiology, Mohanlal Sukhadia University, Udaipur, Rajasthan. She studied Masters in Microbiology from Gurukula Kangri Vishwavidyalaya, Haridwar. She did her doctoral study in Life Science (Biotechnology) at Punjabi University Patiala, Punjab. She has more than fourteen years of research, industrial and teaching experience. She is having two Indian patents on Aflatoxin M1 and *Enteroccoci* in milk to her credit. She has published more than twenty publications in International and National reputed journals along with twelve book chapters. Her current research focuses on applied microbiology and food safety. She has handled two research projects sanctioned by UGC and RUSA, MHRD, Delhi. She has many prestigious awards to his credit like Dr. C.M. Singh Memorial NABS-Best Research Paper Award by NABS, Chennai, India, Certificate of appreciation by DST Rajasthan, Best Paper Award by Indian Dairy Association (20000 Cash Prize), Best Inventor award, by Microbiologists Society India etc.

Dr Rama Kanwar Khangarot earned her doctorate degree in Chemistry from IIT Bombay, Mumbai, India in 2012 under the supervision of Professor Krishna P. Kaliappan. She subsequently held a postdoctoral position in the lab of Professor Barbara Zajc at City College of New York, NY, USA. Then she joined the Department of Chemistry, IISER Bhopal, India, as a research associate, and worked with Professor V. K. Singh. Later, she worked at Manipal University Jaipur, Jaipur, Rajasthan as an Assistant Professor. Afterwards, she has moved to Mohanlal Sukhadia University, Udaipur, Rajasthan, and has been working as an Assistant Professor in the Department of Chemistry. Her research interests focus on the synthesis of biologically active molecules and nanomaterials. Her work has been featured in numerous highly regarded international journals. She has contributed chapters to multiple national and international books as an author.

Dr. Girima Nagda is currently working as Assistant Professor at the Department of Zoology, University College of Science Mohanlal Sukhadia University (MLSU), Udaipur Rajasthan. She completed her M.Sc. in Zoology with specialization in Cell and Molecular Biology in the year 2004 and received Gold Medal for standing first in Mohanlal Sukhadia University. She cleared her CSIR-NET exam in the year 2004. She has been awarded Ph.D. degree by Mohanlal Sukhadia University in the year 2012. She has 15 years of UG and PG teaching experience. Her research areas include Nanotechnology, Toxicology, Biochemistry, Developmental Biology etc. She is handling one RUSA MHRD project on "Pesticide Residue Analysis in Fruits and Vegetables" as a Co-PI. She has published about 14 research/review articles in reputed international and national journals and 8 chapters in edited books.

Nanotechnology
Journey from Laboratory to Society

Namita Ashish Singh
Assistant Professor
Department of Microbiology
Mohanlal Sukhadia University
Udaipur-313001, Rajasthan

Rama Kanwar Khangarot
Assistant Professor
Department of Chemistry
Mohanlal Sukhadia University
Udaipur-313001, Rajasthan

Girima Nagda
Assistant Professor
Department of Zoology
Mohanlal Sukhadia University
Udaipur-313001, Rajasthan

NIPA® GENX ELECTRONIC RESOURCES & SOLUTIONS P. LTD.
New Delhi-110 034

NIPA® GENX ELECTRONIC RESOURCES & SOLUTIONS P. LTD.

101,103, Vikas Surya Plaza, CU Block
L.S.C.Market, Pitam Pura, New Delhi-110 034
Ph : +91 11 27341616, 27341717, 27341718
E-mail: newindiapublishingagency@gmail.com
www: www.nipabooks.com

For customer assistance, please contact
Phone: + 91-11-27 34 17 17
Fax: + 91-11-27 34 16 16

Print ISBN: 978-93-58878-35-6
ebook ISBN: 978-93-58873-85-6

Composed and Designed by NIPA®.

Dedicated
to
Our Family

Preface

Welcome to "Nanotechnology: Journey from Laboratory to Society." In this book, we embark on a comprehensive exploration of the dynamic realm of nanotechnology, where innovation meets societal impact. Nanotechnology, with its ability to manipulate matter at the nanoscale, has emerged as a transformative force, revolutionizing industries and research disciplines across the globe.

The themes within this volume reflect the diverse and dynamic nature of recent advances in nanotechnology. From the synthesis of nanostructured materials to their applications in various fields, including healthcare, energy, environment, sensing, food science, agriculture, and manufacturing, this book delves into the multifaceted dimensions of nanotechnology. Each theme showcases the wide-ranging impact of nanotechnology, fostering innovation and addressing complex challenges in today's world.

The book includes eighteen chapters highlighting various aspects of nanotechnology. As we journey through the pages of this book, we will explore the synthesis and characterization of nanomaterials, their diverse applications, and the intricate interplay between science, technology, and society. Moreover, we will delve into critical discussions surrounding safety concerns, future prospects, and the challenges that accompany the integration of nanotechnology into our lives.

The contributors to this volume, experts in their respective fields, offer insightful perspectives and cutting-edge research, illutminating the pathways from laboratory discovery to real-world applications. Whether you are a seasoned researcher, a student eager to delve into the exciting world of nanotechnology, or a curious reader interested in understanding its societal implications, this book provides a comprehensive and accessible resource.

This book serves as a comprehensive exploration of the latest advancements, trends, and challenges in the realm of nanotechnology. From the fundamental principles of fabrication strategies, encompassing top-down, bottom-up, and hybrid approaches, to the intricate interactions between nanoparticles and biological systems, each chapter delves into a distinct facet of this burgeoning field.

The chapters dedicated to sensing and detection highlight the unparalleled analytical capabilities unlocked by nanomaterials, offering insights into their utilization for chemical detection and disease diagnosis. Moreover, the intersection of nanotechnology with medicine presents groundbreaking opportunities for personalized healthcare, ranging from nanomedicine applications to the detection and treatment of parasitic diseases.

Beyond the realms of healthcare, nanotechnology holds immense promise for enhancing food preservation, packaging, and safety. By harnessing the power of nanoparticles, researchers are paving the way for cleaner and greener approaches to wastewater remediation and environmental remediation, thus mitigating the impact of heavy metals and contaminants on our ecosystems.

Furthermore, this book delves into the sustainable synthesis and diverse environmental applications of zinc oxide nanoparticles, shedding light on their pivotal role in advancing environmental stewardship. From sustainable agriculture to the development of nanobiopesticides, nanotechnology offers innovative solutions to address pressing challenges in food security and agricultural sustainability.

As we navigate through the intricacies of nanotechnology, it is crucial to acknowledge the broader implications of research, including patenting trends and ethical considerations. By fostering interdisciplinary collaboration and dialogue, we can harness the full potential of nanotechnology while ensuring responsible innovation and equitable access to its benefits.

In compiling this book, our aim is to provide researchers, practitioners, and enthusiasts alike with a comprehensive resource that not only elucidates the latest advancements in nanotechnology but also inspires future exploration and innovation. We invite readers to embark on this journey with us, as we unravel the endless possibilities and potential of nanotechnology to shape a brighter, more sustainable future for generations to come.

Editors

Contents

1

Fabrication Strategies for Nanomaterials: Top-down, Bottom-up and Hybrid Approaches

Manisha Khandelwal, Gangotri Pemawat and Rama Kanwar Khangarot*

Department of Chemistry, University College of Science, Mohanlal Sukhadia University Udaipur-313001, Rajasthan, India

Abstract

Nanomaterials are a rapidly growing technological sector with diverse applications. These nanomaterials are categorized into distinct groups according to their shape, size, and materials used in their fabrication. The synthesis of nanomaterials follows different strategies that are tailored to their specific applications. This chapter describes the different approaches used for the synthesis of nanomaterials, including top-down, bottom-up, and hybrid approaches. Furthermore, this chapter discusses the advantages and disadvantages of each fabrication method and explores the diverse applications of nanomaterials across multiple fields.

Keywords: *Nanomaterials, Synthesis methods, Advantages, Disadvantages*

1 Introduction

Nanomaterials, characterized by their unique properties at the nanoscale, have garnered significant attention in recent years for their diverse range of applications across various fields (Barhoum *et al.*, 2022; El-Khawaga *et al.*, 2023). These materials exhibit distinct physical, chemical, and biological properties compared to their bulk counterparts, owing to their high surface area-to-volume ratio and quantum confinement effects (El-Kady *et al.*, 2023). The nanomaterials are categorized into four main types, i.e., zero-dimensional (0-D), one-dimensional (1-D), two-dimensional (2-D), and three-dimensional structures (3-D), according to their size and shape as shown in Figure 1. Zero-

Correspondence author e-mail: ramakanwar@mlsu.ac.in

dimensional contains spherical shapes and one-dimensional includes nanowires, nanotubes, nanorods, and nanofibers. Two-dimensional structures consist of thin films and sheets layered one by one. Three-dimensional structures are arrays of nanotubes, nanocubes, nanowires, etc. (Vaseghi & Nematollahzadeh, 2020). Nanomaterials are also classified based on the materials used for the synthesis. This classification typically includes carbon-based nanomaterials, inorganic-based nanomaterials, organic-based nanomaterials, composite-based nanomaterials, and plant-based nanomaterials (Figure 2) (Singh *et al.*, 2020).

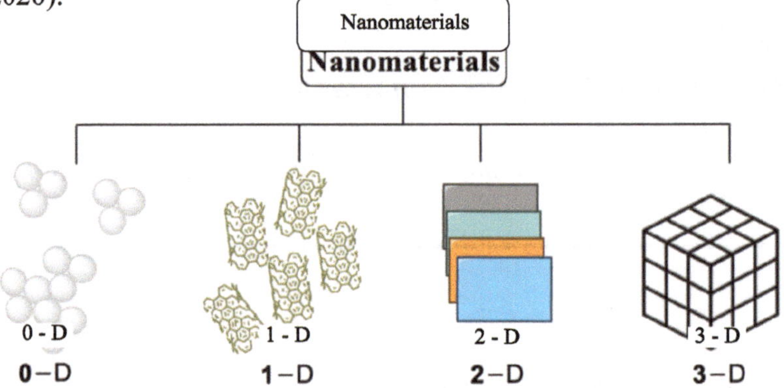

Fig. 1: Classification of nanomaterials based on their dimensions.

The fabrication of these nanomaterials involves precise control over their size, shape, composition, and structure, which can be achieved through a variety of synthesis/fabrication methods (Alshammari *et al.*, 2023; Saleh, 2020). This chapter offers an overview of diverse methods employed in nanomaterials synthesis. The chapter also discusses the advantages and disadvantages related to various methods.

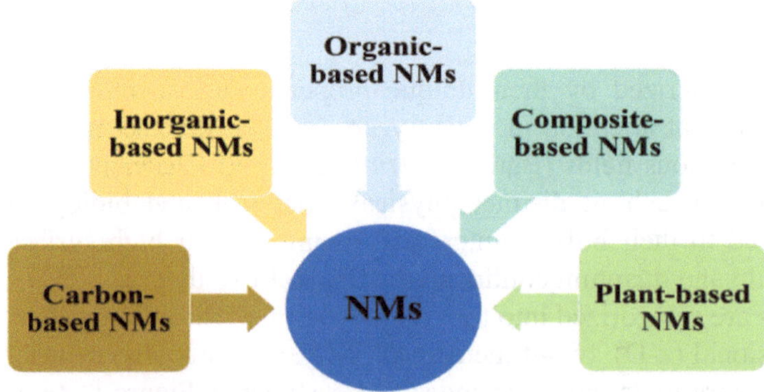

Fig. 2: Classification of nanomaterials based on materials.

2 Methods for fabrication of nanomaterials

The synthesis methods are primarily categorized into two approaches: the top-down approach and the bottom-up approach. The top-down approach consists of various techniques, including high-energy ball milling (De Carvalho *et al.*, 2013; Pentimalli *et al.*, 2015), thermal evaporation (Kar *et al.*, 2006), laser ablation (Shabalina *et al.*, 2022), sputtering (Alexeeva & Fateev, 2016), etc. The bottom-up approach incorporates hydrothermal (Khan, 2020; Rendón-Angeles & Seong, 2023), sol-gel (Bokov *et al.*, 2021), chemical vapour deposition (CVD) (Manawi *et al.*, 2018; Sun *et al.*, 2021), co-precipitation (Khandelwal *et al.*, 2023a), etc. A schematic diagram illustrating the top-down and bottom-up approaches is depicted in Figure 3. Some techniques incorporate a hybrid approach, utilizing a combination of both methodologies, such as in the case of the lithography approach (Colson *et al.*, 2013). In the subsequent section, we provide a detailed description of each method, along with their respective advantages and disadvantages.

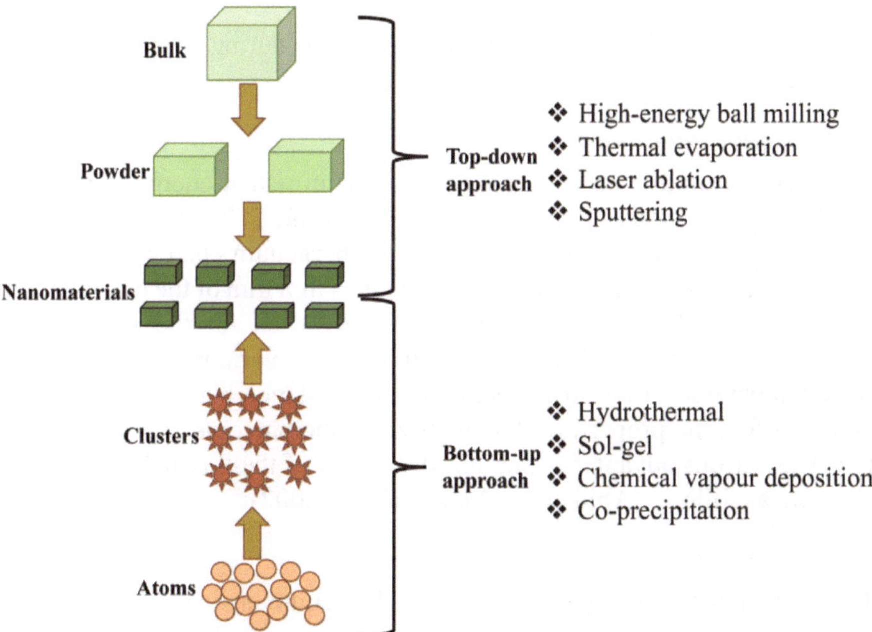

Fig. 3: The top-down and bottom-up approaches for the fabrication of nanomaterials.

2.1 Top-down approaches

2.1.1 High-energy ball milling

High-energy ball milling is a process in which high-energy collisions are used to synthesize the nanomaterials. High-energy collisions are induced by hardened steel balls (Joy *et al.*, 2022). This approach is effective for the large-scale production of nanomaterials from bulk materials. This method is predominantly employed in the fabrication of nanoalloys encompassing elements such as nickel, copper, zinc, aluminium, magnesium, etc. (Charitidis *et al.*, 2014). The advantages and disadvantages of the high-energy ball milling method are listed below (Abid *et al.*, 2022; Delogu *et al.*, 2017; Stolle & Ranu, 2014):

2.1.1.1 Advantages

1. Operational feasibility.
2. Facilitate the controlled and gradual growth of particles.
3. Reliable and safe method.
4. Low maintenance of the equipment.

2.1.1.2 Disadvantages

1. High electricity consumption.
2. Generate substantial noise and high vibrations during processing.
3. Low working efficiency.

2.1.2 Thermal evaporation

Thermal evaporation is an endothermic process in which energy is required to break down the bond of molecules (Bashir *et al.*, 2020; Li *et al.*, 2011). In this process, the sample is heated in the high vacuum chamber, generates vapours of the substrate and then condensed into a thin film of the nanomaterial (Andrews *et al.*, 2010). This method is widely used for the synthesis of inorganic nanomaterials (Zheng *et al.*, 2011). The various factors such as furnace temperature, time of evaporation, and distance between source and substrate, affect the properties of synthesized nanomaterials (S. Wang *et al.*, 2018). The various advantages and disadvantages of the thermal evaporation method are as follows (Abid *et al.*, 2022; Mattox, 2002):

2.1.2.1 Advantages

1. The nanomaterial is fabricated in a high vacuum environment, resulting in the formation of pure material.
2. Utilizes a greener approach as it does not require any solvent.
3. Effective in generating uniform layers.
4. Particularly effective for materials having low melting points.

2.1.2.2 Disadvantages

1. This technique cannot be used for materials that are non-volatile and undergo decomposition when heated.
2. High complexity and cost of the technological process.
3. Difficulty in obtaining homogeneous composite films due to long-term optimization of the material.
4. Several alloy compositions and combinations are challenging to deposit.

2.1.3 Laser ablation

Laser ablation is a top-down method utilized for the synthesis of nanomaterials. In this method, a high-energy pulsed laser beam is radiated onto a target material, leading to the vaporization of the material (Sadrolhosseini *et al.*, 2019). The vapours of the material immediately solvated in the surfactant liquid solution and the formation of nanomaterials occurred (AlMalki *et al.*, 2022). This technique is employed across various fields due to its versatility and ability to produce nanomaterials with unique properties. The ablation rate in laser-based processes is influenced by several key characteristics of the laser system, including fluency, wavelength, repetition rate, and pulse duration (Balachandran *et al.*, 2022). Additionally, factors such as transmission properties, light absorption efficiency of the target material, and chemical composition of any surrounding liquid also play significant roles in determining the rate of ablation. Here are some advantages and disadvantages associated with the laser ablation method (Balachandran *et al.*, 2022; Naser *et al.*, 2019):

2.1.3.1 Advantages

1. Laser ablation allows for precise control over the size, shape, and composition of the synthesized nanomaterial, enabling tailored properties for specific applications.
2. Laser ablation can be applied to a wide range of materials, including metals, semiconductors, and ceramics, offering flexibility in nanomaterial synthesis.
3. Nanomaterials can be synthesized rapidly using laser ablation, making it a time-efficient process compared to some other synthesis methods.
4. One of the significant advantages of laser ablation synthesis in solution is its capability to produce ligand-free noble nanomaterials across various solvents.

2.1.3.2 Disadvantages

1. Laser ablation is often restricted to small-scale production due to challenges in scaling up the process for mass production.

2. Setting up and maintaining the necessary laser equipment can be costly and technically challenging, requiring specialized expertise.
3. Laser ablation consumes significant amounts of energy, especially when operating high-power lasers, leading to increased operational costs.

2.1.4 Sputtering

The sputtering method stands as a prominent technique for the synthesis of nanomaterials (Alfonso *et al.*, 2012). This process involves bombarding a target material with high-energy ions or atoms, causing the ejection of small atoms from the target to deposit onto a substrate, thus forming a thin film of nanomaterial (El-Khawaga *et al.*, 2023). The sputtering method incorporates various approaches, including magnetron sputtering, ion beam sputtering and reactive sputtering, each tailored to specific material requirements and desired outcomes (Liang *et al.*, 2021). The sputtering method is controlled by various parameters such as the energy and mass of a bombarding particle annealing duration and temperature, target and substrate angle, etc. (F. Wang *et al.*, 2020). The benefits and limitations of the sputtering method include the following (Depla, 2021; Kukla, 1997; Lee *et al.*, 2022; Nguyen *et al.*, 2022):

2.1.4.1 Advantages

1. The sputtering method allows for precise control over the deposition rate and thickness of the resulting material, facilitating the fabrication of uniform and tailored nanostructures.
2. The target used for sputtering deposition serves as a reliable and stable source for vapourization.
3. Sputtering operates in a vacuum environment, minimizing contamination and ensuring the high purity of the synthesized nanomaterials.
4. The sputtering technique is compatible with versatile substrates, including silicon, alloys, glass, and flexible polymers, offering flexibility in material selection and substrate integration for diverse applications.
5. Sputtering systems can be readily scaled up for industrial production, making it feasible for large-scale manufacturing of nanomaterials with consistent quality and reproducibility.
6. The use of solid targets enables precise control over the type and composition of the material.

2.1.4.2 Disadvantages

1. Sputtering targets frequently come with a high cost.

2. The sputtering process can generate heat, particularly in high-power systems, leading to substrate heating and potential thermal damage, which may limit its applicability for temperature-sensitive materials or substrates.
3. Reactive sputtering methods necessitate the use of specific gases for deposition, adding complexity to the process and requiring careful gas handling to prevent poisoning of sputtering targets.

2.2 Bottom-up approaches

2.2.1 Hydrothermal

The hydrothermal synthesis of single crystals relies on the solubility of precursor solid materials in hot water under high pressure (Bharti & Bharati, 2017). Crystal growth occurs within a steel reactor vessel known as an autoclave, where a nutrient is supplied alongside water (Lal *et al.*, 2023; Maghsoudi *et al.*, 2023). Temperature gradients are maintained across the growth chamber, with the hotter end facilitating nutrient dissolution and the cooler end promoting seed growth. Through controlled nucleation and growth processes, nanomaterials with specific sizes, shapes, and compositions are manufactured (Gan *et al.*, 2020; Maghsoudi *et al.*, 2023). The benefits and drawbacks of the hydrothermal method are written below (Asim *et al.*, 2014; Byrappa & Adschiri, 2007; Byrappa & Yoshimura, 2012; Ghasaban *et al.*, 2017).

2.2.1.1 Advantages

1. Hydrothermal synthesis allows specific control over nanomaterial properties such as size, shape, and composition by adjusting reaction parameters like temperature, pressure, and precursor concentration.
2. This method applies to a diverse range of materials including metal oxides, sulfides, and semiconductors, facilitating the synthesis of nanoparticles with tailored functionalities for various applications.
3. This method often yields nanomaterials with high yield, making it suitable for both research endeavours and industrial-scale production.
4. The synthesis process can typically be conducted in a single reaction vessel, minimizing complexity and simplifying purification steps.
5. The utilization of water as a solvent in hydrothermal synthesis reduces reliance on organic solvents, aligning with green approaches.

2.2.1.2 Disadvantages

1. Nanoparticles synthesized *via* hydrothermal methods may exhibit tendencies towards aggregation or agglomeration, affecting their dispersibility and uniformity in suspension.

2. Safety concerns may arise during the reaction process.
3. The closed black box hinders the observation of particle growth.

2.2.2 Sol-gel method

The sol-gel process, alternatively referred to as chemical solution deposition is a wet chemical technique. This method is extensively utilized across ceramic engineering and material science disciplines (Kanwal *et al.*, 2023). This technique emerges as a straightforward approach to fabricate nanomaterials with precise control over size, composition, and morphology (Alabada *et al.*, 2023). This method involves the conversion of precursor solutions (sol) into a gel-like network through hydrolysis. The hydrolysis of precursors generates hydroxide ions, which initiate the condensation reaction, leading to the formation of a three-dimensional network of desired nanomaterials suspended in the solution (Mandal *et al.*, 2023). Subsequent ageing and drying steps promote the gelation of the sol, transforming it into a solid gel matrix. Additional thermal treatment, such as calcination or annealing, may be employed to remove residual solvent and organic components, resulting in the formation of pure nanomaterials (Mahalingam & Sivaraju, 2023). The advantages and disadvantages of the sol-gel method are as follows (Aegerter & Mennig, 2013; Dislich, 1986; Hench & West, 1990; Mackenzie, 1988; Ullattil & Periyat, 2017):

2.2.2.1 Advantages

1. The sol-gel method enables precise control over the composition, morphology, and structure of nanomaterials by adjusting precursor concentration, solvent composition, and processing parameters, facilitating the synthesis of custom-tailored materials for specific applications.
2. This technique offers excellent homogeneity and uniform distribution of nanomaterials within the gel matrix, ensuring consistent material properties and performance.
3. Sol-gel method is compatible with various materials, including oxides, sulfides, nitrides, ceramics, composites and hybrid organic-inorganic compounds, providing versatility in material selection and synthesis.
4. The synthesis process typically operates at relatively low temperatures, making it suitable for temperature-sensitive materials and reducing energy consumption compared to high-temperature techniques.
5. Sol-gel synthesis can be easily scaled up for industrial production, offering a cost-effective and scalable approach for producing nanomaterials in large quantities with consistent quality.

2.2.2.2 Disadvantages

1. Sol-gel synthesis often requires prolonged processing time due to the slow kinetics of hydrolysis and condensation reactions, potentially extending the overall synthesis duration.
2. The sol-gel process involves complicated chemical reactions, including hydrolysis, condensation, and gelation, which may require careful optimization and control to achieve desired material properties and avoid undesirable side reactions.
3. The use of organic solvents in sol-gel synthesis raises concerns regarding solvent handling, waste disposal, and environmental impact, necessitating proper waste management practices.
4. In some cases, the sol-gel method may result in the formation of porous structures within the nanomaterials, which can affect material properties such as mechanical strength, density, and surface area.

2.2.3 Chemical vapour deposition

The chemical vapour deposition (CVD) method acts as a pivotal technique in the domain of nanomaterial synthesis, offering precise control over the growth of thin films, nanowires, and other nanostructures with tailored properties (Ahmed *et al.*, 2023; Kumar *et al.*, 2023). During this procedure, a substrate undergoes the deposition of a thin film of nanomaterials layer by layer through chemical reactions involving gas-phase precursors (Roniboss *et al.*, 2023). Various types of CVD techniques exist, including thermal CVD, plasma-enhanced CVD (PECVD), and atomic layer deposition (ALD), each offering unique advantages and capabilities (Asim *et al.*, 2014). The advantages and limitations of the CVD method are listed below (Bahri *et al.*, 2023; Jones & Hitchman, 2009; Morosanu, 2016; Sun *et al.*, 2021; Yan & Xu, 2010):

2.2.3.1 Advantages

1. CVD allows precise control over nanomaterial growth, enabling the deposition of thin films and nanostructures with uniform thickness, composition, and crystallinity.
2. This method applies to a wide range of materials, including metals, semiconductors, oxides, and carbon-based materials, offering versatility in material selection and synthesis.
3. CVD processes can be easily scaled up for industrial production, making them suitable for large-scale manufacturing of nanomaterials with consistent quality and reproducibility.

4. CVD typically operates in a vacuum or controlled atmosphere, minimizing contamination and ensuring high purity of synthesized nanomaterials, critical for applications requiring pristine material characteristics.
5. CVD enables the deposition of nanomaterials onto complex and three-dimensional substrates with excellent conformal coverage, allowing for the fabrication of nanodevices and coatings with tailored properties.

2.2.3.2 Disadvantages

1. CVD systems often require sophisticated equipment and expertise to operate, including vacuum chambers, gas handling systems, and temperature control mechanisms, leading to higher initial investment and maintenance costs.
2. Many CVD processes operate at elevated temperatures, which may limit the choice of substrates and materials, and pose challenges for temperature-sensitive applications or substrates.
3. Some precursor gases used in CVD processes may be hazardous or toxic, requiring careful handling, storage, and disposal procedures to ensure safety and environmental compliance.
4. Achieving uniform film thickness and minimizing defects across large-area substrates can be challenging in CVD, particularly for complex nanostructures or heterogeneous materials.

2.2.4 Co-precipitation

The co-precipitation method is a widely utilized technique for the synthesis of nanomaterials, offering simplicity, versatility, and tunability in the manufacturing of nanomaterials with tailored properties (Gholizadeh *et al.*, 2023). In this method, nanomaterials are formed through the simultaneous precipitation of multiple precursor ions from a solution. The process typically involves dissolving precursor salts in a solvent, followed by the addition of a precipitating agent or alteration of solution conditions such as pH, temperature, and concentration to induce precipitation (Apostoluk *et al.*, 2023). Hydroxides, chlorides, carbonates, and oxalates are commonly used as precipitating agents (Abid *et al.*, 2022). Following precipitation, the resulting particles are aged to increase their size and then collected *via* filtration or centrifugation. Further purification steps involving washes with ethanol, distilled water, or other solvents are necessary to remove contaminants and obtain high-purity nanomaterials (Khandelwal *et al.*, 2023b). Post-treatments such as annealing, sintering, or calcination are employed to achieve desired crystal structures and morphologies. The advantages and disadvantages of the co-precipitation method are listed below (Darwish *et al.*, 2019; Hou *et al.*, 2017; Rahmani *et al.*, 2018; Theiss *et al.*, 2016):

2.2.4.1 Advantages

1. The co-precipitation method allows for the synthesis of various types of nanomaterials, including oxides, hydroxides, sulfides, and carbonates, among others.
2. This method is relatively simple and scalable, making it suitable for large-scale production of nanomaterials.
3. By controlling parameters such as precursor concentration, reaction temperature, pH, and reaction time, researchers can finely tune the size, morphology, composition, and crystallinity of the resulting nanomaterials.
4. Co-precipitation can be performed at low temperatures, reducing energy consumption and minimizing the formation of undesirable by-products.
5. Cost-effectiveness method

2.2.4.2 Disadvantages

1. Agglomeration of particles occurs, which can affect the uniformity and performance of the nanomaterials.
2. Impurities from the precursor salts or reaction environment may be incorporated into the synthesized nanomaterials, affecting their purity and properties.

2.3 Hybrid approach

2.3.1 Lithography

This technique represents a hybrid approach, integrating the principle of both top-down and bottom-up methods (Colson *et al.*, 2013). This technique acts as a valuable tool for the manufacturing of nanostructures which contain complex patterns (Y. Wang *et al.*, 2019). The fabrication of nanomaterials using lithography involves the use of a focused beam of light or electrons. Lithography can be broadly divided into two types: masked and maskless lithography (E. Sharma *et al.*, 2022). Masked lithography involves transferring nanopatterns across a large surface area using a designated mask or template. On the other hand, maskless lithography contains nanopattern writing without the need for a mask or template. The lithography method has evolved into a versatile technique with numerous advanced modifications, including electron beam lithography, ion beam lithography, X-ray lithography, extreme ultraviolet lithography, dip-pen nanolithography, nanoimprint lithography, soft lithography, atomic layer deposition lithography, coherent lithography, and more (E. Sharma *et al.*, 2022). Some advantages and shortcomings of the lithography methods are listed below (Gatzen *et al.*, 2015; Hasan & Luo, 2018, 2018; Lawson & Robinson, 2016; Saeedeh Ebad-Sichani *et al.*, 2023; Sebastian *et al.*, 2020; Thompson, 1983; Wu *et al.*, 2016):

2.3.1.1 Advantages

1. Lithography techniques enable the fabrication of nanostructures with extremely high precision, down to the sub-10 nanometer scale, facilitating the creation of intricate patterns and devices.
2. The lithography technique allows for precise control over the shape, size, and arrangement of nanostructures, enabling tailored design and optimization for specific applications such as electronics, photonics, and biotechnology.
3. Some lithography methods, such as photolithography and nanoimprint lithography, enable parallel processing, allowing for the simultaneous fabrication of multiple nanostructures on a single substrate, thereby increasing efficiency.

2.3.1.2 Disadvantages

1. Mask lithography techniques require the use of masks or templates to define the desired patterns, requiring additional fabrication steps and costs, and potentially introducing alignment errors or defects during the patterning process.
2. Lithography processes often involve multiple steps and require sophisticated equipment, expertise, and cleanroom facilities, leading to higher operational costs and technical challenges, particularly for small-scale research or production facilities.
3. Certain lithography methods may have limitations in substrate compatibility, such as restrictions on substrate size, type, or surface properties, which may constrain the range of applications or materials that can be utilized.

3 Applications and Future Perspective

Nanomaterial synthesis holds immense potential for a wide range of applications across various fields. The nanomaterials find application in a wide array of fields including sensors (Ghodselahi *et al.*, 2015; Khangarot *et al.*, 2022), supercapacitors (Nagamuthu *et al.*, 2013), water purification (Khandelwal *et al.*, 2023.; Pinto *et al.*, 2020), catalysis (Gawande *et al.*, 2015; Jiang *et al.*, 2005;

N. Sharma *et al.*, 2015), photothermal therapies (Fazal *et al.*, 2014), energy storage (Devi *et al.*, 2021), drug delivery (Liu *et al.*, 2022), coatings (Bechtold *et al.*, 2020; Rai *et al.*, 2010), agriculture (Srujana & Bhagat, 2022), cosmetics (Arroyo *et al.*, 2020), etc. Figure 4 represents the versatile applications of nanomaterials. Despite extensive research into the synthesis of nanomaterials, numerous challenges remain unresolved. Firstly, achieving precise control over the size, shape, and composition of nanomaterials remains a difficult task. Secondly, ensuring the stability and dispersibility of nanomaterials poses a significant challenge. Thirdly, the development of environmentally sustainable synthesis routes lacking defects, agglomeration, and suitable for large-scale production presents a significant challenge. Hence, there is immense potential for the advancement of novel methodologies for nanomaterials fabrication, capable of addressing all challenges associated with nanomaterials synthesis.

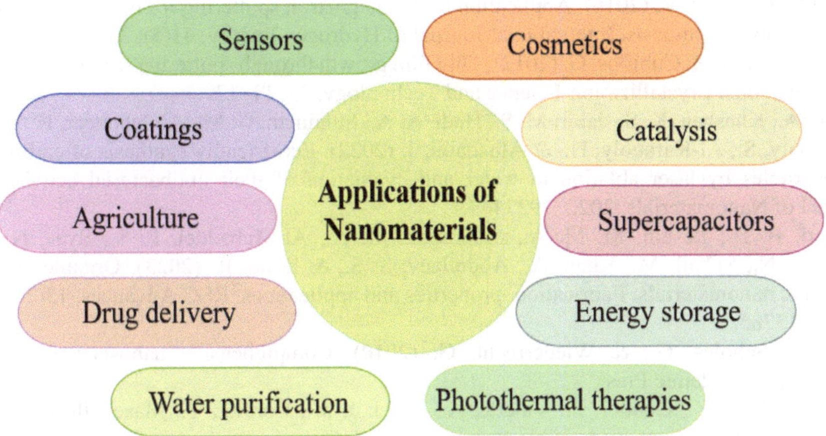

Fig. 4: Versatile applications of the nanomaterials in different fields.

4 Conclusion

In summary, the methods and techniques outlined in this chapter enable the synthesis of a diverse array of nanomaterials. However, a critical aspect of nanoparticle synthesis is the identification of optimal conditions to enhance reproducibility and yield. Fabrication of nanomaterials is accomplished by various top-down and bottom-up approaches, each having advantages and shortcomings. The appropriate method for nanomaterial synthesis should be selected based on specific requirements and criteria. With continuous research efforts and collaborative initiatives, there are exciting prospects for the development of nanomaterial-based technologies that will shape the world in profound ways.

References

Abid, N., Khan, A. M., Shujait, S., Chaudhary, K., Ikram, M., Imran, M., Haider, J., Khan, M., Khan, Q., & Maqbool, M. (2022). Synthesis of nanomaterials using various top-down and bottom-up approaches, influencing factors, advantages, and disadvantages: A review. Advances in Colloid and Interface Science, 300, 102597.

Aegerter, M. A., & Mennig, M. (2013). Sol-gel technologies for glass producers and users. Springer Science & Business Media.

Ahmed, H. R., Aziz, K. H. H., Agha, N. N., Mustafa, F. S., & Hinder, S. J. (2023). Iron-loaded carbon black prepared via chemical vapor deposition as an efficient peroxydisulfate activator for the removal of rhodamine B from water. RSC Advances, 13(37), 26252–26266.

Alabada, R., Kadhim, M. M., sabri Abbas, Z., Rheima, A. M., Altimari, U. S., Dawood, A. H., jawad al-bayati, A. dhari, Talib Abed, Z., Saeed Radhi, R., Salam Jaber, A., Hachim, S. K., Ali, F. K., Mahmoud, Z. H., & Kianfar, E. (2023). Investigation of effective parameters in the production of alumina gel through the sol-gel method. Case Studies in Chemical and Environmental Engineering, 8, 100405.

Alexeeva, O., & Fateev, V. (2016). Application of the magnetron sputtering for nanostructured electrocatalysts synthesis. International Journal of Hydrogen Energy, 41(5), 3373–3386.

Alfonso, E., Olaya, J., & Cubillos, G. (2012). Thin film growth through sputtering technique and its applications. Crystallization-Science and Technology, 23, 11–12.

AlMalki, F. A., Khashan, K. S., Jabir, M. S., Hadi, A. A., Sulaiman, G. M., Abdulameer, F. A., Albukhaty, S., Al-Karagoly, H., & Albaqami, J. (2022). Eco-friendly synthesis of carbon nanoparticles by laser ablation in water and evaluation of their antibacterial activity. Journal of Nanomaterials, 2022, 7927447.

Alshammari, B. H., Lashin, M. M. A., Mahmood, M. A., Al-Mubaddel, F. S., Ilyas, N., Rahman, N., Sohail, M., Khan, A., Abdullaev, S. S., & Khan, R. (2023). Organic and inorganic nanomaterials: Fabrication, properties and applications. RSC Advances, 13(20), 13735–13785.

Andrews, D., Scholes, G., & Wiederrecht, G. (2010). Comprehensive nanoscience and technology. Academic Press.

Apostoluk, A., Zhu, Y., Gautier, P., Valette, A., Bluet, J.-M., Cornier, T., Masenelli, B., & Daniele, S. (2023). Improved visible emission from ZnO nanoparticles synthesized via the co-precipitation method. Materials, 16(15).

Arroyo, G. V., Madrid, A. T., Gavilanes, A. F., Naranjo, B., Debut, A., Arias, M. T., & Angulo, Y. (2020). Green synthesis of silver nanoparticles for application in cosmetics. Journal of Environmental Science and Health, Part A, 55(11), 1304–1320.

Asim, N., Ahmadi, S., Alghoul, M. A., Hammadi, F. Y., Saeedfar, K., & Sopian, K. (2014). Research and development aspects on chemical preparation techniques of photoanodes for dye sensitized solar cells. International Journal of Photoenergy, 2014, 518156.

Bahri, M., Gebre, S. H., Elaguech, M. A., Dajan, F. T., Sendeku, M. G., Tlili, C., & Wang, D. (2023). Recent advances in chemical vapour deposition techniques for graphene-based nanoarchitectures: From synthesis to contemporary applications. Coordination Chemistry Reviews, 475, 214910.

Balachandran, A., Sreenilayam, S. P., Madanan, K., Thomas, S., & Brabazon, D. (2022). Nanoparticle production via laser ablation synthesis in solution method and printed electronic application—A brief review. Results in Engineering, 16, 100646.

Barhoum, A., García-Betancourt, M. L., Jeevanandam, J., Hussien, E. A., Mekkawy, S. A., Mostafa, M., Omran, M. M., S. Abdalla, M., & Bechelany, M. (2022). Review on natural, incidental, bioinspired, and engineered nanomaterials: History, definitions, classifications, synthesis, properties, market, toxicities, risks, and regulations. Nanomaterials, 12(2).

Bashir, A., Awan, T. I., Tehseen, A., Tahir, M. B., & Ijaz, M. (2020). Chapter 3—Interfaces and surfaces. In T. I. Awan, A. Bashir, & A. Tehseen (Eds.), Chemistry of Nanomaterials (pp. 51–87). Elsevier.

Bechtold, M., Valério, A., Ulson de Souza, A., De Oliveira, D., Franco, C., Serafim, R., & Guelli U. Souza, S. (2020). Synthesis and application of silver nanoparticles as biocidal agent in polyurethane coating. Journal of Coatings Technology and Research, 17, 613–620.

Bharti, D. B., & Bharati, A. V. (2017). Synthesis of ZnO nanoparticles using a hydrothermal method and a study its optical activity. Luminescence, 32(3), 317–320.

Bokov, D., Turki Jalil, A., Chupradit, S., Suksatan, W., Javed Ansari, M., Shewael, I. H., Valiev, G. H., & Kianfar, E. (2021). Nanomaterial by sol-gel method: Synthesis and application. Advances in Materials Science and Engineering, 2021, 1–21.

Byrappa, K., & Adschiri, T. (2007). Hydrothermal technology for nanotechnology. Progress in Crystal Growth and Characterization of Materials, 53(2), 117–166.

Byrappa, K., & Yoshimura, M. (2012). Handbook of hydrothermal technology. William Andrew.

Charitidis, C. A., Georgiou, P., Koklioti, M. A., Trompeta, A.-F., & Markakis, V. (2014). Manufacturing nanomaterials: From research to industry. Manufacturing Review, 1, 11.

Colson, P., Henrist, C., & Cloots, R. (2013). Nanosphere lithography: A powerful method for the controlled manufacturing of nanomaterials. Journal of Nanomaterials, 2013, 948510.

Darwish, M. S., Kim, H., Lee, H., Ryu, C., Lee, J. Y., & Yoon, J. (2019). Synthesis of magnetic ferrite nanoparticles with high hyperthermia performance via a controlled co-precipitation method. Nanomaterials, 9(8), 1176.

De Carvalho, J., De Medeiros, S., Morales, M., Dantas, A., & Carriço, A. (2013). Synthesis of magnetite nanoparticles by high energy ball milling. Applied Surface Science, 275, 84–87.

Delogu, F., Gorrasi, G., & Sorrentino, A. (2017). Fabrication of polymer nanocomposites via ball milling: Present status and future perspectives. Progress in Materials Science, 86, 75–126.

Depla, D. (2021). Sputter deposition with powder targets: An overview. Vacuum, 184, 109892.

Devi, N., Sahoo, S., Kumar, R., & Singh, R. K. (2021). A review of the microwave-assisted synthesis of carbon nanomaterials, metal oxides/hydroxides and their composites for energy storage applications. Nanoscale, 13(27), 11679–11711.

Dislich, H. (1986). Sol-gel: Science, processes and products. Journal of Non-Crystalline Solids, 80(1–3), 115–121.

El-Kady, M. M., Ansari, I., Arora, C., Rai, N., Soni, S., Verma, D. K., Singh, P., & Mahmoud, A. E. D. (2023). Nanomaterials: A comprehensive review of applications, toxicity, impact, and fate to environment. Journal of Molecular Liquids, 370, 121046.

El-Khawaga, A. M., Zidan, A., & El-Mageed, A. I. A. A. (2023). Preparation methods of different nanomaterials for various potential applications: A review. Journal of Molecular Structure, 1281, 135148.

Fazal, S., Jayasree, A., Sasidharan, S., Koyakutty, M., Nair, S. V., & Menon, D. (2014). Green synthesis of anisotropic gold nanoparticles for photothermal therapy of cancer. ACS Applied Materials & Interfaces, 6(11), 8080–8089.

Gan, Y. X., Jayatissa, A. H., Yu, Z., Chen, X., & Li, M. (2020). Hydrothermal synthesis of nanomaterials. Journal of Nanomaterials, 2020, 8917013.

Gatzen, H. H., Saile, V., & Leuthold, J. (2015). Lithography. In H. H. Gatzen, V. Saile, & J. Leuthold (Eds.), Micro and Nano Fabrication: Tools and Processes (pp. 313–395). Springer Berlin Heidelberg.

Gawande, M. B., Goswami, A., Asefa, T., Guo, H., Biradar, A. V., Peng, D.-L., Zboril, R., & Varma, R. S. (2015). Core–shell nanoparticles: Synthesis and applications in catalysis and electrocatalysis. Chemical Society Reviews, 44(21), 7540–7590.

Ghasaban, S., Atai, M., & Imani, M. (2017). Simple mass production of zinc oxide nanostructures via low-temperature hydrothermal synthesis. Materials Research Express, 4(3), 035010.

Ghodselahi, T., Aghababaie, N., Mobasheri, H., Salimi, K. Z., Pasha, M. A., & Vesaghi, M. (2015). Fabrication and characterization and biosensor application of gold nanoparticles on the carbon nanotubes. Applied Surface Science, 355, 1175–1179.

Gholizadeh, Z., Aliannezhadi, M., Ghominejad, M., & Tehrani, F. S. (2023). High specific surface area γ-Al2O3 nanoparticles synthesized by facile and low-cost co-precipitation method. Scientific Reports, 13(1), 6131.

Hasan, R. Md. M., & Luo, X. (2018). Promising lithography techniques for next-generation logic devices. Nanomanufacturing and Metrology, 1(2), 67–81.

Hench, L. L., & West, J. K. (1990). The sol-gel process. Chemical Reviews, 90(1), 33–72.

Hou, P., Zhang, H., Zi, Z., Zhang, L., & Xu, X. (2017). Core–shell and concentration-gradient cathodes prepared via co-precipitation reaction for advanced lithium-ion batteries. Journal of Materials Chemistry A, 5(9), 4254–4279.

Jiang, L., Sun, G., Zhou, Z., Sun, S., Wang, Q., Yan, S., Li, H., Tian, J., Guo, J., & Zhou, B. (2005). Size-controllable synthesis of monodispersed SnO2 nanoparticles and application in electrocatalysts. The Journal of Physical Chemistry B, 109(18), 8774–8778.

Jones, A. C., & Hitchman, M. L. (2009). Chemical vapour deposition: Precursors, processes and applications. Royal society of chemistry.

Joy, J., Krishnamoorthy, A., Tanna, A., Kamathe, V., Nagar, R., & Srinivasan, S. (2022). Recent developments on the synthesis of nanocomposite materials via ball milling approach for energy storage applications. Applied Sciences, 12(18), 9312.

Kanwal, S., Tahir Khan, M., Tirth, V., Algahtani, A., Al-Mughanam, T., & Zaman, A. (2023). Room-temperature ferromagnetism in Mn-doped ZnO nanoparticles synthesized by the sol–gel method. ACS Omega, 8(31), 28749–28757.

Kar, S., Ghoshal, T., & Chaudhuri, S. (2006). Simple thermal evaporation route to synthesize Zn and Cd metal nanowires. Chemical Physics Letters, 419(1–3), 174–178.

Khan, F. A. (2020). Synthesis of nanomaterials: Methods & technology. Applications of Nanomaterials in Human Health, 15–21.

Khandelwal, M., Choudhary, S., Kumawat, A., Misra, K. P., Rathore, D. S., & Khangarot, R. K. (2023a). Asterarcys quadricellulare algae-mediated copper oxide nanoparticles as a robust and recyclable catalyst for the degradation of noxious dyes from wastewater. RSC Advances, 13(40), 28179–28196.

Khandelwal, M., Choudhary, S., Kumawat, A., Misra, K. P., Rathore, D. S., & Khangarot, R. K. (2023b). Superior photocatalytic degradation of water pollutant dyes and strong antimicrobial activity in Coelastrella terrestris algal-mediated biosynthesized stable copper hydroxide nanoparticles. Biomass Conversion and Biorefinery, 1–18.

Khandelwal, M., Pemawat, G., Meena, K., & Khangarot, R. K. (2023). Chapter 2—Applications of copper-based polymer nanocomposites for the removal of organic dyes from wastewater. In Green Technology for the Sustainable Development (pp. 18–32). Blue Rose Publishers.

Khangarot, R. K., Khandelwal, M., & Singh, R. (2022). Copper-based polymer nanocomposites: application as sensors. In Metal Nanocomposites for Energy and Environmental Applications (pp. 489–508). Springer.

Kukla, R. (1997). Magnetron sputtering on large scale substrates: An overview on the state of the art. Surface and Coatings Technology, 93(1), 1–6.

Kumar, R., Goel, N., Jarwal, D. K., Hu, Y., Zhang, J., & Kumar, M. (2023). Strategic review on chemical vapor deposition technology-derived 2D material nanostructures for room-temperature gas sensors. Journal of Materials Chemistry C, 11(3), 774–801.

Lal, M., Sharma, P., & Ram, C. (2023). Optical, structural properties and photocatalytic potential of Nd-ZnO nanoparticles synthesized by hydrothermal method. Results in Optics, 10, 100371.

Lawson, R. A., & Robinson, A. P. G. (2016). Chapter 1—Overview of materials and processes for lithography. In A. Robinson & R. Lawson (Eds.), Frontiers of Nanoscience (Vol. 11, pp. 1–90). Elsevier.

Lee, S., Joung, Y.-H., Yoon, Y.-K., & Choi, W. (2022). Preparation of a ZnO nanostructure as the anode material using RF magnetron sputtering system. Nanomaterials, 12(2).

Li, Y., Peng, R., Xiu, X., Zheng, X., Zhang, X., & Zhai, G. (2011). Growth of SnO2 nanoparticles via thermal evaporation method. Superlattices and Microstructures, 50(5), 511–516.

Liang, J., Liu, Q., Li, T., Luo, Y., Lu, S., Shi, X., Zhang, F., Asiri, A. M., & Sun, X. (2021). Magnetron sputtering enabled sustainable synthesis of nanomaterials for energy electrocatalysis. Green Chemistry, 23(8), 2834–2867.

Liu, Y., Yang, G., Hui, Y., Ranaweera, S., & Zhao, C. (2022). Microfluidic nanoparticles for drug delivery. Small, 18(36), 2106580.

Mackenzie, J. D. (1988). Applications of the sol-gel process. Journal of Non-Crystalline Solids, 100(1–3), 162–168.

Maghsoudi, S., Abdullah Mirzaie, R., Ghalkhani, M., & Anaraki Firooz, A. (2023). Synthesis of Ag and Mn/ZnO nanoparticles using a hydrothermal method—A brief study and their role in the electrocatalytic oxidation of glucose in alkaline media. Journal of Physics and Chemistry of Solids, 176, 111269.

Mahalingam, V., & Sivaraju, M. (2023). Microwave-assisted sol-gel synthesis of silica nanoparticles using rice husk as a precursor for corrosion protection application. Silicon, 15(4), 1967–1975.

Manawi, Y. M., Ihsanullah, Samara, A., Al-Ansari, T., & Atieh, M. A. (2018). A review of carbon nanomaterials' synthesis via the chemical vapor deposition (CVD) method. Materials, 11(5), 822.

Mandal, S., Ali, S. I., & Mandal, A. C. (2023). Investigation of structural, optical and photoluminescence properties of the sol–gel synthesized powder ZnS nanoparticles. Applied Physics A, 129(3), 219.

Mattox, D. M. (2002). Physical vapor deposition (PVD) processes. Metal Finishing, 100, 394–408.

Morosanu, C. E. (2016). Thin films by chemical vapour deposition (Vol. 7). Elsevier.

Nagamuthu, S., Vijayakumar, S., & Muralidharan, G. (2013). Synthesis of Mn3O4/amorphous carbon nanoparticles as electrode material for high performance supercapacitor applications. Energy & Fuels, 27(6), 3508–3515.

Naser, H., Alghoul, M. A., Hossain, M. K., Asim, N., Abdullah, M. F., Ali, M. S., Alzubi, F. G., & Amin, N. (2019). The role of laser ablation technique parameters in synthesis of nanoparticles from different target types. Journal of Nanoparticle Research, 21(11), 249.

Nguyen, M. T., Deng, L., & Yonezawa, T. (2022). Control of nanoparticles synthesized via vacuum sputter deposition onto liquids: A review. Soft Matter, 18(1), 19–47.

Pentimalli, M., Bellusci, M., & Padella, F. (2015). High□energy ball milling as a general tool for nanomaterials synthesis and processing. Handbook of Mechanical Nanostructuring, 663–679.

Pinto, M., Ramalho, P., Moreira, N., Gonçalves, A., Nunes, O., Pereira, M., & Soares, O. (2020). Application of magnetic nanoparticles for water purification. Environmental Advances, 2, 100010.

Rahmani, M., Mirzaee, O., Tajally, M., & Loghman-Estarki, M. R. (2018). A comparative study of synthesis and spark plasma sintering of YAG nano powders by different co-precipitation methods. Ceramics International, 44(9), 10035–10046.

Rai, A., Prabhune, A., & Perry, C. C. (2010). Antibiotic mediated synthesis of gold nanoparticles with potent antimicrobial activity and their application in antimicrobial coatings. Journal of Materials Chemistry, 20(32), 6789–6798.

Rendón-Angeles, J. C., & Seong, G. (2023). Editorial for the Special Issue: "Hydrothermal Synthesis of Nanoparticles." Nanomaterials, 13(9), 1463.

Roniboss, A., Sindhu, S., Kennedy, L. J., & Arockiasamy, S. (2023). Synthesis and thermal properties of two novel cobalt (II) Schiff's base complexes as precursors for coating cobalt oxide (Co3O4) thin film by a plasma enhanced metallo-organic chemical vapour deposition. Journal of Molecular Structure, 1272, 134189.

Sadrolhosseini, A. R., Mahdi, M. A., Alizadeh, F., & Rashid, S. A. (2019). Laser ablation technique for synthesis of metal nanoparticle in liquid. Laser Technology and Its Applications, 63–83.

Saeedeh Ebad-Sichani, Paria Sharafi-Badr, Payam Hayati, & Mohammad Jaafar Soltanian-Fard. (2023). Nano-Fabrication Methods. In Dipti Ranjan Sahu (Ed.), Nanofabrication Techniques (p. Ch. 1). IntechOpen.

Saleh, T. A. (2020). Nanomaterials: Classification, properties, and environmental toxicities. Environmental Technology & Innovation, 20, 101067.

Sebastian, E. M., Jain, S. K., Purohit, R., Dhakad, S. K., & Rana, R. S. (2020). Nanolithography and its current advancements. 10th International Conference of Materials Processing and Characterization, 26, 2351–2356.

Shabalina, A. V., Svetlichnyi, V. A., & Kulinich, S. A. (2022). Green laser ablation-based synthesis of functional nanomaterials for generation, storage, and detection of hydrogen. Current Opinion in Green and Sustainable Chemistry, 33, 100566.

Sharma, E., Rathi, R., Misharwal, J., Sinhmar, B., Kumari, S., Dalal, J., & Kumar, A. (2022). Evolution in lithography techniques: Microlithography to nanolithography. Nanomaterials, 12(16).

Sharma, N., Ojha, H., Bharadwaj, A., Pathak, D. P., & Sharma, R. K. (2015). Preparation and catalytic applications of nanomaterials: A review. Rsc Advances, 5(66), 53381–53403.

Singh, V., Yadav, P., & Mishra, V. (2020). Recent Advances on Classification, Properties, Synthesis, and Characterization of Nanomaterials. In Green Synthesis of Nanomaterials for Bioenergy Applications (pp. 83–97). John Wiley & Sons Ltd.

Srujana, S., & Bhagat, D. (2022). Chemical-based synthesis of ZnO nanoparticles and their applications in agriculture. Nanotechnology for Environmental Engineering, 7(1), 269–275.

Stolle, A., & Ranu, B. (2014). Ball milling towards green synthesis: Applications, projects, challenges. Royal Society of Chemistry.

Sun, L., Yuan, G., Gao, L., Yang, J., Chhowalla, M., Gharahcheshmeh, M. H., Gleason, K. K., Choi, Y. S., Hong, B. H., & Liu, Z. (2021). Chemical vapour deposition. Nature Reviews Methods Primers, 1(1), 1–20.

Theiss, F. L., Ayoko, G. A., & Frost, R. L. (2016). Synthesis of layered double hydroxides containing Mg2+, Zn2+, Ca2+ and Al3+ layer cations by co-precipitation methods—A review. Applied Surface Science, 383, 200–213.

Thompson, L. F. (1983). An introduction to lithography. In Introduction to Microlithography (Vol. 219, pp. 1–13). American Chemical Society.

Ullattil, S. G., & Periyat, P. (2017). Sol-gel synthesis of titanium dioxide. Sol-Gel Materials for Energy, Environment and Electronic Applications, 271–283.

Vaseghi, Z., & Nematollahzadeh, A. (2020). Nanomaterials. In Green Synthesis ofNanomaterials for Bioenergy Applications (pp. 23–82).

Wang, F., Zhao, H., Liang, J., Li, T., Luo, Y., Lu, S., Shi, X., Zheng, B., Du, J., & Sun, X. (2020). Magnetron sputtering enabled synthesis of nanostructured materials for electrochemical energy storage. Journal of Materials Chemistry A, 8(39), 20260–20285.

Wang, S., Li, X., Wu, J., Wen, W., & Qi, Y. (2018). Fabrication of efficient metal halide perovskite solar cells by vacuum thermal evaporation: A progress review. Current Opinion in Electrochemistry, 11, 130–140.

Wang, Y., Pan, J.-A., Wu, H., & Talapin, D. V. (2019). Direct wavelength-selective optical and electron-beam lithography of functional inorganic nanomaterials. ACS Nano, 13(12), 13917–13931.

Wu, D., S Rajput, N., & Luo, X. (2016). Nanoimprint lithography-the past, the present and the future. Current Nanoscience, 12(6), 712–724.

Yan, X.-T., & Xu, Y. (2010). Chemical vapour deposition: An integrated engineering design for advanced materials. Springer Science & Business Media.

Zheng, J., Jiang, Q., & Lian, J. (2011). Synthesis and optical properties of flower-like ZnO nanorods by thermal evaporation method. Applied Surface Science, 257(11), 5083–5087.

2

Metal Nanostructures in Surface-Enhanced Raman Spectroscopy A Promising Approach for Chemical Detection

***Priya Jasrotia*[1*], *Tanuj Kumar*[1] *and Jyoti*[2]**

[1]*Department of Nanoscience and Materials, Central University of Jammu-181143 Jammu, India*

[2]*Department of Physics, Government College for Women, Gurawara, Rewari-123035 India*

Abstract

Surface-Enhanced Raman Spectroscopy (SERS) may detect and identify chemical and biological species ultrasensitively. Through localized surface plasmon resonance (LSPR) effects, metal nanostructures boost analyte Raman signals. This chapter covers current advances and essential principles in metal nanostructures for SERS chemical detection. It discusses SERS fundamentals and enhancing processes, such as metal nanostructure size, shape, and composition. Top-down and bottom-up processes including lithography, chemical synthesis, and self-assembly are used to create repeatable and controllable metal nanostructures. The complex interactions between light, metal nanostructures, and analyte molecules are best understood by combining theoretical modeling and experimental observations. Metal nanostructures are shown to be successful in SERS-based chemical sensing for environmental monitoring and biological diagnostics in case studies. Functionalization with molecular linkers and integration with microfluidic technologies can improve SERS substrate stability, sensitivity, and selectivity. The chapter finishes with a look ahead at nanomaterial fabrication, device downsizing, and integration with complementary analytical methods. This chapter is useful for academics and practitioners interested in using metal nanostructures to improve SERS-based chemical detection methods.

Keywords: *Nanostructures, Surface-Enhanced Raman Spectroscopy (SERS), Electromagnetic, Chemical, Sensing.*

---

* *Correspondence author e-mail: priyajasrotia996@gmail.com*

1. Fundamentals of Surface Enhanced Raman Scattering: Principles and Applications

Surface-enhanced Raman scattering (SERS) stands as a widely adopted sensing technique, wherein the inelastic light scattering of molecules (illustrated in Figure 1) experiences substantial amplification upon their adsorption onto textured metal surfaces, notably silver or gold nanoparticles (NPs). This augmentation can achieve magnification factors surpassing 10^8, enabling the realm of single molecule (SM) SERS (Langer *et al.*, 2019). Since its discovery over 40 years ago, the scientific community has gradually nurtured an interest in SERS, spurring the emergence of numerous spectroscopic methodologies leveraging the intensified local fields generated by plasmon excitation within the NPs. These methodologies extend to optical phenomena including fluorescence and nonlinear optics.

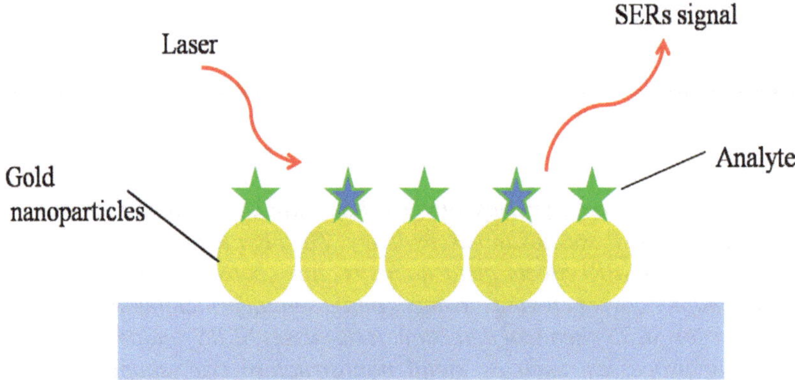

Fig. 1: SERS on gold/silver nanoparticles surface.

Utilizing the phenomenon of inelastic scattering of laser light, surface-enhanced Raman spectroscopy (SERS) emerges as a powerful emission technique that elicits frequency-shifted spectrum peaks based on molecular vibrational modes. In order to amplify Raman signals, analyte molecules necessitate adherence to roughened surfaces typically composed of silver, copper, or gold, serving as SERS substrates. The resulting SERS spectra boast narrow peaks and exceptional resolution, facilitating the simultaneous analysis of multiple components. SERS offers advantages such as streamlined sample handling, rapid analysis, prompt analyte identification, and the convenience of portable Raman spectrometers (Allakhverdiev *et al.*, 2024).

Nonetheless, SERS is not devoid of limitations. Challenges include the requirement for intimate contact between analyte and enhancing surface, signal degradation over time, limited selectivity, single-use substrates, and fluctuations in SERS signals within substrates. Despite these drawbacks, SERS amalgamates the benefits of Raman spectroscopy, enabling the simultaneous detection of numerous analytes, analysis of biological fluids, molecule detection, non-destructive analysis, minimal sample preparation, and on-site analysis with portable instruments.

Traditionally associated with metals like copper, gold, and particularly silver, SERS has recently witnessed explorations into the utilization of other metals and materials (Mosier-Boss, 2017). While generalization to alternative metals and materials has been under scrutiny since the inception of SERS, only in recent years have well-characterized studies with consistent enhancements emerged. Notably, contemporary research reveals the detectability of SERS for metals such as rhodium and platinum, which hold significance in the chemical industry due to their catalytic roles. Despite limited research, metals like ruthenium and aluminum remain intriguing due to their technological applications. The correlation between these findings and SERS enhancement mechanisms presents an intriguing area for further investigation.

In terms of signal strength, Raman scattering and fluorescence present competitive alternatives, exemplified by the landmark discovery of single-molecule SERS in 1997 (Lin *et al.*, 2023). The evolution of SERS has been propelled by advancements in Raman instrumentation and nanofabrication over the past two decades. While silver substrates are commonly employed in SERS, copper and gold substrates also offer enhancement capabilities. Various gold-deposited metal films, including silver island films and silver-coated beads, have been developed to generate roughened surfaces conducive to SERS.

Successful identification of chemical analytes through SERS hinges on meeting the following prerequisites:

a) Substrate Quality: The efficacy of enhancement, reproducibility, durability, and lifespan critically rely on the quality of the substrate. Optimal SERS performance necessitates a roughened surface.

b) Analyte Absorption: Efficient adhesion of the analyte to the substrate surface is imperative. To ensure clear signals, the analyte should possess a higher SERS cross-section than potential interferents.

c) Excitation Regulation: Precise control of excitation intensity is essential to mitigate surface photochemistry, which could interfere with SERS data integrity.

d) Averaging for Quantitative Measurements: Achieving precise quantitative measurements is best accomplished through the averaging of numerous events. This can be attained by controlling the number of active sites analyzed, as well as the volume and duration of interrogation.

e) Standardization: Incorporating a standard reference material is strongly recommended to monitor any variations stemming from substrate discrepancies. This standardization process enhances the repeatability and reliability of quantitative SERS readings.

2. The Basics of SERS Electromagnetic Enhancement

The enhancement of Raman intensities correlates with the intensity of the incident field and the derivative of polarizability. Surface-Enhanced Raman Spectroscopy (SERS) commonly operates through two primary mechanisms: one focuses on enhancing field intensity via plasmon resonance excitation, while the other accentuates polarizability through chemical effects such as charge-transfer excited states (Zhang, Wheeler, Schwartzberg, & Shi, 2014). The former method embodies the electromagnetic enhancement mechanism, whereby the enhancement factor E for each molecule is roughly determined.

$$\mathrm{E} = |\boldsymbol{E}(\omega)|^2|\boldsymbol{E}(\omega')|^2,$$

The symbol E(ω) denotes the local electric-field enhancement factor at the incident frequency ω, while E(ω') represents a comparable factor at the Stokes-shifted frequency ω'. In the context of single-molecule SERS (SMSERS), the maximal enhancement Emax holds significance over E, which is typically averaged across the particle surface area where molecules can adsorb to generate the reported enhancement factor in traditional SERS. It's crucial to recognize the substantial disparity between these two enhancement estimations, as E_{max} can be orders of magnitude greater than E.

Furthermore, it's common practice to estimate E by assuming equivalence between E(ω) and E(ω'), leading to $E = |E(\omega)|^4$. This assumption leverages the fact that the plasmon width often exceeds the Stokes shift. However, this approximation tends to overestimate the enhancement factor by a factor of three or more in SERS experiments involving isolated homogeneous particles.

3. SERS's Fundamental Enhancement Mechanism

As previously mentioned, coinage metals such as Ag, Au, and Cu are commonly employed as substrates in Surface-Enhanced Raman Spectroscopy (SERS) due to their optical absorptions within the visible spectrum (Gaur *et al.*, 2021). Among these metals, Ag and Au are particularly favored for SERS research owing to their well-defined nanostructures, robust plasmonic effects, and relative chemical inertness. The electromagnetic enhancement mechanism, one of the two predominant enhancement mechanisms in SERS, plays a significant role in elucidating the theoretical foundation of SERS. By employing an excitation wavelength resonant with the plasmon absorption profile of specific nanoparticles, surface plasmons are activated. This activation generates a potent electromagnetic field with an evanescent nature on the metal surface. Given that Raman intensity is proportionate to the square of the incident electromagnetic field amplitude, molecules in close proximity to a metallic surface experience a notable enhancement in their Raman modes.

The Raman scattered light, emitted in all directions and averaged for all molecule orientations, is collected and analyzed using a microscope objective, exploiting the emission characteristics akin to a Hertzian dipole. For Chemical Enhancement (CE) to manifest, there must be a chemical bond between the SERS substrate and the probe molecule. CE encompasses three main contributions to the chemical mechanism (Yuan *et al.*, 2017):

1. The incident light aligns with an electronic transition within the molecule, leading to a resonance Raman (RR) effect, contributing significantly (10^3–10^6).
2. Alternatively, the incident light can induce a charge-transfer (CT) effect when it resonates with a metal–molecule or molecule–metal transition.
3. The ground-state orbital overlap between the metal and the molecule results in a nonresonant chemical (CHEM) effect, accounting for a lesser contribution (≤ 10–10^4).

Figure 2A illustrates the interaction between incident photons and molecules, which leads to the generation of emission photons when exposed to laser irradiation. Figure 2B illustrates the energy differences resulting from the vibration and rotation states of the molecules. Only the Raman signal of molecules adsorbed on or very close to the surface of the nanostructures can be amplified, since SERS enhancement decreased almost exponentially with the distance between the interesting point and nanostructures (Figure 2D).

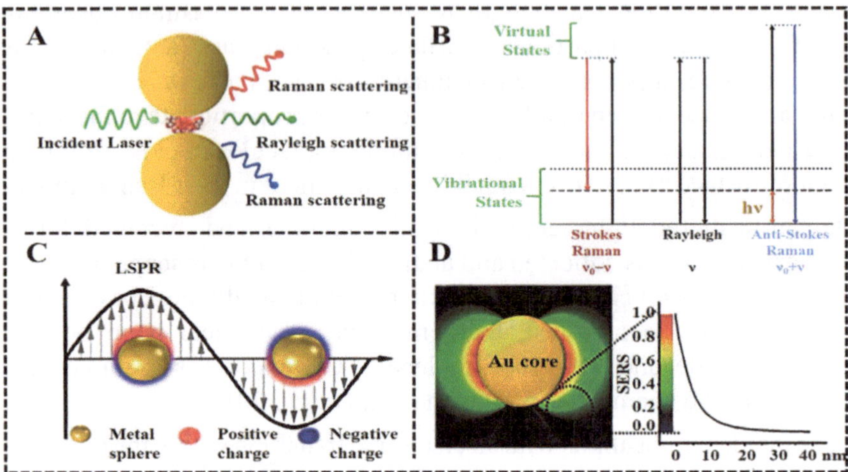

Fig. 2: (A) Diagram showing how a molecule situated between two metallic nanoparticles scatters light using Rayleigh and Raman techniques (Long, Ju, Yang, & Li, 2022). **(B)** Quantum energy represented by the Jablonski diagram (Long *et al.*, 2022). **(C)** Illustrations depicting the phenomenon of localized surface plasmon resonance (LSPR) (Cui, Zhang, Yang, Zhang, & Zhu, 2019). **(D)** Simulation of the electromagnetic field distribution and its impact on the surface-enhanced Raman scattering (SERS) enhancement as a function of distance from the nanoparticle surface using the finite difference time domain method (Cui *et al.*, 2019).

4. Electromagnetic and Chemical Enhancements

In 1973, Martin Fleischmann, Patrick J. Hendra, and A. James McQuillan from the University of Southampton's UK Department of Chemistry observed Surface-Enhanced Raman Spectroscopy (SERS) from pyridine adsorbed on electrochemically roughened silver (Fleischmann, Hendra, & McQuillan, 1974). This groundbreaking discovery has garnered over 7,000 citations and earned the University of Southampton a National Chemical Landmark plaque from the Royal Society of Chemistry in commemoration of its 40th anniversary. Despite numerous attempts at research, unanswered questions persist, underscoring the complexity of the Surface-Enhanced Raman Scattering (SERS) phenomenon.

In 1977, two separate research teams made notable discoveries independently, each offering an explanation for the amplified signal observed. Albrecht and Creighton (Albrecht & Creighton, 1977) proposed a charge-transfer effect, while Jeanmaire and Richard Van Duyne (Jeanmaire & Van Duyne, 1977) suggested an electromagnetic effect. The electromagnetic enhancement process, which is widely accepted and acknowledged, stands as the most prevalent theory. According to this theory, the Localized Surface Plasmon Resonance (LSPR) effect generates local fields on the surface of noble metal nanoparticles, leading to the amplification of Raman signals. With enhancement factors ranging from 10^6 to 10^8 times, this effect has been established as the primary mechanism driving Surface-Enhanced Raman Spectroscopy (SERS) improvement. Figure 3 illustrates the electromagnetic and chemical enhancement mechanism of SERS.

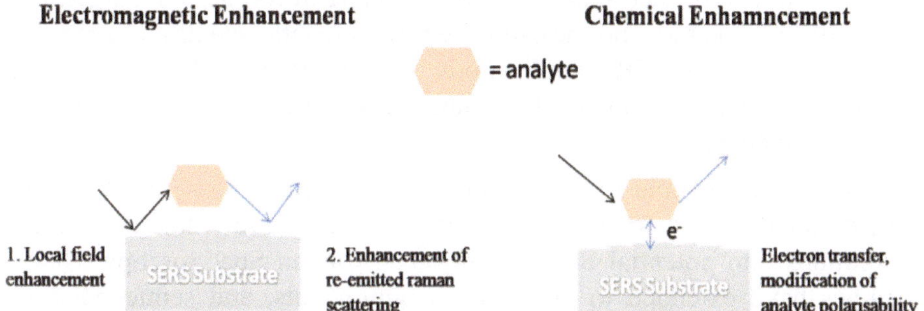

Fig. 3: SERS mechanisms of electromagnetic and chemical enhancement.

The term "plasmon" refers to the coherent oscillations of delocalized electrons in noble metals, explicable through classical physics concepts. Surface plasmons manifest as collective electron oscillations relative to the stationary positive ions in the metal, occurring within nanoparticles with diameters equal to or greater than the incident light wavelength used to excite them. While these oscillations are typically dampened by the metallic lattice or defects, they become less attenuated when the metal dimensions are smaller than the incident wavelength, allowing them to respond to light. This observation underscores that the effect's impact varies depending on the bulk material size, resulting in either a limited intensity increase or negligible effects in vast quantities of precious metals (Tan, Wei, Shang, & Yang, 2021).

On thin-film precious metal surfaces, there is a notable rise in electron density, leading to the creation of a potent electromagnetic field near the surface, known as the Surface Plasmon Resonance (SPR) effect. This electric field strengthens the polarization of Raman scattering at the analyte's surface, thereby amplifying Raman emission intensity. The Localized Surface Plasmon Resonance (LSPR) or Surface Plasmon Resonance (SPR) effect of the SERS substrate primarily drives this electric field. Consequently, there has been considerable interest in employing precious metal nanomaterials exhibiting notable LSPR effects for SERS. Signal amplification is most pronounced in regions with fissures, gaps, or sharp features in plasmonic materials, where the strongest electromagnetic field is concentrated. Enhancement factors as high as 10^{10}–10^{11} times have been observed on well-tuned surfaces.

For Surface-Enhanced Raman Spectroscopy (SERS) to operate effectively, molecules must interact with the surfaces of plasmonic nanostructures, typically composed of copper (Cu), silver (Ag), or gold (Au) in conventional substrates. Gold and silver are commonly favored for SERS substrates due to their air stability, whereas copper exhibits higher reactivity. All three metals exhibit Localized Surface Plasmon Resonances (LSPRs) across a significant portion of the visible and near-infrared wavelength spectrum, rendering them well-suited for SERS applications. This is particularly advantageous considering that the majority of Raman measurements occur within these wavelength ranges.

A material is deemed plasmonic if its dielectric constant possesses a modest, positive imaginary component alongside a negative real component. Explorations into potential Surface-Enhanced Raman Spectroscopy (SERS) substrates have extended to graphene, quantum dots, and semiconductors, aiming to uncover novel materials and technologies. These investigations reveal that enhancements observed in such substrates stem solely from chemical origins, arising from resonances with charge transfer bands, inter-band transitions, and π-π stacking interactions. However, these substrates are not anticipated to yield enhancement factors exceeding 10^3 due to the absence of substantial electromagnetic enhancement. The emergence of infrared plasmonic material graphene holds promise for the future viability of SERS on graphene.

Charge-transfer states offer a pathway for resonant excitation akin to resonance Raman spectroscopy, augmenting the likelihood of a Raman transition. This mechanism is contingent upon the analyte and is site-specific. Chemical amplification is not feasible unless the molecule promptly adheres to a roughened surface. Understanding electromagnetic enhancement necessitates consideration of the size, structure, and composition of nanoscale roughness features, which dictate the resonance frequency of conduction electrons in metallic nanostructures. Upon excitation of localized surface plasmon resonances (LSPRs), large electromagnetic fields are generated at the surface of roughness features, selectively absorbing and scattering resonant electromagnetic radiation. The average enhancement factor from electromagnetic enhancement is estimated at 10,000, contingent upon these electromagnetic fields containing Raman-active compounds. Extensive theoretical modeling has thoroughly investigated how noble-metal nanoparticles may amplify electromagnetic fields, often employing a quasistatic method to describe the interaction with the incoming electromagnetic field, as exemplified by the simple case of a single sphere.

$$E^2 \propto E_0^2 \left| \frac{\in_m - \in_0}{\in_m - 2\in_0} \right|^2$$

In this equation, the electric field strength at the surface of the sphere is denoted by E, the incident field strength is represented by E_o, the wavelength-dependent dielectric constant of the metal sphere is indicated by $\square_m$, and the dielectric constant of the surrounding local environment is symbolized by $\square_o$. This relationship illustrates that when the condition □m= −2$\square_0$ is satisfied, there is a significant increase in the electric field magnitude at the sphere's surface. This heightened field enhancement is achieved for silver and gold at specific wavelengths within the visible and near-infrared range, meeting the criteria for localized surface plasmon resonance (LSPR). Similarly, the nanoparticle's extinction and scattering cross sections can be linked. By optimizing these cross sections at resonance wavelengths, one can spectroscopically discern the activation of LSPR (Petryayeva & Krull, 2011).

Surface-Enhanced Raman Spectroscopy (SERS) exhibits another intrinsic enhancement mechanism wherein the Raman signal is amplified through electron transfer between the analyte and the SERS-enhanced substrate via chemical bonds. This process, known as the Chemical Mechanism (CM), involves the resonance of charge-transfer electrons in various states with the incident wavelength, thereby amplifying the Raman signal of analyte molecular oscillations. However, these mechanisms display a strong dependence on the Raman excitation wavelength, posing challenges in identifying and confirming their contribution to signal enhancement. Conversely, electromagnetic phenomena play a significant role in intensifying the Surface-Enhanced Raman Scattering (SERS) mechanism due to the substantial electromagnetic field generated by plasmons. Evidence supports the significant amplification of electromagnetic waves owing to the direct proportionality between the enhancement coefficient arising from the local fields of metal nanoparticles and the fourth power of the electromagnetic field (Ding, You, Tian, & Moskovits, 2017). Surface-enhanced Raman spectroscopy (SERS) has been extensively explored across various domains, particularly in sensor development aimed at identifying biological molecules, proteins, DNA, chemical substances, and cancer, among other applications.

5. Experimental Techniques for Analyzing SERS Advancements

i. **Single Molecule SERS:** The transition from analyzing bulk samples to detecting single molecules in SERS was achieved approximately 15 years ago. Single molecule SERS (SMSERS) offers several advantages over single molecule fluorescence, including reduced sample bleaching and the provision of precise chemical information resembling a fingerprint (Jain, 2021).

ii. **UV SERS:** SERS can be employed for resonantly detecting various biological substances, such as DNA bases and protein residues, in the UV frequency range. While most metals exhibit stronger enhancements in the visible or near-infrared range, activating SERS in the UV range enables resonant detection of substrates like gold (Au) and silver (Ag), commonly utilized in SERS (NIR). Biological molecules often possess electronic resonances in the deep ultraviolet (DUV) region, making the use of DUV light particularly advantageous for their investigation. DUV light enhances both SERS and electronic resonance (Jha, 2013).

iii. **Ultrafast and Stimulated SERS:** Current SERS research primarily focuses on integrating ultrafast spectroscopic techniques with SERS benefits. Femtosecond stimulated Raman spectroscopy (FSRS) provides detailed structural information within a femtosecond timeframe through coherent and stimulated vibration. Surface-enhanced femtosecond stimulated Raman spectroscopy (SE-FSRS) has investigated various photo-driven processes to examine their structural dynamics, achieving enhancement factors between 10^4 and 10^6 in recent experiments with SERS nanotags (Frontiera, Henry, Gruenke, & Van Duyne, 2011).

iv. **Tip-Enhanced Raman Spectroscopy (TERS):** TERS employs a sharp metallic tip illuminated by laser light to amplify the electromagnetic field. By approaching the sample of interest with the tip, TERS generates a localized zone of SERS enhancement, enabling nanoscale spatial resolution for structural and compositional analysis. Atomic force microscopes (AFMs) or scanning tunneling microscopes (STMs) are commonly integrated with optical excitation and collecting optics for TERS investigations (Bhushan & Marti, 2017).

6. Utilization of Electromagnetic and Chemical Enhancements in Surface-Enhanced Raman Spectroscopy (SERS)

The strength of SERS lies in its ability to identify chemical species and provide structural details across various scientific domains, including electrochemistry, polymer and materials research, biochemistry and biosensing, and catalysis (Schlücker, 2014). Enhanced surface Raman scattering has garnered increasing attention for developing innovative techniques and tools for biomedical purposes, aiming for efficient diagnostics and affordable methods capable of detecting various disease-associated biomarkers at the point-of-care. To compute hyper-Raman spectra, a comparable theory is formulated at the model A level. In the context of SERS and SEHRS, enhancements on the scale of

10^4 and 10^9 are achieved by incorporating both chemical and electromagnetic mechanisms at the model A level. Metal nanoparticles are typically integrated into SERS platforms to generate the necessary electromagnetic enhancements. Natelson *et al.* outline the intricate plasmonic physics involved in these extended systems and highlight recent advancements in the field, such as tip-enhanced Raman spectroscopy (TERS), mechanical break junctions, and electromigrated junctions .Natelson *et al.* (Natelson, Li, & Herzog, 2013) also detail the process of conducting in situ gap-enhanced Raman measurements of molecular-scale junctions while subjecting these structures to electronic transport. Strong localized increases in electromagnetic fields are observed in individual Al nanostructures, leading to significant SERS activity. Lay *et al.* explore aluminum nanostructures for flexible molecular security label micropatterning, demonstrating robust visible-range surface plasmon activity (Lay *et al.*, 2018). Their work underscores the adaptability of Al-based structures in current, cost-effective chemical nano-analytics and forgery protection.

According to Wang *et al.*, graphene proves to be a valuable component of SERS-active structures, especially in biosensor applications, and can serve as a fundamental element for SERS processes (P. Wang, Liang, Zhang, Schroeder, & Xie, 2013). A biocompatible graphene-based SERS platform with a local enhancement factor exceeding 10^{10} facilitates a high density of hot spots. SERS analysis of volatile organic compounds (VOCs) is a compelling method for fast disease screening; for example, aldehydes in exhaled breath from humans have been utilized as markers for lung cancer. However, two significant barriers to VOC-based diagnostics are the low affinity of gaseous molecules for SERS substrates and the weak Raman scattering. For the SERS detection of gaseous aldehydes, Qiao and colleagues employed an Au@ZIF-8 3D structure made by depositing a ZIF-8 layer over self-assembled gold superparticles (GSPs) (Figure 4A) (Qiao *et al.*, 2018). The local heating effect has a major impact on the stability and functionality of nanoscale electronic devices, particularly at current-carrying molecule junctions with low heat capacity. Based on the non-equilibrium occupancy of the vibrational modes, surface emissivity reflectance spectroscopy (SERS) can be a spectroscopic method for heat monitoring since the effective temperature at the molecular junction is connected to the inelastic scattering of electrons and can be reflected by molecular vibrations. Ioffe and colleagues used a delicate design to illustrate this idea, utilizing a self-assembled monolayer of 4,40-biphenyldithiol molecules as linkers between two neighboring sides of an Ag nanostructure that was "on-edge" (Figure 4B) (Ioffe *et al.*, 2008). In a sealed transparent Raman cell configuration, a monolayer of Au nanoparticles (36.7 ± 2.8 nm) self-assembled on a Ni current collector was enclosed as the plasmonic SERS substrate. Yang and colleagues reported a SERS detection of the electrolyte solvent structure at the solid-liquid interface (Figure 4C) (Yang *et al.*, 2018).

A plasmonic metal-semiconductor hybrid was shown to exhibit the PIERS effect, according to Glass *et al.* The enhancement was caused by surface oxygen vacancies (VO) that developed on semiconductor surfaces after they were pre-irradiated with high-energy photons. On the other hand, the assessment of atomic VO dynamics on the surface region of metal oxides is made possible by such a chemical improvement in SERS (Figure 4D) (Glass *et al.*, 2019).

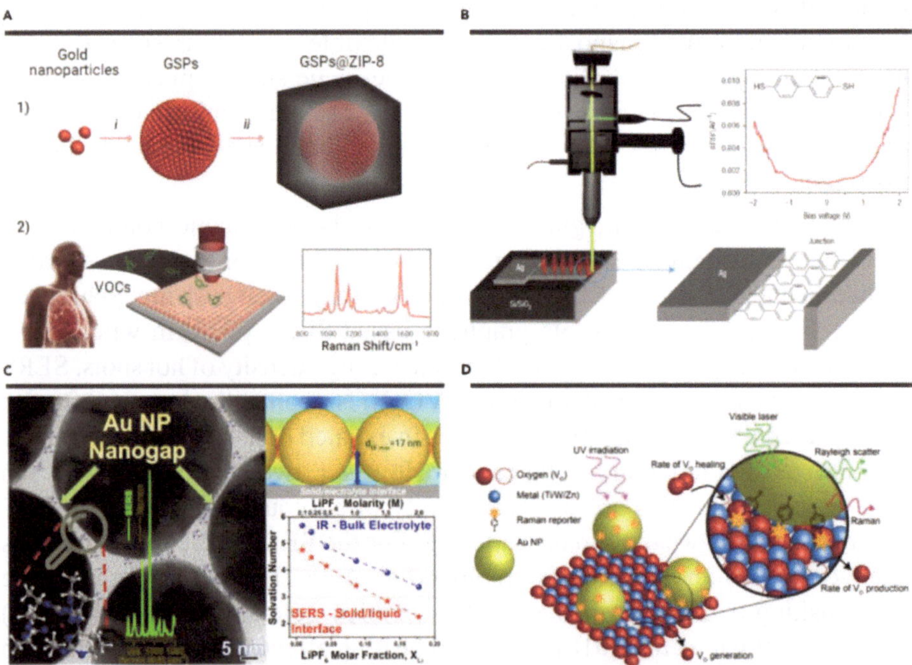

Fig. 4: Common Uses of SERS **(A)** Au@ZIF-8 3D Structure for the detection of VOCs linked to cancer. **(B)** Diagram showing a "on-edge" connection with a Raman setup. **(C)** SERS identification of the solid-liquid interface electrolyte solvation structure. **(D)** SERS tracking metal-oxide semiconductor surface oxygen vacancy dynamics produced by photons (Qiao *et al.*, 2018) (Ioffe *et al.*, 2008) (Yang *et al.*, 2018) (Glass *et al.*, 2019).

7. Challenges and Future Directions in SERS Enhancement Mechanisms

Despite the substantial expansion of SERS applications in biosensing over the past decade, its widespread adoption as a routine clinical diagnostic tool has not kept pace with this rapid growth. Qian *et al.* introduce a novel class of colloidal metal nanoparticles capable of enhancing surface-enhanced Raman scattering

(SERS) efficiency by up to 10^{14}–10^{15} fold. Notably, understanding of enhancing mechanisms has significantly progressed, with conclusive evidence supporting the single-molecule origin of fluctuating SERS signals. Wang *et al.* provide a comprehensive analysis of interconnected SERS devices, elucidating coupling enhancement processes, materials, and techniques (H. Wang *et al.*, 2021). Despite recent advancements in equipment and substrate design improving the utility, repeatability, and quantitative capabilities of SERS, certain challenges persist. Intermolecular interactions, surface selection criteria, and the solubility and accessibility of target molecules are explored in relation to developments in quantitative SERS detection by Norton *et al.* (Norton, Phan, Gibbons, & Haes, 2022). The field of SERS holds promise for both real-world applications and laboratory research challenges. Future research in SERS enhancement techniques may focus on integrating new nanomaterials with tailored characteristics to achieve enhanced electromagnetic and chemical effects. This may involve the development of sophisticated nanostructures such as hierarchical nanostructures, core-shell nanoparticles, or nanorods engineered for optimal SERS performance. Advancements in theoretical modeling tools will be essential for gaining a comprehensive understanding and predicting SERS enhancement mechanisms. Future studies may involve the creation of advanced computer models capable of accurately simulating interactions between plasmonic nanostructures, electromagnetic fields, and analyte molecules, providing insights into complex enhancement processes. Surface engineering techniques offer opportunities to modify surface characteristics and enhance SERS performance for specific applications. These techniques may include functionalizing SERS substrates with specific ligands, polymers, or biomolecules. Research into cutting-edge surface modification techniques could enhance the stability, repeatability, and biocompatibility of SERS substrates, leading to innovations in drug delivery, medical diagnostics, and biological sensing (Das, Managò, Mangini, & De Luca, 2021). Moving forward, the primary objectives for SERS enhancement mechanisms should be to deepen fundamental knowledge, explore new applications, and address real-world challenges, enabling SERS to realize its full potential as a versatile and powerful analytical tool.

8. Conclusion

In conclusion, the comprehensive exploration of Surface-Enhanced Raman Scattering (SERS) has elucidated its foundational principles, including chemical and electromagnetic enhancements, elucidating their complementary functions and applications across diverse fields. This discourse has underscored the significance of experimental approaches for fabricating and characterizing substrates, offering fresh insights into the mechanisms driving SERS enhancement. Despite the remarkable promise of SERS in materials science, environmental monitoring, and biomedical diagnostics, challenges pertaining to repeatability and standardization persist. Realizing the full potential of SERS as a versatile analytical tool with far-reaching implications for both research and industry hinges on overcoming these hurdles and fostering interdisciplinary collaboration to deepen fundamental understanding.

References

Albrecht, M. G., & Creighton, J. A. (1977). Anomalously intense Raman spectra of pyridine at a silver electrode. Journal of the American Chemical Society, 99(15), 5215-5217.

Allakhverdiev, E. S., Kossalbayev, B. D., Sadvakasova, A. K., Bauenova, M. O., Belkozhayev, A. M., Rodnenkov, O. V., & Allakhverdiev, S. I. (2024). Spectral insights: Navigating the frontiers of biomedical and microbiological exploration with Raman spectroscopy. Journal of Photochemistry and Photobiology B: Biology, 112870.

Bhushan, B., & Marti, O. (2017). Scanning probe microscopy—principle of operation, instrumentation, and probes. Nanotribology and Nanomechanics: An Introduction, 33-93.

Cui, L., Zhang, D., Yang, K., Zhang, X., & Zhu, Y.-G. (2019). Perspective on surface-enhanced Raman spectroscopic investigation of microbial world. Analytical Chemistry, 91(24), 15345-15354.

Das, G. M., Managò, S., Mangini, M., & De Luca, A. C. (2021). Biosensing using SERS active gold nanostructures. Nanomaterials, 11(10), 2679.

Ding, S.-Y., You, E.-M., Tian, Z.-Q., & Moskovits, M. (2017). Electromagnetic theories of surface-enhanced Raman spectroscopy. Chemical Society Reviews, 46(13), 4042-4076.

Fleischmann, M., Hendra, P. J., & McQuillan, A. J. (1974). Raman spectra of pyridine adsorbed at a silver electrode. Chemical physics letters, 26(2), 163-166.

Frontiera, R. R., Henry, A.-I., Gruenke, N. L., & Van Duyne, R. P. (2011). Surface-enhanced femtosecond stimulated Raman spectroscopy. The Journal of Physical Chemistry Letters, 2(10), 1199-1203.

Gaur, R., Manikandan, P., Manikandan, D., Umapathy, S., Padhy, H. M., Maaza, M., & Elayaperumal, M. (2021). Noble metal ion embedded nanocomposite glass materials for optical functionality of UV–visible surface plasmon resonance (SPR) surface-enhanced raman scattering (SERS) X-ray and electron microscopic studies: An overview. Plasmonics, 16, 1461-1493.

Glass, D., Cortés, E., Ben□Jaber, S., Brick, T., Peveler, W. J., Blackman, C. S., & Maier, S. A. (2019). Dynamics of photo□induced surface oxygen vacancies in metal□oxide semiconductors studied under ambient conditions. Advanced Science, 6(22), 1901841.

Ioffe, Z., Shamai, T., Ophir, A., Noy, G., Yutsis, I., Kfir, K., & Selzer, Y. (2008). Detection of heating in current-carrying molecular junctions by Raman scattering. Nature Nanotechnology, 3(12), 727-732.

Jain, S. (2021). Surface Enhanced Raman Spectroscopy: Biosensing and Diagnostic Technique for Healthcare Applications: Bentham Science Publishers.

Jeanmaire, D. L., & Van Duyne, R. P. (1977). Surface Raman spectroelectrochemistry: Part I. Heterocyclic, aromatic, and aliphatic amines adsorbed on the anodized silver electrode. Journal of electroanalytical chemistry and interfacial electrochemistry, 84(1), 1-20.

Jha, S. K. (2013). Deep-ultraviolet surface-enhanced Raman scattering and fluorescence using aluminum nanostructures. ETH Zurich.

Langer, J., Jimenez de Aberasturi, D., Aizpurua, J., Alvarez-Puebla, R. A., Auguié, B., Baumberg, J. J., & Brolo, A. G. (2019). Present and future of surface-enhanced Raman scattering. ACS nano, 14(1), 28-117.

Lay, C. L., Koh, C. S. L., Wang, J., Lee, Y. H., Jiang, R., Yang, Y., & Ling, X. Y. (2018). Aluminum nanostructures with strong visible-range SERS activity for versatile micropatterning of molecular security labels. Nanoscale, 10(2), 575-581.

Lin, C., Li, Y., Peng, Y., Zhao, S., Xu, M., Zhang, L., & Yang, Y. (2023). Recent development of surface-enhanced Raman scattering for biosensing. Journal of Nanobiotechnology, 21(1), 149.

Long, L., Ju, W., Yang, H.-Y., & Li, Z. (2022). Dimensional design for surface-enhanced raman spectroscopy. ACS Materials Au, 2(5), 552-575.

Mosier-Boss, P. A. (2017). Review of SERS substrates for chemical sensing. Nanomaterials, 7(6), 142.

Natelson, D., Li, Y., & Herzog, J. B. (2013). Nanogap structures: combining enhanced Raman spectroscopy and electronic transport. Physical Chemistry Chemical Physics, 15(15), 5262-5275.

Norton, R. D., Phan, H. T., Gibbons, S. N., & Haes, A. J. (2022). Quantitative surface-enhanced spectroscopy. Annual review of physical chemistry, 73, 141-162.

Petryayeva, E., & Krull, U. J. (2011). Localized surface plasmon resonance: Nanostructures, bioassays and biosensing—A review. Analytica Chimica Acta, 706(1), 8-24.

Qiao, X., Su, B., Liu, C., Song, Q., Luo, D., Mo, G., & Wang, T. (2018). Selective surface enhanced Raman scattering for quantitative detection of lung cancer biomarkers in superparticle@ MOF structure. Advanced Materials, 30(5), 1702275.

Schlücker, S. (2014). Surface□Enhanced raman spectroscopy: Concepts and chemical applications. Angewandte Chemie International Edition, 53(19), 4756-4795.

Tan, L. L., Wei, M., Shang, L., & Yang, Y. W. (2021). Cucurbiturils□mediated noble metal nanoparticles for applications in sensing, sers, theranostics, and catalysis. Advanced Functional Materials, 31(1), 2007277.

Wang, H., Liu, Y., Rao, G., Wang, Y., Du, X., Hu, A., & Xiong, J. (2021). Coupling enhancement mechanisms, materials, and strategies for surface-enhanced Raman scattering devices. Analyst, 146(16), 5008-5032.

Wang, P., Liang, O., Zhang, W., Schroeder, T., & Xie, Y. H. (2013). Ultra□sensitive graphene□ plasmonic hybrid platform for label- free detection. Advanced Materials, 25(35), 4918-4924.

Yang, G., Ivanov, I. N., Ruther, R. E., Sacci, R. L., Subjakova, V., Hallinan, D. T., & Nanda, J. (2018). Electrolyte solvation structure at solid–liquid interface probed by nanogap surface-enhanced Raman spectroscopy. ACS Nano, 12(10), 10159-10170.

Yuan, Y., Panwar, N., Yap, S. H. K., Wu, Q., Zeng, S., Xu, J., & Yong, K.-T. (2017). SERS-based ultrasensitive sensing platform: An insight into design and practical applications. Coordination Chemistry Reviews, 337, 1-33.

Zhang, J. Z., Wheeler, D. A., Schwartzberg, A. M., & Shi, J. (2014). Basics and practice of surface enhanced Raman scattering (SERS) and tip enhanced Raman scattering (TERS). Biomedical Spectroscopy and Imaging, 3(2), 121-159.

3

Nanomaterials for Sensing and Detection

***Riddhi Garg*[1], *Vandana Jhalora*[1], *Harshita Sachdeva*[2] and *Renu Bist*[1*]**

[1]Department of Zoology, Centre of Advanced Studies, University of Rajasthan Jaipur-302004

[2]Department of Chemistry, Centre of Advanced Studies, University of Rajasthan Jaipur-302004

Abstract

Nanomaterials have emerged as promising candidates for revolutionizing sensing and detection technologies, offering unprecedented capabilities in sensitivity, selectivity, and miniaturization. The current chapter provides a concise overview of recent advancements in the utilization of nanomaterials for sensing and detection applications. It encompasses a diverse range of nanomaterials, including nanoparticles, nanotubes, and nanowires, each exhibiting unique properties that enable enhanced performance in sensing applications. The integration of nanomaterials into sensing platforms has demonstrated remarkable progress in various fields, such as environmental monitoring, healthcare diagnostics & security applications. The exceptional surface-to-volume ratios and distinctive electronic, optical, and catalytic properties of nanomaterials play pivotal roles in enhancing sensor performance. Moreover, the design and engineering of nanomaterial-based sensors have facilitated the development of highly efficient and selective detection systems. Besides highlighting the key challenges, the future prospects in the field, have also emphasized the need for continued research to address issues such as stability, reproducibility, and scalability of nanomaterials in sensing and detection. As nanomaterial-based sensing technologies continue to evolve, their widespread implementation is anticipated to significantly impact diverse industries, paving the way for innovative solutions in real-time monitoring, disease diagnosis, and advanced security systems.

Keywords: *Sensing, nanomaterials, detection, nanowires, diagnosis, nano-tubes*

---

* *Corresponding author e-mail: renu_bisht22@yahoo.co.in*

1. Introduction

1.1 Overview of nanomaterials (NMs) and their unique properties

NMs, characterized by their nanoscale dimensions ranging from 1 to 100 nm in at least one dimension, represent a diverse and impactful class of materials. They exhibit unique physical, chemical, and biological properties that differ from their bulk counterparts, primarily due to quantum effects and increased surface area-to-volume ratios (Bayda *et al.*, 2020). These materials encompass various categories, including carbon-based NMs such as graphene, carbon nanotubes, and fullerenes, renowned for their exceptional mechanical, electrical, and thermal properties (Díez-Pascual, 2021). Metal-based NMs, such as nanoparticles (NPs) and nanowires, offer catalytic, optical, and magnetic functionalities, while semiconductor NMs like quantum dots and nanowires provide size-dependent optical and electronic properties suitable for diverse applications (Sweet *et al.*, 2012). Other NMs include nanocomposites (Camargo *et al.*, 2009), nanoporous materials (Y. Xu *et al.*, 2021), and biomaterials (Ajmal *et al.*, 2022), each with distinct applications spanning electronics, energy, catalysis, biomedical engineering, and environmental remediation.

1.2 Rationale for utilizing NMs in sensing and detection applications

The unique properties of NPs, driven by quantum effects, distinguish them from bulk materials, making them highly influential in the development of advanced sensing and detection technologies (Vo-Dinh *et al.*, 2001). Sensing involves detecting changes in physical or chemical properties relative to the surrounding environment, while detection refers to the output provided by the sensing mechanism. They are classified on the basis of the type of transducer such as piezoelectric (Tressler *et al.*, 1998), optical (Lepore & Delfino, 2022), colorimetry (F.-Q. Yang & Ge, 2023), electrochemical (Baranwal *et al.*, 2022), etc. and on the basis of type of receptors such as aptamers (McConnell *et al.*, 2020), antibodies (Byrne *et al.*, 2009), molecularly imprinted polymers (Yano & Karube, 1999), etc. Sensors find applications in diverse fields, from environmental monitoring to medical diagnostics. Understanding sensor classifications aids in developing innovative technologies to address societal needs.

These processes play significant roles across various fields such as filter technology, safety, surveillance, automotive, manufacturing, biotechnology, and tissue engineering. The discovery of new NMs and innovations in existing ones drive the development of new sensors. Fabricating NMs into nano-sensors often alters their physical, chemical, or biological properties, directly impacting sensor performance. Thus, understanding the relationship between these changes and sensor performance is crucial in the field of NMs. Key characteristics and benefits of NMs-based sensors include lower power consumption for portable systems, reduced sensor costs, smaller dimensions and weight, batch processing capabilities similar to semiconductor devices, heightened sensitivity, lowered detection limits, improved signal-to-noise ratio, faster response times, and the potential for self-powering through mechanical loads or solar radiation exposure (Malik *et al.*, 2023).

1.3 Importance of Sensing and Detection

Sensing and detection technologies are integral components across a spectrum of fields, serving critical roles in healthcare, environmental monitoring, security, and beyond (Malik *et al.*, 2023). These technologies provide the means to identify, measure, and analyze diverse phenomena with precision and efficiency, enabling a wide range of applications. In healthcare, they facilitate early disease diagnosis by detecting biomarkers and abnormal physiological conditions, thereby improving treatment outcomes and patient care (Haleem *et al.*, 2021). In environmental monitoring, sensing and detection systems play a vital role in assessing air and water quality, pollution levels, and climate change indicators, supporting efforts to safeguard ecosystems and public health (Huang *et al.*, 2023). Additionally, they ensure food safety and quality control by detecting contaminants and allergens in the food supply chain (Thakur & Ragavan, 2013). Moreover, they contribute to industrial process monitoring, research and development, driving innovation and efficiency across various sectors (Katey *et al.*, 2023). As technology continues to advance, the importance of sensing and detection in addressing emerging challenges and advancing societal well-being cannot be overstated.

2. Nanomaterials for Sensing

2.1 Carbon-Based Nanomaterials

The rising demand for sensors based on carbon nanomaterials (CNMs) across various sectors such as chemical, medical, automotive, food safety, and energy is driven by their advantageous intrinsic properties. These properties include chemical resistance, electrical conductivity, optical clarity, flame retardance, thermal stability, surface appearance, decreased water permeability, and dimensional stability among others. CNMs come in various forms, ranging from zero-dimensional fullerenes to three-dimensional structures like graphene, carbon nanotubes (CNTs), nanodiamonds (NDs), carbon quantum dots (CDs), graphene quantum dots (GQDs), carbon nanofibers (CNFs), carbon nanohorns (CNHs), carbon nano-onions (CNOs), and carbon black (CB) (Speranza, 2021). In recent years, these inherent characteristics of CNMs have found increasing applications in advanced sensor technologies. However, despite their promise, CNM-based sensors face practical challenges, particularly in terms of cost-effective processing for large-scale manufacturing of high-purity nanomaterials. The existing market prices for CNMs used in sensor fabrication remain high, limiting their commercial applications in various sectors.

2.1.1 Fullerenes

Fullerenes, also known as buckyballs, represent a fascinating class of carbon-based molecules characterized by a hollow, cage-like structure composed entirely of carbon atoms. Its cage-like structure provides a large surface area, offering ample opportunities for molecular interactions as it enhances their sensitivity to analytes, enabling the detection of even trace amounts of target molecules (Krätschmer, 2011). Additionally, fullerenes possess exceptional electron-accepting and electron-donating capabilities, allowing them to undergo reversible electron transfer processes with various analytes (Afreen *et al.*, 2015). This property forms the basis for their utilization in electrochemical sensing platforms, where they serve as redox-active components for signal transduction. Fullerenes find applications across a diverse range of sensing and detection platforms, encompassing environmental monitoring, biomedical diagnostics, chemical analysis, and beyond (Uygun *et al.*, 2020).

Functionalized fullerenes serve as excellent electron mediators and conductive substrates in electrochemical glucose biosensors. By modifying electrode surfaces with functionalized fullerenes, researchers have achieved improved electron transfer kinetics and enhanced sensitivity for glucose detection (Gavalas & Chaniotakis, 2000). Additionally, functionalized fullerenes can also facilitate the immobilization of glucose oxidase or other glucose recognition elements, enabling selective and rapid detection of glucose levels in biological samples (Chuang & Shih, 2001). Furthermore, fullerenes have also been utilized in DNA biosensors for the detection of DNA hybridization and sequence-specific recognition (You *et al.*, 2020). Their potential extends to neurotransmitter biosensors, where they demonstrate efficacy in monitoring neurotransmitter levels in biological fluids (Pattanaik *et al.*, 2023).

2.1.2 Graphene

Graphene, with its atomically thin structure forming a two-dimensional honeycomb lattice of carbon atoms, serves as the foundational element for a diverse array of carbon-based materials such as graphite, carbon nanotubes, and fullerenes. Graphene has been widely used in the development of gas sensors that are capable of detecting several gases. For instance, graphene oxide (GO) along with the metal nanoparticles has facilitated the selective detection of ammonia gas, playing a vital role in environmental monitoring and industrial safety protocols (Tang *et al.*, 2021). Moreover, graphene field-effect transistor (GFET) biosensors have emerged for the rapid and label-free detection of DNA sequences, employing DNA probes for applications in genomics and medical diagnostics (Sengupta & Hussain, 2021). Beyond this, graphene thermistors have found integration into electronic devices, contributing to precise temperature monitoring and efficient thermal management, thereby enhancing device performance and reliability. Additionally, graphene photodetectors have been deployed to detect ultraviolet (UV) light, enabling critical applications in environmental monitoring, UV dosimetry, and biomedical imaging (Phukan & Sahu, 2020). Furthermore, graphene-based pressure sensors have been seamlessly integrated into electronic skins utilized in robotics and prosthetics, facilitating tactile sensing and precise object manipulation (K.-Y. Chen *et al.*, 2023). These examples underscore the versatility and transformative potential of graphene across a spectrum of sensing applications, driving innovation and progress in various fields.

2.1.3 Carbon nanotubes

(CNT): These are cylindrical structures composed of carbon atoms arranged in a hexagonal lattice. They come in two main types: single-walled carbon nanotubes (SWCNTs) and multi-walled carbon nanotubes (MWCNTs) (Solorio-Rodriguez *et al.*, 2023). SWCNTs consist of a single layer of carbon atoms rolled seamlessly into a cylinder, typically with diameters ranging from 0.4 to 3 nm (Kharlamova & Kramberger, 2021). They possess exceptional electrical conductivity, mechanical strength, and unique optical properties. On the other hand, MWCNTs comprise multiple layers of graphene rolled concentrically, resembling nested cylinders, with diameters typically ranging from 2 to 100 nm (Ahn *et al.*, 2011). MWCNTs retain the high strength and electrical conductivity of SWCNTs while offering enhanced mechanical flexibility.

SWCNTs functionalized with gold (Au) and titanium dioxide (TiO_2) NPs have been utilized for detecting nitrogen dioxide (NO_2) gas (Fort *et al.*, 2019), while flexible gas sensors based on CNT/graphene hybrid films have demonstrated high sensitivity and rapid response times for ammonia gas detection, indicating their potential for real-time environmental monitoring (Bisht *et al.*, 2015). Additionally, label-free electrochemical biosensors employing vertically aligned CNT arrays have shown promise for glucose detection, offering high sensitivity, selectivity, and stability, particularly suitable for clinical diagnostics and diabetes monitoring (Y. Lin *et al.*, 2003). Furthermore, highly sensitive electrochemical sensors based on three-dimensional porous CNT networks have been developed for detecting heavy metal ions in water (Shirsat & Hianik, 2023). CNT-based electrochemical biosensors have also been employed for sensitive and selective detection of cancer biomarkers (Sengupta & Hussain, 2023). Moreover, CNTs integrated into photonic devices have shown tunable photoluminescence properties, enabling sensitive detection of biomolecules and environmental analytes in optical sensing applications (Watahiki *et al.*, 2012). These examples highlight the versatility of CNTs across various sensing applications, owing to their exceptional properties and structural characteristics.

2.2 Metal-Based Nanomaterials

Metal-based nanoparticles are nanoscale particles composed primarily of metal atoms. These nanoparticles exhibit unique physical, chemical, and optical properties due to their small size and high surface area-to-volume ratio.

2.2.1 Gold NPs (AuNPs)

They have a broad spectrum of applications and can be synthesized using diverse components. They have proven utility in drug delivery (Khandelia *et al.*, 2015), biological and chemical sensing (Alex & Tiwari, 2015), cancer treatment (Sahoo *et al.*, 2014), and imaging (Borse *et al.*, 2017). As colorimetric sensors, AuNPs are adept at detecting heavy metals including Cr^{3+} (Shellaiah & Sun, 2023), Cd^{2+} (Bhamore *et al.*, 2021), As^{3+} (J. Li *et al.*, 2020), Zn^{2+} and Ni^{2+} (Azzam *et al.*, 2014), Pb^{2+} (Sengan *et al.*, 2020), Hg^{2+} (Amanulla *et al.*, 2019). They exhibit strong surface plasmon resonance (SPR) effects, making them ideal for optical sensing applications (Ferrari, 2023). Functionalized AuNPs are often used as labels or probes for the detection of biomolecules (Tiwari *et al.*, 2011), DNA (Rink *et al.*, 2010), and other analytes through changes in their optical properties. AuNPs can serve as excellent electrode modifiers due to their high conductivity and catalytic activity (De Bortoli *et al.*, 2023). Modified electrodes with AuNPs enhance the electron transfer rate and sensitivity of electrochemical sensors, enabling the detection of various analytes such as heavy metals (Kusuma *et al.*, 2023), glucose (Cao *et al.* 2019), and neurotransmitters (Qin *et al.*, 2021).

2.2.2 Silver NPs (AgNPs)

They exhibit strong surface-enhanced Raman scattering (SERS) effects allowing for ultrasensitive detection of analytes through their vibrational signatures (J. Lu *et al.*, 2019). AgNPs-based SERS sensors are widely used for chemical analysis, bioimaging, and environmental monitoring. They can induce color changes in solutions due to their localized surface plasmon resonance (LSPR) effects (Adrianto *et al.*, 2022). Moreover, they enable the detection of specific analytes through colorimetric changes, offering a simple and cost-effective sensing platform for various applications. Furthermore, AgNPs demonstrate antimicrobial properties, which have been exploited in biosensing applications for the detection of pathogens and infectious diseases (Karn-orachai *et al.*, 2016). Functionalized AgNPs can selectively bind to target pathogens, leading to changes in their optical or electrical properties that can be measured for detection purposes. This makes AgNP-based biosensors valuable tools for rapid and sensitive detection of bacterial, viral, and fungal pathogens in clinical, food safety, and environmental monitoring applications.

2.2.3 *Platinum NPs (PtNPs)*

They are commonly used as sensing materials in gas sensors due to their high catalytic activity towards gases such as hydrogen (Loof *et al.*, 2021), carbon monoxide (Alvi *et al.*, 2014), and other volatile organic compounds (VOCs). PtNPs-based gas sensors offer rapid response times and high sensitivity, making them valuable for environmental monitoring and industrial safety applications. PtNPs modified electrodes hold significant promise in the development of glucose sensors due to their exceptional catalytic activity and biocompatibility (Ngun *et al.*, 2022). These sensors play a crucial role in monitoring glucose levels, which is essential for managing diabetes and other metabolic disorders.

2.2.4 Palladium nanoparticles (PdNPs)

These NPs are widely employed in hydrogen sensors due to their ability to absorb and desorb hydrogen gas reversibly (Graham, 1997; Ndaya *et al.*, 2019). PdNPs-based sensors exhibit high sensitivity and fast response times, making them suitable for hydrogen detection in fuel cells, automotive systems, and industrial processes. PdNPs-based catalytic sensors enable the detection of organic compounds, and pollutants through changes in reaction kinetics or gas concentration. They are also useful in antibacterial applications, hydrogen storage, anticancer treatment, fuel cells and photothermal therapy (Yurkov *et al.*, 2022).

2.2.5 Metal oxide nanoparticles

Iron oxide NPs such as magnetite (Fe_3O_4), are utilized in magnetic sensors for detecting magnetic fields and measuring magnetic properties (Kolhatkar *et al.*, 2017). Fe_3O_4 nanoparticles-based sensors find applications in magnetic resonance imaging (MRI), magnetic resonance sensors, and magnetic field mapping. Titanium dioxide nanoparticles are utilized in photocatalytic sensors for the detection and degradation of organic pollutants and hazardous substances under light irradiation (Bai & Zhou, 2014). TiO_2 NPs-based sensors offer high efficiency and selectivity in environmental monitoring and wastewater treatment applications.

2.2.6 Metal-Organic Frameworks (MOFs)

MOFs are crystalline materials composed of metal ions or clusters coordinated with organic ligands. These materials have a high surface area and tunable porosity, making them suitable for immobilizing sensing elements and enhancing analyte capture (W. Li *et al.*, 2023). MOFs can be functionalized with chromogenic ligands or nanoparticles to enable colorimetric detection of specific analytes through changes in the optical properties of the material (Hu *et al.*, 2023).

2.3 Semiconductor Nanomaterials

Semiconductor nanomaterials have garnered significant interest in biosensing applications due to their unique optical, electrical, and structural properties (Naresh & Lee, 2021). These materials offer several advantages for biosensing, including high sensitivity, tunable bandgap, biocompatibility, and ease of functionalization.

2.3.1 Quantum Dots (QDs)

They are semiconductor nanocrystals known for their unique optical properties, including size-tunable fluorescence and exceptional photostability (Ying Lim *et al.*, 2015). Widely employed as fluorescent probes in biosensing applications, QDs offer remarkable sensitivity and multiplexing capabilities. They serve as fluorescent labels or probes for detecting various biomolecules, including proteins, DNA, and viruses, enabling sensitive and multiplexed detection in biological samples (Rizvi *et al.*, 2010; Sapsford *et al.*, 2006). Additionally, QDs are utilized as contrast agents for cellular imaging, providing bright and photostable fluorescence for high-resolution imaging of cellular structures and dynamic processes within living cells (Jin & Hildebrandt, 2012).

Further more, in drug delivery systems, surface-functionalized QDs can be conjugated with therapeutic agents and targeting ligands for precise delivery to diseased tissues or cells, facilitating personalized medicine approaches (Badıllı *et al.*, 2020). In optical sensor platforms, QDs serve as sensitive probes with advantages such as high sensitivity, rapid response times, and multiplexed detection capabilities, making them valuable for applications in environmental monitoring, healthcare diagnostics, and food safety (Freeman & Willner, 2012). Moreover, QDs find utility in non-invasive *in vivo* imaging, enabling visualization of molecular and cellular events in real-time for studying disease mechanisms and evaluating therapeutic interventions (Z.-B. Li *et al.*, 2007). Additionally, QDs can serve as photosensitizers in photodynamic therapy, offering enhanced light absorption and efficient generation of reactive oxygen species for improved therapeutic outcomes in cancer treatment (Samia *et al.*, 2003).

2.3.2 Nanowires (NWs)

Semiconductor nanowires represent a class of one-dimensional nanostructures characterized by their high aspect ratios and large surface-to-volume ratios. These unique structural features endow nanowires with excellent electrical properties, making them highly suitable for a variety of sensing applications (W. Lu & Lieber, 2006). One prominent use of nanowires lies in electrical biosensing, where functionalized nanowires serve as sensitive transducers for label-free detection of biomolecules such as DNA, proteins, and enzymes (Y. Wang *et al.*, 2013). This approach enables rapid and real-time detection with exceptional sensitivity, facilitating advancements in medical diagnostics and biotechnology. In addition to biosensing, nanowires find extensive application in gas sensing, particularly for the detection of toxic gases, pollutants, and environmental contaminants (Galstyan *et al.*, 2022). Nanowire-based gas sensors offer several advantages, including rapid response times, low detection limits, and compatibility with miniaturization for portable and wearable devices (X. Chen *et al.*, 2013). Their high sensitivity and selectivity make them invaluable tools for environmental monitoring, industrial safety, and healthcare diagnostics.

Further more, nanowires serve as versatile sensors for temperature, strain, and pressure detection (Menzel *et al.*, 2011). Temperature sensors based on nanowires rely on their temperature-dependent electrical resistance, allowing for precise and accurate temperature measurements across a wide range of applications. Nanowires can also be utilized as strain and pressure sensors, monitoring mechanical deformations and stresses in various materials and structures (Mai *et al.*, 2012; D. Wang *et al.*, 2020). Changes in electrical resistance or piezoelectric properties of nanowires in response to mechanical strain or pressure enable the detection and quantification of these physical parameters, enabling applications in structural health monitoring, robotics, and wearable technology (C. Pan *et al.*, 2019).

2.3.3 Nanoporous Silicon (Si)

Nanoporous silicon is a porous form of silicon with a high surface area, making it suitable for biomolecule immobilization and detection. Nanoporous Si biosensors utilize the change in electrical conductivity, optical properties, or surface functionalization upon biomolecule binding for sensitive detection of analytes such as DNA, proteins, and viruses (Boehringer *et al.*, 2012). These sensors offer high sensitivity, rapid response times, and compatibility with microfabrication techniques. Si nanoporous semiconductors functionalized with metal oxides have been employed for gas sensing applications (Urbiztondo *et al.*, 2009). Furthermore, Si nanoporous semiconductors integrated with fluorescent dyes or quantum dots have been developed for chemical sensing applications (DeLouise & Ouyang, 2009). These hybrid nanomaterials exhibit changes in fluorescence properties upon exposure to target analytes, enabling sensitive and real-time detection of chemical species in environmental or industrial settings. Silica nanoporous semiconductors incorporated into photonic structures, such as photonic crystals or waveguides, have been explored for optical sensing applications (Pacholski, 2013).

2.3.4 Nanoparticle-Decorated Semiconductor Surfaces

Semiconductor surfaces functionalized with metal or metal oxide nanoparticles offer enhanced sensitivity and selectivity in biosensing applications (Pescaglini& Iacopino, 2015). Metal nanoparticles can serve as catalytic sites for signal amplification or as platforms for immobilizing biomolecules. Metal oxide nanoparticles, such as zinc oxide or titanium dioxide, can provide enhanced photocatalytic properties for photoelectrochemical biosensing applications. Nanostructures that combine metal NPs with semiconductor nanowires offer great potential for the development of plasmon-enhanced optoelectronics and sensing applications (Pescaglini & Iacopino, 2015). The silver vanadate ($AgVO_3$) nanoribbons decorated with AgNPs has shown efficient plasmonic photocatalytic activity (Wei *et al.*, 2014). The integration of AuNPs with tungsten trioxide (WO_3) photoelectrodes represents a novel approach to enhance the photoelectrochemical properties of WO_3-based devices (F. Xu *et al.*, 2015).

3. Applications of Nanomaterial-Based Sensors

Owing to the quick development of nanomaterials, a variety of nanostructured materials are frequently used to create innovative and effective biosensors because of their special optical, electrical, magnetic, and catalytic properties. Nanomaterials can also be used to enhance the sensitivity and other characteristics of biosensors. Many nanomaterials have been widely used in a variety of application fields to fabricate sensors and biosensors (Ferrag & Kerman, 2020; Y. Liu *et al.*, 2024). This proliferation has solidified nanomaterial-based sensors as a prominent subject of interest, evidenced by the potential growth of scholarly publications within the field. Nanomaterial-based sensors are captivating at the forefront of modern sensing technology, providing unparalleled capabilities to detect and monitor a broad spectrum of analytes with exceptional sensitivity and specificity (A. Chen & Chatterjee,

2013; Maduraiveeran & Jin, 2017; Manikandan *et al.*, 2018). These sensors find extensive use in environmental monitoring, offering sensitive and selective detection of pollutants, heavy metals, and toxic gases in air and water. In healthcare, nanomaterial-based sensors are employed for early disease diagnosis, drug delivery monitoring, and point-of-care testing, facilitating personalized medicine and improving patient outcomes. Their use in food safety ensures the detection of contaminants and pathogens, safeguarding public health and enhancing food quality control measures. Moreover, in industrial settings, these sensors enable precise monitoring of parameters like temperature, pressure, and chemical composition, optimizing processes and enhancing productivity. In homeland security, these sensors are utilized for detecting explosives, chemical warfare agents, and biological threats, bolstering security measures and emergency response capabilities. The versatility, sensitivity, and miniaturization potential of nanomaterial-based sensors continue to drive innovation across various sectors (Ferrag & Kerman, 2020; A. Kumar, 2018), promising transformative solutions to complex challenges in society and industry alike.

3.1 Environmental Monitoring (Table 1)

The field of nanosensor research for environmental pollutants is expanding quickly and novel and inventive techniques of combining nanomaterials and recognition agents are always being developed. An exciting new technology known as nanomaterial enabled sensors allows for the precise detection of environmental toxins at nanomolar to sub-picomolar levels (Aragay *et al.*, 2012; Willner & Vikesland, 2018). The use of carbon nanotubes, gold nanoparticles, silicon nanowires, and quantum dots in nanomaterial-based sensors presents an affordable substitute for conventional detection techniques such as SPR, HPLC, GC-MS, and ICP-MS (Banerjee *et al.*, 2024; Su *et al.*, 2012). These sensors identify industrial chemicals, hazardous gases, insecticides, and toxic metal ions in environmental samples with exceptional sensitivity and selectivity (Zor *et al.*, 2024). By using the extensive surface area, reactivity, catalytic effectiveness, and adsorption capacity of nanomaterials, they offer straightforward but efficient methods for regular environmental monitoring, improving the detection of trace pollutants (Su *et al.*, 2012). In this article, we explore recent advancements in 2024, utilizing various nanomaterial classes to fabricate devices and demonstrate their application in detecting traces of environmental contamination (Table 1).

3.2 Biomedical Sensing (Table 2)

Nanomaterials have gained significant attraction in the biomedical industry due to their small size and exceptional physicochemical properties, particularly in imaging and medication delivery. Research has highlighted the effectiveness of carbon-based nanomaterials like graphene oxide (GO), nanotubes, as well as gold and silver nanoparticles (NPs) in applications ranging from tissue engineering to cancer treatment and intelligent biosensors (Mukherjee *et al.*, 2024). Moreover, various nanostructures have been pivotal in the development of electrochemical sensing and biosensing techniques for a wide array of biological and biomedical purposes (Maduraiveeran *et al.*, 2018). With their immense potential, nanomaterials serve as versatile diagnostic and therapeutic platforms for both in vivo and in vitro medical applications. Notably, the advancement of nanomaterial-based biosensors holds promise for the early detection and diagnosis of disease-related biomarkers (Alemu & Alemu, 2022; A. Chen & Chatterjee, 2013). Furthermore, recent developments in the healthcare industry have seen the emergence of various nanomaterial-based sensors for detecting serum, blood, protein, and nucleic acid biomarkers, as well as addressing specific health concerns such as COVID-19, single molecule detection, diabetes, cancer, neural sensing, and skin sensing, as summarized in Table 2.

Table 1. Detection of analytes using different nano-sensing materials in environmental monitoring

Sensing material	Type of sensor	Detected analyte	Response	Reference
Polyvinyl alcohol/yttrium oxide (PVA/Y_2O_3) nanocomposite films	Chemiresistive sensor	CO_2 gas detection	Sensing response (S) of 92.72%, optimized low temperature of 100 °C and a fast response/recovery time of □15/11 s	(Y. R. Kumar *et al.*, 2024)
Flower-shaped photoactive Cu_3SnS_4 nanocrystals based on protein corona-induced aggregation effect (PCIAE)	Portable photo-electrochemical sensor	Polystyrene microplastics in aquatic environment	Linear range of 0.5–500 μg mL^{-1}, detection limit of 0.06 μg mL^{-1}, and limit of quantification of 0.14 μg mL^{-1}	(Xiao *et al.*, 2024)
Fe_2O_3@C-800, Co/CoO@C-600, and $CoFe_2O_4$@C-600 MOFs on carbon electrodes	Magnetic electrochemical sensor	Heavy metals like lead ions (Pb^{2+})	Linear detection range of 0.001–60 μM, detection limit of 0.0043 μM and sensitivity of 22.22 μA μM·cm^{-2}	(Y. Zhang *et al.*, 2024)
N-doped carbon dots with Europium-based metal organic frameworks (N-CDs@Eu-MOF)	Ratiometric imprinted fluorescence sensor	Pesticide malathion in lettuce, tap water, and soil samples	Detection range of 1–10 μM and limit of detection (LOD) of 0.05 μM	(L. Yang *et al.*, 2024)
$K_3Fe(CN)_6$ regulated silver nanoparticles (Ag NPs@Apt)	ratiometric electrochemical biosensor	Marine toxin-saxitoxin	Detection range of 0.04 to 0.15 μM and limit of detection (LOD) be 1 nM	(Zeng *et al.*, 2024)

Table 2. Recent developments in sensors using various nanomaterials in biomedical diagnostics

Nanomaterial used	Type of sensor	Metabolite detected	Sensor response	Reference
CuCo2O4-metal-organic frameworks (MOFs) and AuNPs (gold nanoparticles)	Nanocomposites as dual mode colorimetric/ Electrochemical sensor with peroxidase activity	Dopamine (DA)	Low detection limit- 0.07μM and linear detection range- 0-1mM	(Yi *et al.*, 2024)
Cu-MOF heterostructure	Electrochemical sensor	Glutathione	Detection limit and dynamic detection range- 0.1-20μM and sensitivity- 0.0437μA μM-1	(Kaimal *et al.*, 2024)
Screen-printed carbon electrodes (SPCEs) modified with sea urchin-like tungsten oxide ($SUWO_3$) nanostructures	Nanocomposites as electrochemical sensor with catalytic activity	Dopamine (DA) and Uric Acid (UA) in synthetic urine	DA (LOD- 51.4nM and sensitivity- 127 μA mM^{-1} $cm^{-2)}$ and UA (LOD- 253 nM and sensitivity- 55.9 μA mM^{-1} cm^{-2})	(Oğuz *et al.*, 2024)
Polypyrrole (PPy), V2O5 and TiO2 polymeric nanocomposite surface modified with Polyvinylpyrrolidone/ CuO nanorods (NRs) and Fe2O3 NRs	Conductive polymer nanocomposite as electrochemical sensor with conductive activity	Creatinine in blood serum	Linear detection range- 10 nmol L^{-1} –1.3 mmol L^{-1}	(Dehghan-Manshadi *et al.*, 2024)
Single red upconversion nanoparticles with core shell structure	Fluorescent sensor on enzyme-controlled cascade signal amplification (ECSAm)	Glucose and glutathione (GSH)	Detection limit for glucose and GSH as 0.03 μM and 0.075 μM respectively	(T. Zhao *et al.*, 2024)
Cadmium sulfide–gold (CdS–Au) nanocomposite	Photoelectrochemical (PEC) sensor	Ascorbic acid in blood samples	Liner detection range- 0.01 μM–200 μM, detection limit- 0.2 nM and sensitivity- 642.9 μA mM^{-1} cm^{-2}	(H. Zhang *et al.*, 2024)
Liquid metal nanoparticles (gallium-indium alloy) on polyamide substrate	Wearable flexible strain sensor	Human motion and strain sense	Real time monitoring of facial expressions and vocal cord vibrations	(Zhu *et al.*, 2024)

Cu-MOF nanosheets and BiVO4 NRs	Nanorod as electrogenerated Chemiluminescence sensor	KRAS gene in cancer colorectal cancer samples	Linear detection range- 0.1 pM–1 nM and detection limit- 0.02 fM.	(X. Zhang *et al.*, 2024)
Gold nanoparticle-DNA Pendulum (GDP) and reduced graphene modified electrode	Electrochemical aptamer-based sensor	Vascular endothelial growth factor (VEGF) in PBS and artificial urine	Real time monitoring of VEGF with detection range- 13 fM to 130 nM,	(Y. Wang *et al.*, 2024)
Tungsten disulfide Quantum Dots (WS_2 QDs) /CoOOH nanosheet	Nanosheet as ratiometric fluorescence sensor	α-glucosidase in human serum	Low detection limit of α-Glu and α-Glu inhibitor- 0.009 U/mL and 0.07 μM respectively	(Zhai *et al.*, 2024)
Modified electrodes of MXene with sodium ascorbate (NaAsc) nanosheets and cadmium selenide (CdSe) with zinc sulphide (ZnS) QDs	Nanosheet and QDs as ratiometric molecular imprinted electrochemiluminescence sensor	Uric acid in human serum	Optical and electrical signal detected range- 1×10^{-10} to 1×10^{-4} mol/L and LOD- 18.13 pmol/L	(M. Liu *et al.*, 2024)
Tetrahedral DNA nanostructures (TDNs) fixed on surface of gold electrode (GE)	Impedence based electrochemical aptamer sensor	Doxorubicin (DOX) in serum sample	relative standard deviation (RSD) range- 3.6 to 5.9% and detection limit (LOD)- 3.0 nM	(Ou *et al.*, 2024)
Molecularly imprinted polymeric nanoparticles (nanoMIPs) and self-assembled thioctic acid (SAM-TA) monolayer on gold electrodes	Chronoamperometry based Impedimetric sensor	Trypsin in saliva	Detection range- 3 to 46 ng mL−1 and LOD- 1.06 ± 0.5 ng mL−1	(Di Masi *et al.*, 2024)
Iron oxide-graphene (Fe3O4–GO) nanocomposite	Wearable polymethylmethacrylate (PMMA) substrated microneedle electrochemical sensor	Dopamine in interstitial fluid	Detection range- 3 to 32mM Sensitivity- 0.01528 mA mM−1 and LOD- 0.6 mM	(M. R. *et al.*, 2024)

Luminescent CDs and copper nanocomposites	Fluorescent sensor	Creatinine in urine	Linear detection range and limit of detection as 10^{-5} to 0.1 mg·dL^{-1} and 5.1 × 10^{-16} mg·dL^{-1} respectively.	(Bhatt *et al.*, 2024)
Photoelectric nanocomposite (COF) with Single walled carbon nanotubes (SWCNT)	Photoelectrochemical (PEC) sensor	Trace ATP detection	Detection range- 10 fmol L^{-1} to 10 nmol L^{-1} and detection limit- low as 2.75 fmol L^{-1}	(Xi *et al.*, 2024)
Bismuth vanadate ($BiVO_4$) nanocomposite with molecularly imprinted poly (acrylic acid) (MIP-AA) on carbon NTs	Photoelectrochemical (PEC) sensor	Dopamine in urine	Limit of detection (LOD)- 2.9 nmol L^{-1} and linear range- 9.7 to 150 nmol L^{-1}	(Gorla *et al.*, 2024)
Water soluble silver nanoclusters (AgNCs@PEI)	Dual colorimetric and fluorometric sensor	Ascorbic acid	Detection limit- 10^3 µM and dynamic detection range- 1-10^5 µM	(Raj *et al.*, 2024)
Gold nanoparticles (AuNPs) and copper oxide (CuO) nanoparticles on screen-printed carbon electrode (Au/CuO/SPCE)	Nonenzymatic electrochemical sensor.	Glucose in saliva	Linear detection range- 2 mM to 397 mM and sensitivity- 236.70 mAmM−1 cm−2	(Ali *et al.*, 2024)
Reduced graphene oxide (RGO) and CoNi-MOF as bimetallic metal organic framework	Molecularly imprinted electrochemical sensor	Hippuric acid (HA) in urine	Detection range- 2–800 nM and detection limit- 0.97 nM.	(Gao *et al.*, 2024)
Bird's nest CDs (B-CDs)	Fluorescent sensor	Vitamin B12 in serum	The linear range- 0 □ 100 µM and detection limit- 0.24 µM.	(J. Chen *et al.*, 2024)
Zinc oxide NRs arrays on Gallium Nitride substrates with nanogold films (ZnO/Au/GaN) heterojunction	Photoelectrochemical (PEC) sensor	alpha-fetoprotein (AFP) in serum	Linear range- 0.080–5.0 ng/mL and low detection limit- 0.027 ng/mL	(J. Li *et al.*, 2024)

3.3 Food Safety and Quality Control

The detection of different harmful compounds in foods has become a focus of study in the field of food safety due to the development of numerous chemical or biological sensing devices based on different operating principles in recent studies (Pan *et al.*, 2019; Scognamiglio *et al.*, 2014). Addressing emerging challenges in food safety necessitates the adoption of innovative approaches, prompting the development of novel, user-friendly, rapid, portable, and miniaturized, in-field detection methods with streamlined sample preparation procedures, low-cost implementation, and timely analytical capabilities, exemplified by advanced sensor technologies. At various stages of the food supply chain, from the producer to the consumer, the sensing would facilitate the control and access to the food's quality outside of the laboratory (Shruti *et al.*, 2024). A significant body of research is being published every day on the use of nanomaterials-based sensors in the field of food safety screening. Some of the recent researches are reviewed like utilization of advanced carbon-based nanomaterials, including Carbon Dots (CDs), doped CDs, and functionalized CDs, for sensitive Tartrazine detection (Chaudhari *et al.*, 2024), molecularly imprinted Copper MOFs (CuMOF) with reduced Graphene Oxide (GO) for bisphenol A in milk (Zhou *et al.*, 2024), silver nanowires (Au-NWs) with Molybednum disulphide (MoS2) for thiabendazole determination in food (Zheng *et al.*, 2024), using electrochemical sensor, Citric Acid CDs (CA-CDs) for ascorbic acid in juice (Du *et al.*, 2024; Z. Lin *et al.*, 2024), Europium based CDs (Eu-CDs) for tetracycline in seafood aquaculture (J. Zhang *et al.*, 2024), Eu-MOFs with N doped CDs for malathione in water and soil samples (L. Yang *et al.*, 2024) N@CDs for allura red dye in beverages (Salman, 2024), N/S CDs for lead ion sensing in food samples (Aboobakri *et al.*, 2024) and upconversion nanoparticles (UCNPs) for chlorpyrifos pesticide (X. Zhao *et al.*, 2024) using paper-based dual-emission ratiometric fluorescence sensor, carbon nanotubes (CNTs) and cadmium selenium with zinc selenium quantum dots (CdSe@ZnSe QDs) (Wu *et al.*, 2024) and zinc sulphide QDs (ZnS-QDs) (Yilmaz *et al.*, 2024) for melamine in milk using electrochemiluminescence sensor. Hence, the seamless integration of these advanced technologies, alongside parallel improvements in the analytical capabilities of nanoparticle-based devices, is poised to propel the evolution of food safety and quality sensors into the next phase.

3.4 Industrial and Security Applications

Sensitive, precise, and dependable explosives detection is desperately needed to meet the demands of environmental preservation and global security. The necessity to swiftly identify biological threat agents, coupled with the inadequacies of current detection techniques, underscores the imperative for ongoing innovation in sensor development. Nanomaterial-based biosensors represent a promising avenue, demonstrating heightened sensitivity and accuracy compared to conventional biothreat detection methods. Moreover, these sensors achieve these advancements while minimizing sample volume, reducing preparation time, and lowering assay costs (Rowland *et al.*, 2016). Counterfeiting presents formidable challenges with profound economic, security, and health implications worldwide. This underground industry, estimated in the billions, demands advanced solutions to combat its increasing sophistication. It introduces a novel approach utilizing a range of security ink formulations, incorporating lanthanide-doped luminescent nanomaterials (P. Kumar *et al.*, 2016), QDs (both semiconductor and carbon-based), metal-organic frameworks, porous metal-oxides based gas sensors (Barreca *et al.*, 2018; Galstyan, 2017) and plasmonic nanomaterials. Because of their unique chemical, optical, and electrical characteristics as well as their simplicity of synthesis and functionalization, colloidal QDs provide intriguing solutions. The academic field of explosive detection has seen a notable increase in research efforts as a result (De Iacovo *et al.*, 2024). Furthermore, it provides insights into the historical background, fundamental concepts, synthesis methods, and characterization techniques of luminescent nanomaterials.

4. Challenges and Future Prespectives

Nanomaterial-based sensors hold immense promise for revolutionizing various fields, including environmental monitoring, healthcare, and security. However, they also face several challenges and offer intriguing future perspectives. Achieving consistent sensor performance, especially in complex environmental analysis, poses significant challenges. Sensor signals must remain robust in fluctuating conditions and extreme temperatures, crucial for deployment in isolated and harsh environments.

Metal oxide gas sensors have seen sensitivity and specificity advancements through heterostructure integration, promoting catalytic reactions and enhancing gas adsorption rates. However, maintaining sensor consistency over time, particularly in varying temperature environments, is essential. Nanomaterials offer exceptional sensitivity but raise concerns about longevity due to nanoparticle agglomeration and detachment. Integrating sol-gel technology and ceramics with nanomaterials is increasingly common to enhance sensor durability. Nature-inspired approaches hold promise for overcoming current sensor limitations, such as insufficient sensitivity and instability. Yet, reproducing consistent results with nanomaterial-based sensors remains challenging due to unpredictable nanomaterial placement. Achieving precise and controlled nanomaterial arrangements is a pressing challenge. Despite their extensive use in biosensors, nanomaterials have shortcomings, but their large surface areas enable the attachment of signal-amplifying reporter molecules, like enzymes and fluorescent tags, enhancing detection sensitivity. Combining nanotechnology with ultra-sensitive detection methods, such as digital PCR and single-molecule analysis, offers further sensitivity improvements.

Addressing these challenges requires multidisciplinary collaborations between scientists, engineers, policymakers, and stakeholders. By overcoming these hurdles and embracing emerging opportunities, nanotechnology sensing can significantly impact healthcare, environmental protection, safety, and quality of life in the future.

5. Conclusion

Nanomaterials offer significant potential for enhancing sensing and detection applications. Throughout this chapter the diverse properties of nanomaterials like carbon nanotubes, quantum dots, and graphene, which enable highly sensitive detection across various domains were discussed. Their large surface area-to-volume ratios, customizable properties, and compatibility make them ideal for sensor development. Leveraging these unique traits, researchers can create innovative sensing devices capable of detecting diverse targets with exceptional precision. As our understanding of nanomaterial synthesis and integration deepens, the future promises even more sophisticated sensing technologies. These advancements hold promise for transformative changes in fields like healthcare, environmental monitoring, and security.

References

Aboobakri, E., Heidari, T., & Jahani, M. (2024). Determination of lead (II) in food samples using a functionalized paper-based fluorescent sensor modified by carbon dots. Luminescence, 39 (2), e4690. https://doi.org/10.1002/bio.4690

Adrianto, N., Panre, A. M., Istiqomah, N. I., Riswan, M., Apriliani, F., & Suharyadi, E. (2022). Localized surface plasmon resonance properties of green synthesized silver nanoparticles. Nano-Structures & Nano-Objects, 31, 100895. https://doi.org/10.1016/j.nanoso.2022.100895

Afreen, S., Muthoosamy, K., Manickam, S., & Hashim, U. (2015). Functionalized fullerene (C60) as a potential nanomediator in the fabrication of highly sensitive biosensors. Biosensors and Bioelectronics, 63, 354–364. https://doi.org/10.1016/j.bios.2014.07.044

Ahn, K., Kim, S.-M., & Yu, I. J. (2011). Multi-Walled Carbon Nanotube (MWCNT) Dispersion and Aerosolization with Hot Water Atomization without Addition of Any Surfactant. Safety and Health at Work, 2 (1), 65–69. https://doi.org/10.5491/SHAW.2011.2.1.65

Ajmal, S., Athar Hashmi, F., & Imran, I. (2022). Recent progress in development and applications of biomaterials. Materials Today: Proceedings, 62, 385–391. https://doi.org/10.1016/j.matpr.2022.04.233

Alex, S., & Tiwari, A. (2015). Functionalized Gold Nanoparticles: Synthesis, Properties and Applications—A Review. Journal of Nanoscience and Nanotechnology, 15 (3), 1869–1894. https://doi.org/10.1166/jnn.2015.9718

Ali, M. Y., Abdulrahman, H. B., Ting, W.-T., & Howlader, M. M. R. (2024). Green synthesized gold nanoparticles and CuO-based nonenzymatic sensor for saliva glucose monitoring. RSC Advances, 14 (1), 577–588. https://doi.org/10.1039/D3RA05644A

Alvi, mohammad akhtar, Al-Hartomy, O., & Al-Ghamdi, A. (2014). Carbon mono-oxide gas sensing properties of multi-walled carbon nanotubes decorated with platinum nanoparticles based film sensors. Digest Journal of Nanomaterials and Biostructures, 9, 35–43.

Amanulla, B., Perumal, K. N., & Ramaraj, S. K. (2019). Chitosan functionalized gold nanoparticles assembled on sulphur doped graphitic carbon nitride as a new platform for colorimetric detection of trace Hg2+. Sensors and Actuators B: Chemical, 281, 281–287. https://doi.org/10.1016/j.snb.2018.10.039

Aragay, G., Pino, F., & Merkoçi, A. (2012). Nanomaterials for sensing and destroying pesticides. Chemical Reviews, 112 (10), 5317–5338. https://doi.org/10.1021/cr300020c

Azzam, E. M. S., El-farargy, A. F. M., & Abd-Elaal, Ali. A. (2014). Enhancement the detection of Ni2+ and Zn2+ ions using nanostructure of synthesized dithiol surfactants with gold nanoparticles. Journal of Industrial and Engineering Chemistry, 20 (5), 3905–3912. https://doi.org/10.1016/j.jiec.2013.12.097

Badıllı, U., Mollarasouli, F., Bakirhan, N. K., Ozkan, Y., & Ozkan, S. A. (2020). Role of quantum dots in pharmaceutical and biomedical analysis, and its application in drug delivery. TrAC Trends in Analytical Chemistry, 131, 116013. https://doi.org/10.1016/j.trac.2020.116013

Bai, J., & Zhou, B. (2014). Titanium Dioxide Nanomaterials for Sensor Applications. Chemical Reviews, 114 (19), 10131–10176. https://doi.org/10.1021/cr400625j

Banerjee, D., Adhikary, S., Bhattacharya, S., Chakraborty, A., Dutta, S., Chatterjee, S., Ganguly, A., Nanda, S., & Rajak, P. (2024). Breaking boundaries: Artificial intelligence for pesticide detection and eco-friendly degradation. Environmental Research, 241, 117601. https://doi.org/10.1016/j.envres.2023.117601

Baranwal, J., Barse, B., Gatto, G., Broncova, G., & Kumar, A. (2022). Electrochemical Sensors and Their Applications: A Review. Chemosensors, 10 (9), Article 9. https://doi.org/10.3390/chemosensors10090363

Barreca, D., Gasparotto, A., Gri, F., Comini, E., & Maccato, C. (2018). Plasma-Assisted Growth of β-MnO2 Nanosystems as Gas Sensors for Safety and Food Industry Applications. Advanced Materials Interfaces, 5 (23), 1800792. https://doi.org/10.1002/admi.201800792

Bayda, S., Adeel, M., Tuccinardi, T., Cordani, M., & Rizzolio, F. (2020). The History of Nanoscience and Nanotechnology: From Chemical–Physical Applications to Nanomedicine. Molecules, 25 (1), Article 1. https://doi.org/10.3390/molecules25010112

Bhamore, J. R., Gul, A. R., Kailasa, S. K., Kim, K.-W., Lee, J. S., Park, H., & Park, T. J. (2021). Functionalization of gold nanoparticles using guanidine thiocyanate for sensitive and selective visual detection of Cd2+. Sensors and Actuators B: Chemical, 334, 129685. https://doi.org/10.1016/j.snb.2021.129685

Bhatt, P., Kukkar, D., Yadav, A. K., & Kim, K.-H. (2024). Carbon dot-copper nanocomposite-based fluorescent sensor for detection of creatinine in urine samples of CKD patients. Spectrochimica Acta Part A: Molecular and Biomolecular Spectroscopy, 307, 123666. https://doi.org/10.1016/j.saa.2023.123666

Bisht, A., Chockalingam, S., Panwar, O. S., Kesarwani, A. K., Ishpal, ., Singh, B. P., &

Singh, V. N. (2015). Structural, Field Emission and Ammonia Gas Sensing Properties of Multiwalled Carbon Nanotube-Graphene Like Hybrid Films Deposited by Microwave Plasma Enhanced Chemical Vapor Deposition Technique. Science of Advanced Materials, 7 (7), 1424–1434. https://doi.org/10.1166/sam.2015.2100

Boehringer, M., Artmann, H., & Witt, K. (2012). Porous Silicon in a Semiconductor Manufacturing Environment. Journal of Microelectromechanical Systems, 21 (6), 1375–1381. https://doi.org/10.1109/JMEMS.2012.2205900

Borse, V., Thakur, M., Sengupta, S., & Srivastava, R. (2017). N-doped multi-fluorescent carbon dots for 'turn off-on' silver-biothiol dual sensing and mammalian cell imaging application. Sensors and Actuators B: Chemical, 248, 481–492. https://doi.org/10.1016/j.snb.2017.03.158

Byrne, B., Stack, E., Gilmartin, N., & O'Kennedy, R. (2009). Antibody-Based Sensors: Principles, Problems and Potential for Detection of Pathogens and Associated Toxins. Sensors (Basel, Switzerland), 9 (6), 4407–4445. https://doi.org/10.3390/s90604407

Camargo, P. H. C., Satyanarayana, K. G., & Wypych, F. (2009). Nanocomposites: Synthesis, structure, properties and new application opportunities. Materials Research, 12, 1–39. https://doi.org/10.1590/S1516-14392009000100002

Cao, L., Wang, P., Chen, L., Wu, Y., & Di, J. (2019). A photoelectrochemical glucose sensor based on gold nanoparticles as a mimic enzyme of glucose oxidase. RSC Advances, 9 (27), 15307–15313. https://doi.org/10.1039/C9RA02088H

Chaudhari, S. S., Patil, P. O., Bari, S. B., & Khan, Z. G. (2024). A comprehensive exploration of tartrazine detection in food products: Leveraging fluorescence nanomaterials and electrochemical sensors: Recent progress and future trends. Food Chemistry, 433, 137425. https://doi.org/10.1016/j.foodchem.2023.137425

Chen, A., & Chatterjee, S. (2013). Nanomaterials based electrochemical sensors for biomedical applications. Chemical Society Reviews, 42 (12), 5425–5438. https://doi.org/10.1039/c3cs35518g

Chen, J., Mao, C., Ye, H., Gao, X., & Zhao, L. (2024). Natural biomass carbon Dots-Based fluorescence sensor for high precision detection of vitamin B12 in serum. Spectrochimica Acta Part A: Molecular and Biomolecular Spectroscopy, 305, 123459. https://doi.org/10.1016/j.saa.2023.123459

Chen, K.-Y., Xu, Y.-T., Zhao, Y., Li, J.-K., Wang, X.-P., & Qu, L.-T. (2023). Recent progress in graphene-based wearable piezoresistive sensors: From 1D to 3D device geometries. Nano Materials Science, 5 (3), 247–264. https://doi.org/10.1016/j.nanoms.2021.11.003

Chen, X., Wong, C. K. Y., Yuan, C. A., & Zhang, G. (2013). Nanowire-based gas sensors. Sensors and Actuators B: Chemical, 177, 178–195. https://doi.org/10.1016/j.snb.2012.10.134

Chuang, C.-W., & Shih, J.-S. (2001). Preparation and application of immobilized C60-glucose oxidase enzyme in fullerene C60-coated piezoelectric quartz crystal glucose sensor. Sensors and Actuators B-Chemical - Sensor Actuator B-Chem, 81, 1–8. https://doi.org/10.1016/S0925-4005 (01)00914-5

De Bortoli, L. S., Vanoni, C. R., Jost, C. L., Mezalira, D. Z., & Fredel, M. C. (2023). Stable and ligand-free gold nanoparticles produced by laser ablation as efficient electrocatalysts for electrochemical sensing of dopamine. Journal of Electroanalytical Chemistry, 947, 117744. https://doi.org/10.1016/j.jelechem.2023.117744

De Iacovo, A., Mitri, F., De Santis, S., Giansante, C., & Colace, L. (2024). Colloidal Quantum Dots for Explosive Detection: Trends and Perspectives. ACS Sensors, 9 (2), 555–576. https://doi.org/10.1021/acssensors.3c02097

Dehghan-Manshadi, H., Mazloum-Ardakani, M., & Mozaffari, S. A. (2024). A flexible capacity-metric creatinine sensor based on polygon-shape polyvinylpyrrolidone/CuO and Fe2O3 NRDs electrodeposited on three-dimensional TiO2–V2O5–Polypyrrole nanocomposite. Biosensors and Bioelectronics, 246, 115881. https://doi.org/10.1016/j.bios.2023.115881

DeLouise, L. A., & Ouyang, H. (2009). Photoinduced fluorescence enhancement and energy transfer effects of quantum dots porous silicon. Physica Status Solidi c, 6 (7), 1729–1735. https://doi.org/10.1002/pssc.200881089

Di Masi, S., Costa, M., Canfarotta, F., Guerreiro, A., Hartley, A., Piletsky, S. A., & Malitesta, C. (2024). An impedimetric sensor based on molecularly imprinted nanoparticles for the determination of trypsin in artificial matrices – towards point-of-care diagnostics. Analytical Methods, 16 (5), 742–750. https://doi.org/10.1039/D3AY01762A

Díez-Pascual, A. M. (2021). Carbon-Based Nanomaterials. International Journal of Molecular Sciences, 22 (14), 7726. https://doi.org/10.3390/ijms22147726

Du, Q., Zhao, X., Mei, X., Zhao, Y., Dong, C., & Li, J. (2024). A sensitive sensor based on carbon dots for the determination of Fe^{3+} and ascorbic acid in foods. Analytical Methods, 16 (6), 939–949. https://doi.org/10.1039/D3AY02133E

Ferrag, C., & Kerman, K. (2020). Grand Challenges in Nanomaterial-Based Electrochemical Sensors. Frontiers in Sensors, 1. https://doi.org/10.3389/fsens.2020.583822

Ferrari, E. (2023). Gold Nanoparticle-Based Plasmonic Biosensors. Biosensors, 13 (3), 411. https://doi.org/10.3390/bios13030411

Fort, A., Panzardi, E., Al-Hamry, A., Vignoli, V., Mugnaini, M., Addabbo, T., & Kanoun, O. (2019). Highly Sensitive Detection of NO2 by Au and TiO2 Nanoparticles Decorated SWCNTs Sensors. Sensors (Basel, Switzerland), 20 (1), 12. https://doi.org/10.3390/s20010012

Freeman, R., & Willner, I. (2012). Optical molecular sensing with semiconductor quantum dots (QDs). Chemical Society Reviews, 41 (10), 4067–4085. https://doi.org/10.1039/C2CS15357B

Galstyan, V. (2017). Porous TiO2-Based Gas Sensors for Cyber Chemical Systems to Provide Security and Medical Diagnosis. Sensors, 17 (12), Article 12. https://doi.org/10.3390/s17122947

Galstyan, V., Moumen, A., Kumarage, G. W. C., & Comini, E. (2022). Progress towards chemical gas sensors: Nanowires and 2D semiconductors. Sensors and Actuators B: Chemical, 357, 131466. https://doi.org/10.1016/j.snb.2022.131466

Gao, H., Chai, J., Jin, C., & Tian, M. (2024). Molecularly imprinted electrochemical sensor based on CoNi-MOF/RGO nanocomposites for sensitive detection of the hippuric acid. Analytica Chimica Acta, 1296, 342307. https://doi.org/10.1016/j.aca.2024.342307

Gavalas, V. G., & Chaniotakis, N. A. (2000). [60]Fullerene-mediated amperometric biosensors. Analytica Chimica Acta, 409 (1):131-135.https://doi.org/10.1016/S0003-2670(99)00887-9

Gorla, F. A., Santana Dos Santos, C., De Matos, R., Antigo Medeiros, R., Do Prado Ferreira, M., Pereira Camargo, L., Henrique Dall'Antonia, L., & Tarley, C.R.T. (2024). Development of novel nanocomposite-modified photoelectrochemical sensor based on the association of bismuth vanadate and MWCNT-grafted-molecularly imprinted poly (acrylic acid) for dopamine determination at nanomolar level. Talanta, 266, 125044. https://doi.org/10.1016/j.talanta.2023.125044

Graham, T. (1997). XVIII. On the absorption and dialytic separation of gases by colloid septa. Philosophical Transactions of the Royal Society of London, 156, 399–439. https://doi.org/10.1098/rstl.1866.0018

Haleem, A., Javaid, M., Singh, R. P., Suman, R., & Rab, S. (2021). Biosensors applications in medical field: A brief review. Sensors International, 2, 100100. https://doi.org/10.1016/j.sintl.2021.100100

Hu, J., Chen, S., Liu, Z., Li, J.-R., Huang, J.-H., Jiang, Z., Ou, W., Liao, W.-M., Lu, J., & He, J. (2023). MOF-based colorimetric sensor for rapid and visual readout of trace acetylene. Journal of Materials Chemistry A, 11 (20), 10577–10583. https://doi.org/10.1039/D3TA00582H

Huang, C.-W., Lin, C., Nguyen, M. K., Hussain, A., Bui, X.-T., & Ngo, H. H. (n.d.). A review of biosensor for environmental monitoring: Principle, application, and corresponding achievement of sustainable development goals. Bioengineered, 14 (1), 58–80. https://doi.org/10.1080/21655979.2022.2095089

Jin, Z., & Hildebrandt, N. (2012). Semiconductor quantum dots for in vitro diagnostics and cellular imaging. Trends in Biotechnology, 30 (7), 394–403. https://doi.org/10.1016/j.tibtech.2012.04.005

Kaimal, R., Dube, A., Souwaileh, A. A., Wu, J. J., & Anandan, S. (2024). A copper metal–organic framework-based electrochemical sensor for identification of glutathione in pharmaceutical samples. The Analyst, 149 (3), 947–957. https://doi.org/10.1039/D3AN01714A

Karn-orachai, K., Sakamoto, K., Laocharoensuk, R., Bamrungsap, S., Songsivilai, S., Dharakul, T., & Miki, K. (2016). Extrinsic surface-enhanced Raman scattering detection of influenza A virus enhanced by two-dimensional gold@silver core–shell nanoparticle arrays. RSC Advances, 6 (100), 97791–97799. https://doi.org/10.1039/C6RA17143E

Katey, B., Voiculescu, I., Penkova, A. N., & Untaroiu, A. (Eds.). (2023). A Review of Biosensors and Their Applications. ASME Open Journal of Engineering, 2 (020201). https://doi.org/10.1115/1.4063500

Khandelia, R., Bhandari, S., Pan, U. N., Ghosh, S. S., & Chattopadhyay, A. (2015). Gold Nanocluster Embedded Albumin Nanoparticles for Two-Photon Imaging of Cancer Cells Accompanying Drug Delivery. Small, 11 (33), 4075–4081. https://doi.org/10.1002/smll.201500216

Kharlamova, M. V., & Kramberger, C. (2021). Applications of Filled Single-Walled Carbon Nanotubes: Progress, Challenges, and Perspectives. Nanomaterials, 11 (11), 2863. https://doi.org/10.3390/nano11112863

Kolhatkar, A. G., Chen, Y.-T., Chinwangso, P., Nekrashevich, I., Dannangoda, G. C., Singh, A., Jamison, A. C., Zenasni, O., Rusakova, I. A., Martirosyan, K. S., Litvinov, D., Xu, S., Willson, R. C., & Lee, T. R. (2017). Magnetic Sensing Potential of Fe3O4 Nanocubes Exceeds That of Fe3O4 Nanospheres. ACS Omega, 2 (11), 8010–8019. https://doi.org/10.1021/acsomega.7b01312

Krätschmer, W. (2011). The story of making fullerenes. Nanoscale, 3 (6), 2485–2489. https://doi.org/10.1039/c0nr00925c

Kumar, A. (2018). Nanosensors: Applications and Challenges. 8 (7).

Kumar, Y. R., Thangamani, J. G., Karthik, T. V. K., Deshmukh, K., & Pasha, S. K. K. (2024). A novel flexible CO_2 gas sensor based on polyvinyl alcohol/yttrium oxide nanocomposite films. RSC Advances, 14 (8), 5022–5036. https://doi.org/10.1039/D3RA04257J

Kusuma, S. A. F., Harmonis, J. A., Pratiwi, R., & Hasanah, A. N. (2023). Gold Nanoparticle-Based Colorimetric Sensors: Properties and Application in Detection of Heavy Metals and Biological Molecules. Sensors, 23 (19), Article 19. https://doi.org/10.3390/s23198172

Lepore, M., & Delfino, I. (2022). Optical Sensors Technology and Applications. Sensors (Basel, Switzerland), 22 (20), 7905. https://doi.org/10.3390/s22207905

Li, J., Jiang, Y., Xu, A., Luo, F., Lin, C., Qiu, B., Lin, Z., Jiang, Z., & Wang, J. (2024). ZnO/Au/

GaN heterojunction-based self-powered photoelectrochemical Sensor for alpha-fetoprotein detection. Talanta, 268, 125381. https://doi.org/10.1016/j.talanta.2023.125381

Li, J., Zheng, B., Zheng, Z., Li, Y., & Wang, J. (2020). Highly sensitive and selective colorimetric and SERS dual-mode detection of arsenic (III) based on glutathione functionalized gold nanoparticles. Sensors and Actuators Reports, 2 (1), 100013. https://doi.org/10.1016/j.snr.2020.100013

Li, W., Zhu, Z., Chen, Q., Li, J., & Tu, M. (2023). Device fabrication and sensing mechanism in metal-organic framework-based chemical sensors. Cell Reports Physical Science, 4 (12), 101679. https://doi.org/10.1016/j.xcrp.2023.101679

Li, Z.-B., Cai, W., & Chen, X. (2007). Semiconductor Quantum Dots for In Vivo Imaging. Journal of Nanoscience and Nanotechnology, 7 (8), 2567–2581. https://doi.org/10.1166/jnn.2007.628

Lin, Y., Lu, F., Tu, Y., & Ren, Z. (2003). Glucose Biosensors Based on Carbon Nanotube Nanoelectrode Ensembles. Nano Letters - NANO LETT, 4. https://doi.org/10.1021/nl0347233

Lin, Z., Zeng, Q., Yao, W., Chen, W., Cai, C., Yang, J., Lin, X. & Chen, W. (2024). A fluorescence "turn-on" sensor for ascorbic acid in fruit juice and beverage based on ascorbate oxidase-like activity of citric acid-derived carbon dots. Food Chemistry, 437, 137928. https://doi.org/10.1016/j.foodchem.2023.137928

Liu, M., Wang, Y., Tang, S., Wang, W., Liang, A., & Luo, A. (2024). A ratiometric molecular imprinted electrochemiluminescence sensor based on enhanced luminescence of CdSe@ZnS quantum dots by MXene@NaAsc for detecting uric acid. Bioelectrochemistry, 156, 108610. https://doi.org/10.1016/j.bioelechem.2023.108610

Liu, Y., Xiao, Y., Zhang, Y., Gao, X., Wang, H., Niu, B., & Li, W. (2024). ZnO-rGO-based electrochemical biosensor for the detection of organophosphorus pesticides. Bioelectrochemistry, 156, 108599. https://doi.org/10.1016/j.bioelechem.2023.108599

Loof, D., Thüringer, O., Schowalter, M., Mahr, C., Pranti, A. S., Lang, W., Rosenauer, A., Zielasek, V., Kunz, S., & Bäumer, M. (2021). Synthesis and Characterization of Ligand□Linked Pt Nanoparticles: Tunable, Three□Dimensional, Porous Networks for Catalytic Hydrogen Sensing. ChemistryOpen, 10 (7), 697–712. https://doi.org/10.1002/open.202000344

Lu, J., Cai, Z., Zou, Y., Wu, D., Wang, A., Chang, J., Wang, F., Tian, Z., & Liu, G. (2019). Silver Nanoparticle-Based Surface-Enhanced Raman Spectroscopy for the Rapid and Selective Detection of Trace Tropane Alkaloids in Food. ACS Applied Nano Materials, 2 (10), 6592–6601. https://doi.org/10.1021/acsanm.9b01493

Lu, W., & Lieber, C. M. (2006). Semiconductor nanowires. Journal of Physics D: Applied Physics, 39 (21), R387. https://doi.org/10.1088/0022-3727/39/21/R01

M. R., K., Panicker, L. R., Narayan, R., & Kotagiri, Y. G. (2024). Biopolymer-protected graphene-Fe_3O_4 nanocomposite based wearable microneedle sensor: Toward real-time continuous monitoring of dopamine. RSC Advances, 14 (10), 7131–7141. https://doi.org/10.1039/D4RA00110A

Maduraiveeran, G., & Jin, W. (2017). Nanomaterials based electrochemical sensor and biosensor platforms for environmental applications. Trends in Environmental Analytical Chemistry, 13, 10–23. https://doi.org/10.1016/j.teac.2017.02.001

Mai, W., Liang, Z., Zhang, L., Yu, X., Liu, P., Zhu, H., Cai, X., & Tan, S. (2012). Strain sensing mechanism of the fabricated ZnO nanowire-polymer composite strain sensors. Chemical Physics Letters, 538, 99–101. https://doi.org/10.1016/j.cplett.2012.04.041

Malik, S., Singh, J., Goyat, R., Saharan, Y., Chaudhry, V., Umar, A., Ibrahim, A. A., Akbar, S.,

Ameen, S., & Baskoutas, S. (2023). Nanomaterials-based biosensor and their applications: A review. Heliyon, 9 (9), e19929. https://doi.org/10.1016/j.heliyon.2023.e19929

Manikandan, V. S., Adhikari, B., & Chen, A. (2018). Nanomaterial based electrochemical sensors for the safety and quality control of food and beverages. Analyst, 143 (19), 4537–4554. https://doi.org/10.1039/C8AN00497H

McConnell, E. M., Nguyen, J., & Li, Y. (2020). Aptamer-Based Biosensors for Environmental Monitoring. Frontiers in Chemistry, 8. https://doi.org/10.3389/fchem.2020.00434

Menzel, A., Subannajui, K., Güder, F., Moser, D., Paul, O., & Zacharias, M. (2011). Multifunctional ZnO-Nanowire-Based Sensor. Advanced Functional Materials, 21 (22), 4342–4348. https://doi.org/10.1002/adfm.201101549

Naresh, V., & Lee, N. (2021). A Review on Biosensors and Recent Development of Nanostructured Materials-Enabled Biosensors. Sensors, 21 (4), Article 4. https://doi.org/10.3390/s21041109

Ndaya, C. C., Javahiraly, N., & Brioude, A. (2019). Recent Advances in Palladium Nanoparticles-Based Hydrogen Sensors for Leak Detection. Sensors, 19 (20), Article 20. https://doi.org/10.3390/s19204478

Ngun, M. Y., Mohamad Nor, N., Ramli, N. H., & Abdul Razak, K. (2022). Platinum nanoparticles modified electrode for glucose sensor. Materials Today: Proceedings, 66, 2972–2976. https://doi.org/10.1016/j.matpr.2022.06.569

Oğuz, M., Aykaç, A., & Şen, M. (2024). A disposable sensor based on one-pot synthesized tungsten oxide nanostructure-modified screen printed electrodes for selective detection of dopamine and uric acid. Analytical Sciences, 40 (2), 301–308. https://doi.org/10.1007/s44211-023-00459-6

Ou, D., Yan, H., & Chen, Z. (2024). An impedance labeling free electrochemical aptamer sensor based on tetrahedral DNA nanostructures for doxorubicin determination. Microchimica Acta, 191 (2), 94. https://doi.org/10.1007/s00604-024-06176-9

Pacholski, C. (2013). Photonic Crystal Sensors Based on Porous Silicon. Sensors, 13 (4), Article 4. https://doi.org/10.3390/s130404694

Pan, C., Zhai, J., & Wang, Z. L. (2019). Piezotronics and Piezo-phototronics of Third Generation Semiconductor Nanowires. Chemical Reviews, 119 (15), 9303–9359. https://doi.org/10.1021/acs.chemrev.8b00599

Pan, M., Yin, Z., Liu, K., Du, X., Liu, H., & Wang, S. (2019). Carbon-Based Nanomaterials in Sensors for Food Safety. Nanomaterials, 9 (9). https://doi.org/10.3390/nano9091330

Pattanaik, S., Vishwkarma, A. K., Yadav, T., Shakerzadeh, E., Sahu, D., Chakroborty, S., Tripathi, P. K., Zereffa, E. A., Malviya, J., Barik, A., Sarankar, S. K., Sharma, P., Upadhye, V. J., & Wagadre, S. (2023). In silico investigation on sensing of tyramine by boron and silicon doped C60 fullerenes. Scientific Reports, 13 (1), 22264. https://doi.org/10.1038/s41598-023-49414-5

Pescaglini, A., & Iacopino, D. (2015). Metal nanoparticle–semiconductor nanowire hybrid nanostructures for plasmon-enhanced optoelectronics and sensing. Journal of Materials Chemistry C, 3 (45), 11785–11800. https://doi.org/10.1039/C5TC01820J

Phukan, P., & Sahu, P. P. (2020). High performance UV photodetector based on metal-semiconductor-metal structure using TiO2-rGO composite. Optical Materials, 109, 110330. https://doi.org/10.1016/j.optmat.2020.110330

Qin, X., Zhang, J., Shao, W., Liu, X., Zhang, X., Chen, F., Qin, X., Wang, L., Luo, D., & Qiao, X. (2021). Modification of electrodes with self-assembled, close-packed AuNPs for improved signal reproducibility toward electrochemical detection of dopamine. Electrochemistry Communications, 133, 107161. https://doi.org/10.1016/j.elecom.2021.107161

Raj, R., Mradula, Samanta, P., Singh, R., Sachdev, A., & Mishra, S. (2024). Evaluation of

AgNCs@PEI and their integrated hydrogel for colorimetric and fluorometric detection of ascorbic acid. Analytical Biochemistry, 687, 115433. https://doi.org/10.1016/j.ab.2023.115433

Rink, J. S., McMahon, K. M., Chen, X., Mirkin, C. A., Thaxton, C. S., & Kaufman, D. B. (2010). Transfection of pancreatic islets using polyvalent DNA-functionalized gold nanoparticles. Surgery, 148 (2), 335–345. https://doi.org/10.1016/j.surg.2010.05.013

Rizvi, S. B., Ghaderi, S., Keshtgar, M., & Seifalian, A. M. (2010). Semiconductor quantum dots as fluorescent probes for in vitro and in vivo bio-molecular and cellular imaging. Nano Reviews, 1 (1), 5161. https://doi.org/10.3402/nano.v1i0.5161

Rowland, C. E., Brown, C. W., Delehanty, J. B., & Medintz, I. L. (2016). Nanomaterial-based sensors for the detection of biological threat agents. Materials Today, 19 (8), 464–477. https://doi.org/10.1016/j.mattod.2016.02.018

Sahoo, A. K., Banerjee, S., Ghosh, S. S., & Chattopadhyay, A. (2014). Simultaneous RGB Emitting Au Nanoclusters in Chitosan Nanoparticles for Anticancer Gene Theranostics. ACS Applied Materials & Interfaces, 6 (1), 712–724. https://doi.org/10.1021/am4051266

Salman, B.I. (2024). An Innovative Selective Fluorescence Sensor for Quantification of Hazardous Food Colorant Allura Red in Beverages Using Nitrogen-Doped Carbon Quantum Dots. Journal of Fluorescence, 34 (2), 599–608. https://doi.org/10.1007/s10895-023-03303-2

Samia, A. C. S., Chen, X., & Burda, C. (2003). Semiconductor Quantum Dots for Photodynamic Therapy. Journal of the American Chemical Society, 125 (51), 15736–15737. https://doi.org/10.1021/ja0386905

Sapsford, K. E., Pons, T., Medintz, I. L., & Mattoussi, H. (2006). Biosensing with Luminescent Semiconductor Quantum Dots. Sensors, 6 (8), Article 8. https://doi.org/10.3390/s6080925

Scognamiglio, V., Arduini, F., Palleschi, G., & Rea, G. (2014). Biosensing technology for sustainable food safety. TrAC Trends in Analytical Chemistry, 62, 1–10. https://doi.org/10.1016/j.trac.2014.07.007

Sengan, M., Kamlekar, R. K., & Veerappan, A. (2020). Highly selective rapid colorimetric sensing of Pb2+ ion in water samples and paint based on metal induced aggregation of N-decanoyltromethamine capped gold nanoparticles. Spectrochimica Acta Part A: Molecular and Biomolecular Spectroscopy, 239, 118485. https://doi.org/10.1016/j.saa.2020.118485

Sengupta, J., & Hussain, C. M. (2021). Graphene-based field-effect transistor biosensors for the rapid detection and analysis of viruses: A perspective in view of COVID-19. Carbon Trends, 2, 100011. https://doi.org/10.1016/j.cartre.2020.100011

Sengupta, J., & Hussain, C. M. (2023). CNT and Graphene-Based Transistor Biosensors for Cancer Detection: A Review. Biomolecules, 13 (7), 1024. https://doi.org/10.3390/biom13071024

Shellaiah, M., & Sun, K. W. (2023). Conjugation of cysteamine functionalized nanodiamond to gold nanoparticles for pH enhanced colorimetric detection of Cr3+ ions demonstrated by real water sample analysis. Spectrochimica Acta Part A: Molecular and Biomolecular Spectroscopy, 286, 121962. https://doi.org/10.1016/j.saa.2022.121962

Shirsat, M. D., & Hianik, T. (2023). Electrochemical Detection of Heavy Metal Ions Based on Nanocomposite Materials. Journal of Composites Science, 7 (11), Article 11. https://doi.org/10.3390/jcs7110473

Shruti, A., Bage, N., & Kar, P. (2024). Nanomaterials based sensors for analysis of food safety. Food Chemistry, 433, 137284. https://doi.org/10.1016/j.foodchem.2023.137284

Solorio-Rodriguez, S. A., Williams, A., Poulsen, S. S., Knudsen, K. B., Jensen, K. A., Clausen,

P. A., Danielsen, P. H., Wallin, H., Vogel, U., & Halappanavar, S. (2023). Single-Walled vs. Multi-Walled Carbon Nanotubes: Influence of Physico-Chemical Properties on Toxicogenomics Responses in Mouse Lungs. Nanomaterials, 13 (6), 1059. https://doi.org/10.3390/nano13061059

Speranza, G. (2021). Carbon Nanomaterials: Synthesis, Functionalization and Sensing Applications. Nanomaterials, 11 (4), 967. https://doi.org/10.3390/nano11040967

Su, S., Wu, W., Gao, J., Lu, J., & Fan, C. (2012). Nanomaterials-based sensors for applications in environmental monitoring. Journal of Materials Chemistry, 22 (35), 18101. https://doi.org/10.1039/c2jm33284a

Sweet, M. J., Chessher, A., & Singleton, I. (2012). Chapter Five - Review: Metal-Based Nanoparticles; Size, Function, and Areas for Advancement in Applied Microbiology. In S. Sariaslani & G. M. Gadd (Eds.), Advances in Applied Microbiology (Vol. 80, pp. 113–142). Academic Press. https://doi.org/10.1016/B978-0-12-394381-1.00005-2

Tang, X., Debliquy, M., Lahem, D., Yan, Y., & Raskin, J.-P. (2021). A Review on Functionalized Graphene Sensors for Detection of Ammonia. Sensors, 21 (4), Article 4. https://doi.org/10.3390/s21041443

Thakur, M. S., & Ragavan, K. V. (2013). Biosensors in food processing. Journal of Food Science and Technology, 50 (4), 625–641. https://doi.org/10.1007/s13197-012-0783-z

Tiwari, P. M., Vig, K., Dennis, V. A., & Singh, S. R. (2011). Functionalized Gold Nanoparticles and Their Biomedical Applications. Nanomaterials, 1 (1), 31–63. https://doi.org/10.3390/nano1010031

Tressler, J. F., Alkoy, S., & Newnham, R. E. (1998). Piezoelectric Sensors and Sensor Materials. Journal of Electroceramics, 2 (4), 257–272. https://doi.org/10.1023/A:1009926623551

Urbiztondo, M., Pina, P., & Santamaría, J. (2009). Chapter 15—Gas Sensing with Silicon-Based Nanoporous Solids. In V. Valtchev, S. Mintova, & M. Tsapatsis (Eds.), Ordered Porous Solids (pp. 387–411). Elsevier. https://doi.org/10.1016/B978-0-444-53189-6.00015-9

Uygun, H. D. E., Uygun, Z. O., Uygun, H. D. E., & Uygun, Z. O. (2020). Fullerene Based Sensor and Biosensor Technologies. In Smart Nanosystems for Biomedicine, Optoelectronics and Catalysis. IntechOpen. https://doi.org/10.5772/intechopen.93316

Vo-Dinh, T., Cullum, B. M., & Stokes, D. L. (2001). Nanosensors and biochips: Frontiers in biomolecular diagnostics. Sensors and Actuators B: Chemical, 74 (1), 2–11. https://doi.org/10.1016/S0925-4005 (00)00705-X

Wang, D., Wang, L., & Shen, G. (2020). Nanofiber/nanowires-based flexible and stretchable sensors. Journal of Semiconductors, 41 (4), 041605. https://doi.org/10.1088/1674-4926/41/4/041605

Wang, Y., Duan, H., Yalikun, Y., Cheng, S., & Li, M. (2024). A pendulum-type electrochemical aptamer-based sensor for continuous, real-time and stable detection of proteins. Talanta, 266, 125026. https://doi.org/10.1016/j.talanta.2023.125026

Wang, Y., Wang, T., Da, P., Xu, M., Wu, H., & Zheng, G. (2013). Silicon Nanowires for Biosensing, Energy Storage, and Conversion. Advanced Materials, 25 (37), 5177–5195. https://doi.org/10.1002/adma.201301943

Watahiki, R., Shimada, T., Zhao, P., Chiashi, S., Iwamoto, S., Arakawa, Y., Maruyama, S., & Kato, Y. K. (2012). Enhancement of carbon nanotube photoluminescence by photonic crystal nanocavities. Applied Physics Letters, 101 (14), 141124. https://doi.org/10.1063/1.4757876

Wei, Z., Feng, L., Zhi-Ming, J., Xiao-Bo, S., Peng-Hui, Y., Xue-Ren, W., Cheng, S., Zhan-

Qi, G., & Liang-Sheng, L. (2014). Efficient plasmonic photocatalytic activity on silver-nanoparticle-decorated AgVO3 nanoribbons. Journal of Materials Chemistry A, 2 (33), 13226–13231. https://doi.org/10.1039/C4TA02369B

Willner, M. R., & Vikesland, P. J. (2018). Nanomaterial enabled sensors for environmental contaminants. Journal of Nanobiotechnology, 16 (1), 95. https://doi.org/10.1186/s12951-018-0419-1

Wu, Q., Tian, L., Shan, X., Li, H., Yang, S., Li, C., & Lu, J. (2024). An electrochemiluminescence sensor based on the CNTs and CdSe@ZnSe for determination of melamine in milk samples. Food Chemistry, 430, 137028. https://doi.org/10.1016/j.foodchem.2023.137028

Xi, Z., Xing, J., Yuan, R., & Yuan, Y. (2024). Covalent organic frame based high-performance nanocomposite for construction of ATP sensor. Biosensors and Bioelectronics, 250, 116081. https://doi.org/10.1016/j.bios.2024.116081

Xiao, Z., Zhang, X., Hong, S., Zhang, H., & Zhang, Y. (2024). A platform for microplastic assessment in aquatic environments based on the protein corona-induced aggregation effect. Biosensors and Bioelectronics, 249, 116037. https://doi.org/10.1016/j.bios.2024.116037

Xu, F., Yao, Y., Bai, D., Xu, R., Mei, J., Wu, D., Gao, Z., & Jiang, K. (2015). Au nanoparticle decorated WO3 photoelectrode for enhanced photoelectrochemical properties. RSC Advances, 5 (74), 60339–60344. https://doi.org/10.1039/C5RA06241A

Xu, Y., Xu, C., Huang, Q., Cao, L., Teng, F., Zhao, P., & Jia, M. (2021). Size Effect of Mesoporous Silica Nanoparticles on Pesticide Loading, Release, and Delivery in Cucumber Plants. Applied Sciences, 11 (2), Article 2. https://doi.org/10.3390/app11020575

Yang, F.-Q., & Ge, L. (2023). Colorimetric Sensors: Methods and Applications. Sensors, 23 (24), Article 24. https://doi.org/10.3390/s23249887

Yang, L., Hu, W., Pei, F., Liu, Z., Wang, J., Tong, Z., Mu, X., Du, B., Xia, M., Wang, F., & Liu, B. (2024). A ratiometric fluorescence imprinted sensor based on N-CDs and metal–organic frameworks for visual smart detection of malathion. Food Chemistry, 438, 138068. https://doi.org/10.1016/j.foodchem.2023.138068

Yano, K., & Karube, I. (1999). Molecularly imprinted polymers for biosensor applications. TrAC Trends in Analytical Chemistry, 18 (3), 199–204. https://doi.org/10.1016/S0165-9936 (98)00119-8

Yi, H., Ran, J., Tan, Y., Wang, Z., & Liu, B. (2024). A colorimetric/electrochemical sensor based on coral-like CuCo2O4@AuNPs composites for sensitive dopamine detection. Analytical and Bioanalytical Chemistry, 416 (1), 265–276. https://doi.org/10.1007/s00216-023-05014-w

Yilmaz, H., Ertaş, N., & Basan, H. (2024). Development of a new phosphorescence sensor based on surface molecularly imprinted Mn-doped ZnS quantum dots for detection of melamine in milk products. Spectrochimica Acta Part A: Molecular and Biomolecular Spectroscopy, 309, 123818. https://doi.org/10.1016/j.saa.2023.123818

Ying Lim, S., Shen, W., & Gao, Z. (2015). Carbon quantum dots and their applications. Chemical Society Reviews, 44 (1), 362–381. https://doi.org/10.1039/C4CS00269E

You, H., Mu, Z., Zhao, M., Zhou, J., Yuan, Y., & Bai, L. (2020). Functional fullerene-molybdenum disulfide fabricated electrochemical DNA biosensor for Sul1 detection using enzyme-assisted target recycling and a new signal marker for cascade amplification. Sensors and Actuators B: Chemical, 305, 127483. https://doi.org/10.1016/j.snb.2019.127483

Yurkov, G., Koksharov, Y., Fionov, A., Taratanov, N., Kolesov, V., Kirillov, V., Makeev, M., Mikhalev, P., Ryzhenko, D., & Solodilov, V. (2022). Polymer Nanocomposite Containing Palladium Nanoparticles: Synthesis, Characterization, and Properties. Polymers, 14 (18), Article 18. https://doi.org/10.3390/polym14183795

Zeng, W., Tang, X., Wu, T., Han, B., & Wu, L. (2024). Development of a highly sensitive aptamer-based electrochemical sensor for detecting saxitoxin based on K3Fe (CN)6 regulated silver nanoparticles. Analytica Chimica Acta, 1287, 342134. https://doi.org/10.1016/j.aca.2023.342134

Zhai, Z., Wang, W., Chai, Z., Yuan, Y., Zhu, Q., Ge, J., & Li, Z. (2024). A ratiometric fluorescence platform based on WS2 QDs/CoOOH nanosheet system for α-glucosidase activity detection. Spectrochimica Acta Part A: Molecular and Biomolecular Spectroscopy, 310, 123959. https://doi.org/10.1016/j.saa.2024.123959

Zhang, H., Liang, F., Li, S., Zong, F., & Xu, Y. (2024). A high-performance photoelectrochemical sensor based on CdS–Au composite nanomaterials and localized surface plasmon resonance for ultrasensitive detection of ascorbic acid. Analytical Methods, 16 (8), 1175–1184. https://doi.org/10.1039/D3AY02007J

Zhang, J., Chen, Y., Qi, J., Miao, Q., Deng, D., He, H., Yan, X., & Luo, L. (2024). A paper-based ratiometric fluorescence sensor based on carbon dots modified with Eu $^{3+}$ for the selective detection of tetracycline in seafood aquaculture water. The Analyst, 149 (5), 1571–1578. https://doi.org/10.1039/D3AN02133E

Zhang, X., Wang, P., Liang, Z., Zhong, W., & Ma, Q. (2024). A novel Cu-MOFs nanosheet/BiVO4 nanorod-based ECL sensor for colorectal cancer diagnosis. Talanta, 266, 124952. https://doi.org/10.1016/j.talanta.2023.124952

Zhang, Y., Jin, S., Liu, R., Liu, Z., Gong, L., Zhang, L., Zhao, T., Yin, W., Chen, S., Fa, H., & Niu, L. (2024). A portable magnetic electrochemical sensor for highly efficient Pb (II) detection based on bimetal composites from Fe-on-Co-MOF. Environmental Research, 250, 118499. https://doi.org/10.1016/j.envres.2024.118499

Zhao, T., Wu, D., Zhang, X., & Lyu, H. (2024). A fluorescent sensor based on single band bright red luminescent core-shell UCNPs for the high-sensitivity detection of glucose and glutathione. Analytica Chimica Acta, 1295, 342323. https://doi.org/10.1016/j.aca.2024.342323

Zhao, X., Lu, Y., Li, B., Kong, M., Sun, Y., Li, H., Liu, X., & Lu, G. (2024). Self-ratiometric fluorescent platform based on upconversion nanoparticles for on-site detection of chlorpyrifos. Food Chemistry, 439, 138100. https://doi.org/10.1016/j.foodchem.2023.138100

Zheng, X., Guo, L., Zhu, C., Hu, T., Gong, X., Wu, C., Wang, G., Dong, H., & Hou, Y. (2024). A robust electrochemical sensor based on AgNWs@MoS2 for highly sensitive detection of thiabendazole residues in food samples. Food Chemistry, 433, 137304. https://doi.org/10.1016/j.foodchem.2023.137304

Zhou, B., Sheng, X., Cao, J., Xie, H., Li, X., Huang, L., Yang, M., Zhong, M., & Liu, Y.-N. (2024). A novel electrochemical sensor based on dual-functional MMIP-CuMOFs for both target recognition and signal reporting and its application for sensing bisphenol A in milk. Food Chemistry, 437, 137756. https://doi.org/10.1016/j.foodchem.2023.137756

Zhu, H., Sun, Z., Wang, X., & Xia, H. (2024). A High-Performance Strain Sensor for the Detection of Human Motion and Subtle Strain Based on Liquid Metal Microwire. Nanomaterials, 14 (2), 231. https://doi.org/10.3390/nano14020231

Zor, E., Mollarasouli, F., Karadurmus, L., Ozcelikay, G., & Ozkan, S. A. (2024). Carbon Dots in the Detection of Pathogenic Bacteria and Viruses. Critical Reviews in Analytical Chemistry, 54 (2), 219–246. https://doi.org/10.1080/10408347.2022.2072168

4

Nanomaterials Unleashed Revolutionizing Analytical Capabilities in Sensing and Detection

Sunita Panchawat*, Joohee Pradhan and Deepika Bairagee

Department of Pharmaceutical Sciences, Mohanlal Sukhadia University, Udaipur Rajasthan

Abstract

This section explores how nanoparticles have a significant influence on sensing and detection, explaining how they have revolutionized analytical capabilities. Our trip through many kinds of nanomaterials, such as metal nanoparticles, semiconductor nanomaterials, and carbon-based structures, aims to explore their distinct features and how they might be used for novel sensing methods. The intricacies behind sophisticated sensor technologies are clarified through an examination of plasmonic sensing, quantum dot applications, and conductive nanomaterial responses. In order to improve selectivity, the chapter emphasises surface modifications and biomimetic coatings as key components of strategic functionalization techniques. Applications include environmental monitoring for pollution detection, food safety applications guaranteeing fast pathogen identification, and biomedical sensing for early illness detection.We talk about the difficulties in using nanoparticles, such as scalability and toxicity issues, and we look forward to a day when nanomaterials will continue to push the limits of analytical capabilities. "Nanomaterials Unleashed" is a thorough investigation that emphasises the critical role nanoscale materials play in changing analytical techniques and opening doors for creative solutions across a range of sectors.

Keywords: *Nanomaterials,Plasmonic sensing, Semiconductor nanomaterials, Quantam dot application.*

1. Introduction

Even though the word "nano" initially meant "dwarf" or "very low in height," it now mentions to a billionth of a unit in the international system of

* *Corresponding author e-mail: panchawatsunita@gmail.com*

measurements. A nanometer, for example, is one billionth of a metre. In the last ten years, the prefix "nano" has become extensively utilized in numerous scientific domains and has acquired admiration as a designation in scientific publications (Bayda *et al.*, 2019; Andrews *et al.*, 2019). Since of their special qualities, nanomaterials, whose diameters range from 1 to 100 nm have drawn attention from the scientific community in recent years. Substances with dimensions ranging from 1 to 100 nm in at least one dimension are recognized as nanomaterials, and they vary from their bigger counterparts in several significant ways.They may be divided into groups according to their size, place of origin, structural makeup, pore sizes, and likely toxicity. Nanomaterials differ from solid, liquid, gaseous, and plasma states of matter in that they have special chemical, physical, and biological properties that are dependent on their size (Saxena *et al.*, 20220).

Nanomaterials, which may be separated into five main types, are utilized in a diversity of fields, including food engineering, biomedicine, electronics, energy, food processing, transportation, communications, cosmetics, coatings, materials, and mechanical engineering. Aextensive range of sectors might advantage from nanotechnology, which includes the synthesis, engineering, and use of nanoparticles. The study of solid-state materials' size-dependent properties is called nanoscience, and it focuses on the properties of matter at the nanoscale. Interest in nanotechnology has grown since of its novel uses, which have a result on many dissimilar sectors. When it comes to solving issues like illnesses and global climate change, nanomaterials are necessary. Nanomaterial-based green solutions have shown to be more successful in reducing the dangers associated with climate change (Joudeh *et al.*, 2022; Binns, 2021).

The ability of nanomaterials to detect and manage worldwide pandemics such as Covid-19 and monkeypox demonstrates their promise. Nanomaterials are expected to propel global growth as they continue to progress. Having sufficient knowledge and comprehension of nanomaterials becomes essential for people in a variety of industries. In spite of the large body of study on nanoparticles and materials, most contemporary studies concentrate on the characteristics, properties, applications, and techniques of synthesis of particular nanomaterials. For a holistic understanding and use, a thorough method that articulates the characteristics, processing, and categorization of nanomaterials across several disciplines is required (Bayda *et al.*, 2019; Binns, 2021; Rizwan *et al.*, 2021; Sannino, 2021).

In the field of analytical chemistry, nanomaterials have become ground-breaking instruments that greatly improve sensing and detection capacities.

This chapter inspects how nanoparticles have revolutionised analytical methods by illuminating the singular qualities that permit for formerly unheard-of improvements in sensitivity, selectivity, and real-time monitoring. The combination of analytical sciences and nanotechnology has shapednovelchances for the creation of extremely adaptable and well-organized sensors, opening doors for advancements in a range of industries, from environmental monitoring to healthcare.

1.1. Classification of Nanomaterials

Numerous factors, including as size, origin, structure, and possible uses, may be used to categorise nanomaterials. The following is a broad categorization according to origin, size, and structure (Fig. 1):

Based on the size	**Based on Structure**	**Based on Origin**
• Nanoparticles • Nanostructures (Nanowires, Nanotubes, and Nanorods) • Nanocomposites (Polymer matrix with embeded nanopartices)	• Zero-Dimensional (0D) Nanomaterials (Nanoparticles) • One-Dimensional (1D) Nanomaterilas (Nanotubes, Nanowires) • Two-Dimensional (2D) Nanomaterilas (Graphene, thin films) • Three-Dimensional (3D) Nanomaterials (Aerogels, Nanocompoites)	• Natural Nanomaterilas (Viruses, Proteins, Clay minerals) • Engineered Nanomaterilas (Quantum dots, carbon nanotubes)

Fig. 1: General Classification of Nanomaterials

1.1.1. Based on the size

With diameters between 1 and 100 nanometres, nanomaterials are materials with special characteristics and behaviours that set them apart from their macroscopic counterparts. The word 'nano' itself indicates this size, highlighting how tiny these materials are. Materials frequently exhibit enhanced surface area, quantum effects, and changed physical and chemical properties at the nanoscale. Fundamental components are nanoparticles, which have at least one dimension in the nanoscale. Nanoscale characteristics are possessed by nanostructures, which go beyond the conventional definitions of particles, such nanowires and nanotubes. Thanks to the development of nanoengineering, materials with specific nanomaterial compositions known as nanocomposites can now have customised features and functions. This size-based categorization highlights the importance of the nanoscale in revealing new characteristics that have the potential to transform a number of scientific and technical fields.

1.1.2. Based on the structure

When categorised according to their structural makeup, nanomaterials have a wide range of compositions that significantly influence their characteristics and uses. One common classification makes a distinction between nanomaterials that are zero-dimensional (0D), one-dimensional (1D), two-dimensional (2D), and three-dimensional (3D). Zero-dimensional nanomaterials exhibit distinct quantum effects because they have no dimensionality outside of the nanoscale, such as quantum dots and nanoparticles. One-dimensional nanomaterials, such as nanowires and nanotubes, are characterised by their one-dimensional extension, high aspect ratios, and customised conductivity. Remarkable planar patterns are exhibited by two-dimensional nanomaterials, such graphene and other 2D nanosheets, which find uses in composites and electronics.Three-dimensional nanomaterials combine many dimensions, improving functions and paving the way for new developments in materials science. They are frequently found as complex nanostructures or nanocomposites. Nanomaterials structural categorization offers a methodical framework for comprehending and using their unique properties in a variety of scientific and technical fields.

1.1.3. Based on the Origin

Based on their origin, which can come from a variety of sources and contribute to their distinct qualities, nanomaterials can be categorised. Minerals, viruses, and proteins are examples of naturally occurring nanomaterials that have intrinsic nanoscale dimensions. Conversely, engineered nanomaterials are purposefully created and modified for particular uses through the use of processes including biological synthesis, sol-gel procedures, and chemical vapour deposition. Biomimetic nanomaterials mimic biological processes to develop structures with improved functionality, taking inspiration from nature. Furthermore, inadvertent human activity like combustion or industrial pollutants may produce incidental nanomaterials. The origin-based classification highlights the wide range of nanomaterial sources, each with unique benefits and uses in different scientific and technical fields (Rizwan *et al.*, 2021; Sannino, 2021; Saleh, 2020; Buzea and Pacheco, 2017; Saha *et al.*, 2022; Trotta and Mele, 2019; Darweesh, 2018; Khan, 202; Sudha *et al.*, 2019).

1.2. Synthesis Methods of Nanomaterials

Different techniques can be used to synthesise nanomaterials, each with the aim of achieving certain characteristics and uses. There are two basic strategies that are frequently used: top-down and bottom-up techniques (Fig. 2).

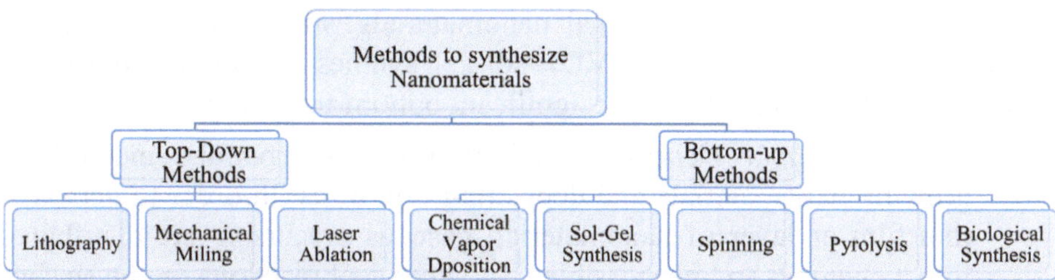

Fig. 2: Methods to synthesize Nanomaterials

1.2.1. Top-Down Methods

Using top-down techniques, bigger structures are shrunk to nanoscale levels, giving perfect control over the final nanomaterials' dimensions. These methods are crucial for applications that demand certain nanoscale characteristics.

Lithography is a well-known top-down method that is often used in the electronics sector. By using light to transmission a pattern from a mask to a photosensitive material, photolithography allows for the formation of nanoscale structures on a substrate. Alike to this, Electron Beam Lithography creates complex nanoscale patterns by using a concentrated electron beam to produce much smaller details.

Mechanical milling, which physically breaks down bulk materials into tiny particles, is another top-down method. This process works especially well for creating nanoparticles and nanocomposites. Larger structures are broken up into nanoscale particles by the mechanical energy imparted by the grinding or milling process. This method is adaptable and works with a variety of materials.

A sophisticated top-down technique called laser ablation uses lasers to remove material from a target while also producing nanoparticles. Particle composition and size may be precisely controlled thanks to the laser ablation process. It works especially well to create metal and semiconductor nanoparticles with special characteristics.

Top-down approaches are appropriate for applications needing particular nanoscale characteristics because they provide high accuracy and control. For some materials and applications, these methods might not be able to achieve uniformity and scalability. Top-down approaches remain essential to the synthesis of nanomaterials and the development of nanotechnology as long as technology progresses.

1.2.2. Bottom-up Methods

Atomic or molecular building blocks are assembled or grown to create the required nanostructure in nanomaterials synthesised bottom-up. These methods provide one control over the final nanomaterials' size, content, and form. Chemical vapour deposition (CVD), sol-gel synthesis, spinning, pyrolysis, and biological synthesis are a raresignificant bottom-up methods.

a. *Chemical Vapor Deposition (CVD):* CVD is a popular method for growing nanomaterials and depositing thin films. In order to deposit a thin film or layer of nanomaterial, gaseous precursors are distributed into a chamber and permissible to do chemical reactions on the surface of a substrate. Graphene layer production, coating creation, and semiconductor manufacture all use CVD.

b. *Sol-Gel Synthesis:* A wet chemical method called sol-gel synthesis is utilized to generate nanomaterials from a solution. Hydrolysis and polycondensation processes take place in a liquid solution of precursors, resulting in a sol that fluctuations into a gel and is then dried to yield the essential nanomaterial. Oxide nanoparticles, such titania or silica, are frequently synthesised using this technique for usage in coatings, sensors, and catalysts.

c. *Spinning:* Nanofibers or nanotubes are made by spinning liquid precursors. Spinning a liquid precursor aims the molecules to align and stretch, resulting in the creation of nanoscale fibres. The procedure of spinning is usually utilised to yield nanofibers for a diversity of uses, like tissue engineering, filtration, and textiles.

d. *Pyrolysis:* In pyrolysis, organic precursors are thermally damaged down to produce nanomaterials. When organic materials are heated in an atmosphere lacking of oxygen, volatile components are free and carbon-based nanoparticles are formed. Carbon nanotubes, graphene, and other carbonaceous nanostructures with uses in electronics and energy storage are shaped via pyrolysis.

e. *Biological Synthesis:* In biological synthesis, the creation of nanomaterials is mediated by living things or their constituent parts. Enzymes, bacteria, and plants are instances of biological organisms that aid in the decrease and maintenance processes that permit the formation of nanoparticles. Developing an eco-friendly process, metal nanoparticles with potential usages in environmental remediation, medicine, and catalysis are synthesised.

Bottom-up methods are important to nanotechnology since they permit nanoparticles to be precisely engineered for specialised uses in a wide range of industries. With the help of these methods, nanomaterial characteristics may be tailored to address a diversity of scientific and technological difficulties

(Baig *et al.*, 2021; Tulinski and Jurczyk, 2017; Varghese *et al.*, 2019; Huston *et al.*, 2021; Kolahalam *et al.*, 2019; Bhagyaraj *et al.*, 2018; Abid *et al.*, 2022; Ahmed; 2020).

1.3. Properties of Nanomaterials

Distinguished from their bulk counterparts by their distinct qualities, nanomaterials are defined by their nanoscale size. These unique features result from surface area expansion, modified electrical and mechanical behaviours, and quantum phenomena. Key characteristics of nanoparticles consist of:

a. *Size-Dependent Properties:* Size has a substantialinfluence on the characteristics of nanomaterials, with quantum effects becoming more projecting at the nanoscale. The electrical and optical actions are enhanced by quantum confinement, giving rise to novel phenomena like the size-dependent tunable fluorescence of quantum dots.

b. *Increased Surface Area:* Oncematched to bulk materials, nanomaterials have knowingly larger surface area-to-volume ratios. Their upgraded reactivity due to their larger surface area qualifies them for usage in medication delivery, sensors, and catalysis.

c. *Optical Properties:* The size-dependent optical characteristics of quantum dots and other nanomaterials include increased emission and absorption. Numerous industries, counting imaging, displays, and optoelectronics, use this characteristic.

d. *Mechanical Properties:* Better-quality mechanical properties such as increasing rigidity and tensile strength are regularlyexhibited by nanoparticles. For instance, the exceptional strength of carbon nanotubes makes them valuable for materials that need reinforcement.

e. *Thermal Conductivity:* Nanomaterials can displayamplified thermal conductivity due to their better-quality phonon transport. This property is vital for applications involving temperature management, nanoelectronics, and thermoelectric devices.

f. *Electrical Conductivity:* Definite nanoparticles and graphene, for instance, display unexpected electrical conductivity. This characteristic novelties application in nanoelectronics, conductive coatings, and electronics.

g. *Chemical Reactivity:* Nanomaterials are active catalysts used in energy conversion, industrial operations, and environmental remediation because of their enormous surface area, which increases chemical reactivity.

h. *Magnetic Properties:* Size-dependent magnetic presences are uncovered by nanoscale magnetic materials, such as nanoparticles. Magnetic storage, biological applications, and magnetic resonance imaging (MRI) all make usage of these materials.

i. *Biocompatibility:* Definite nanomaterials show biocompatibility, particularly those intended for usage in biomedical applications. For applications in nanomedicine like as imaging, medication administration, and others, this feature is crucial.

j. *Tunable Properties:* A lot of nanomaterials have features that can be tuned by changing their surface modifications, composition, or size. When customising nanomaterials for particular uses, this tunability is beneficial (Wu *et al.*, 2020; Lue, 2007; Čitaković, 2019; Asha and Narain, 2020; Gubicza, 2017).

Comprehending and using these distinct characteristics facilitates the creation of pioneering technologies in several fields, including as electronics, medical, energy, and materials research. Investigating and modifying these characteristics becomes more important as nanotechnology develops in order to create customised nanomaterials with improved functionality and performance.

2. Nanomaterials in Sensing Platforms

This segmentdiscovers the extensivediversity of nanomaterials and their applications in sensing platforms, comprising nanoparticles, nanotubes, nanowires, and nanocomposites. It discourses the special physicochemical qualities of nanomaterials, like their high surface-to-volume ratio and amplified reactivity, and displays how these qualities might be utilized to improve sensor performance. The flexibility and usefulness of these materials in analytical chemistry are established by case studies and instances of nanomaterial-based sensors for recognising certain analytes.

2.1. Graphene-Based DNA Sensors

DNA sensors have been produced for genetic detection by taking use of graphene's remarkable electrical conductivity. High sensitivity DNA hybridization episodediscovery is likely using graphene-based field-effect transistor (FET) sensors, which may find usage in medical diagnostics (Wu *et al.*, 2018; Shahdeo *et al.*, 2020).

2.2. Quantum Dot-Based Biosensors

Biosensors using quantum dots (QDs) have been utilized to classify biomolecules. For instance, the sensitive detection of proteins is made likely by CdSe/ZnS QDs functionalized with certain ligands. The watching and discovery of diseases may be affected by these QD-based biosensors (Ma *et al.*, 2018; Zhang *et al.*, 2022).

2.3. Carbon Nanotube Gas Sensors

Gas sensors engagement single-walled carbon nanotubes (SWCNTs) to classify gases like NH3 and NO2. Functionalization of SWCNTs recovers their sensitivity and selectivity, which makes them valuable for industrial safety and environmental monitoring (Verma *et al.*, 2022).

2.4. Gold Nanoparticle-Based Colorimetric Sensors

For a diversity of analytes, colorimetric sensors use gold nanoparticles (AuNPs). For instance, when target molecules are extant, AuNPs functionalized with definite ligands maycollective and alteration colour visibly. This impression is utilized to classify infections, ions, and tiny compounds (Kusuma *et al.*, 2023).

2.5. Magnetic Nanoparticle-Based Immunoassays

In immunoassays, magnetic nanoparticles (MNPs) functionalized with antibodies have been utilized to notice certain proteins. By successfullyunravelling and concentrating target analytes, these magnetic immunoassays recover sensitivity (Nishat *et al.*, 2019).

2.6. 2D Material-Based Gas Sensors

Gas molecules can be detected using sensors built on 2D materials such as graphene and transition metal dichalcogenides (TMDs). Discriminating detection is made promising by the electrical conductivity alterations that TMDs, such MoS2, knowledge when bare to definite gases (Zhang *et al.*, 2019; Buckley *et al.*, 2020; Kumar *et al.*, 2020; Choi and Kim, 2018).

2.7. PlasmonicNanosensors for Chemical Detection

Plasmonic nanoparticles are utilized for chemical finding in surface-enhanced Raman spectroscopy (SERS), like gold nanorods. Chemical trace levels can be recognised thanks to the improved electromagnetic fields everywhere the nanoparticles (Brahmkhatri *et al.*, 2021).

2.8. Polymeric Nanocomposite Sensors for Environmental Monitoring

Environmental sensors usage polymeric nanocomposites that include nanoparticles like metal oxides. These sensors support environmental monitoring creativities by identifyingimpurities in the air and water (Tajik *et al.*, 2021; Toto *et al.*, 2022).

2.9. Nanowire-Based FET Sensors for Biological Molecules

FET sensors centred on silicon nanowires have been planned to categorise biological substances like DNA and proteins. Label-free acknowledgement is completedpossible by the binding of target molecules, which alters the electrical conductance of nanowires (Zhang *et al.*, 2021).

2.10. Magnetic Resonance Imaging (MRI) Contrast Agents

In magnetic resonance imaging (MRI) for biological purposes, superparamagnetic iron oxide nanoparticles (SPIONs) are activated as dissimilarity agents. Functionalized SPIONs proposalbetter-quality imaging capabilities by focusing on definite tissues or cells (Cheng *et al.*, 2013).

These case studies determine the wide range of usages for sensors founded on nanomaterials in numerous fields. The flexibility and sensitivity of nanomaterial-based sensors provision the expansion of analytical chemistry and sensing technologies in a diversity of applications, counting environmental monitoring and medical diagnostics.

3. Advanced Nanomaterials for Sensing Applications

This article inspects the influence that cutting-edge nanomaterials, like graphene, quantum dots, metal-organic frameworks (MOFs), and 2D materials, have in growing the abilities of sensing. The modified design of nanomaterials is given singular consideration in order to get the essential sensing characteristics, which permits for the unmatched precision of the discovery of trace amounts of chemicals. To demonstrate the adaptability of these materials in diverse analytical settings, the integration of nanomaterials into transduction platforms, such as electrochemical, optical, and mass-based sensors, is explored (Wang *et al.*, 2019).

Nanomaterials have transformed sensing technologies by providing improved characteristics for analysis and detection across a range of domains. The needs of sensitive, fast, and selective sensing applications across several domains have been satisfied by the development of advanced nanomaterials. The following are significantgrowths in nanomaterials for sensing applications:

3.1. Graphene and 2D Materials

Two-dimensional (2D) materials such as transition metal dichalcogenides (TMDs) and graphene, a single sheet of carbon atoms, have extraordinary mechanical, optical, and electrical capabilities. Due to their countlesscompassion and rapid reaction times, these materials are utilized in sensors to detect gases, biomolecules, and environmental contaminants (Wu *et al.*, 2018;Shahdeo *et al.*, 2020; Buckley *et al.*, 2020; Kumar *et al.*, 2020; Choi *et al.*, 2018).

3.2. Quantum Dots (QDs)

Founded on their size, semiconductor nanocrystals recognised as quantum dots deliver optical features that may be adjusted. They are utilized in fluorescence-based sensors to classify infections, ions, and biomolecules with accuracy. QDs permit for multiplexed sensing applications and recovercompassion (Ma *et al.*, 2018; Zhang *et al.*, 2022).

3.3. Metal Nanoparticles

Silver (AgNPs) and gold (AuNPs) are binaryinstances of metal nanoparticles that have singularplasmonicfeatures. By altering localised superficialplasmon resonance (LSPR), plasmonicnanosensors take use of these features to deliver label-free detection that permits for rapid and sensitive sensing (Jamkhande *et al.*, 2019).

3.4. Carbon Nanotubes (CNTs)

Both mechanical strength and electrical conductivity are extraordinary qualities of carbon nanotubes. Environmental monitoring, biosensing, and gas sensing all make usage of CNT-based sensors. Real-time detection and better-quality selectivity are made conceivable by functionalized CNTs (Salah *et al.*, 2021).

3.5. Metal-Organic Frameworks (MOFs)

High surface area and adaptable architectures are providing by metal-organic frameworks. Specific molecules are captured using MOF-based sensors, which are used for gas adsorption and detection. MOFs' porous nature improves sensing capabilities (Masoomi *et al.*, 2019).

3.6. Nanocomposites

Several nanoparticles are combined in nanocomposites to enhance sensor performance in a synergistic way. For instance, graphene-based nanocomposites increase electrical conductivity and expand the range of applications, while polymer-nanoparticle composites give improved mechanical qualities (Omanović-Mikličanin *et al.*, 2020).

3.7. Nanowires and Nanorods

Electronic sensors use nanowires and nanorods, such as semiconductor nanowires. Sensitivity is increased by their high aspect ratio and surface reactivity, which facilitate effective charge transfer and interaction with target analytes (Khanal and Zubarev 2020).

3.8. Magnetic Nanoparticles

In magnetic resonance imaging (MRI) for biomedical purposes, magnetic nanoparticles—such as superparamagnetic iron oxide nanoparticles (SPIONs) act as contrast agents. Targeted sensing and imaging are made possible by functionalized SPIONs (Cheng *et al.*, 2013).

3.9. PlasmonicNanorulers

Based on plasmonics principles, plasmonicnanorulers estimate distance precisely at the nanoscale. Applications for these nanorulers include the detection of molecular interactions and the investigation of biological processes (Dahiya *et al.*, 2020).

3.10. 2D Metal-Organic Frameworks (2D MOFs)

Two-dimensional metal-organic frameworks have distinct topologies and a high porosity. They have benefits in terms of stability and selectivity and are used in gas sensing and separation (Zheng *et al.*, 2020).

These cutting-edge nanoparticles demonstrate how sensing technologies are always evolving, opening new avenues for medical advancements, environmental monitoring, and industrial uses. Nanomaterials' customised qualities aid in the creation of very effective and adaptable sensing systems.

4. Challenges and Opportunities

The issues of scalability, repeatability, and possible toxicity that arise from the extensive use of nanomaterials in sensing technologies are discussed in this section. At the same time, it looks at how synthesis method developments, standardisation, and multidisciplinary partnerships might help overcome these obstacles. The chapter emphasises that in order to guarantee the sustainability and moral use of these technologies, responsible nanomaterial design and execution are essential. Although sensors based on nanomaterials have shown to be remarkably capable, they also provide possibilities and obstacles that will determine how far they may be used in different applications. Here, we examine the difficulties posed by sensors made of nanomaterials:

- ***Standardization and Reproducibility:*** One challenge in the production of nanomaterial sensors is achieving repeatability and standardisation. Differences in synthesis techniques and material properties might result in discrepancies in sensor performance. The broad usage of nanomaterial sensors is incomplete by their absence of standardisation, which makes it problematic to integrate them seamlessly into applied applications.
- ***Toxicity and Biocompatibility:*** Concerns over the effects of some nanomaterials on the environment and human health have been raised by their potential toxicity. Biocompatibility must be guaranteed, particularly for environmental and biomedical sensing applications. Restrictions on the safe deployment of nanomaterial sensors in sensitive locations and regulatory permissions might be caused by toxicity problems.

- ***Scalability and Cost:*** One major problem is to increase the output of nanomaterials sensors while keeping them affordable. Methods used in large-scale synthesis could not be the same as those used in laboratories. Limited scalability may prohibit scale manufacture, which would keep nanomaterial sensors from becoming widely available and affordable.
- ***Selectivity and Interference:*** It is difficult to achieve good selectivity in nanomaterial sensors, particularly in intricate sample matrices. False-positive or false-negative conclusions might be produced by interference from other substances. Discrimination is a crucial factor that regulates how reliable and accurate nanomaterial sensors are in real-world applications.

Here, we explore the opportunities associated with nanomaterials sensors:

- ***Multifunctional Nanomaterials:*** The development of multifunctional nanomaterials with a range of characteristics has potential for the development of adaptable sensors. Enhancing sensor capabilities involves the integration of functions including imaging, sensing, and catalysis. The range of applications is expanded by multifunctional nanomaterials, which allow for sensors with many functions and improved performance.
- ***Integration with Internet of Things (IoT):*** Real-time data transmission and analysis is enabled by the integration of nanomaterials sensors with IoT systems. This makes it easier to monitor and operate sensing equipment remotely. Smart sensing solutions for a range of applications are made possible by IoT integration, which improves the use and accessibility of nanomaterials sensors.
- ***Advancements in Nanofabrication Techniques:*** The characteristics and architectures of nanomaterials can be precisely controlled because to ongoing improvements in nanofabrication techniques. Advanced lithography and 3D printing are two methods that improve sensor design. The creation of high-performance nanomaterial sensors with customised features is facilitated by improved manufacturing techniques.
- ***Emerging Materials and Nanocomposites:*** The materials palette for sensor development is expanded by the investigation of new nanomaterials and nanocomposites. Novel combinations of nanomaterials provide special synergy. New materials are being developed to help create sensors that are more stable, selective, and sensitive.
- ***Biomedical and Environmental Applications:*** Environmental monitoring and biological diagnostics both heavily rely on nanomaterial

sensors. There are chances to deal with particular issues in resource management, pollution detection, and healthcare. Advances in early illness diagnosis, personalised therapy, and environmentally friendly behaviours are made possible by nanosensors.

The way nanomaterials sensors are developed in the future will depend on how well these obstacles are handled and how new opportunities are taken advantage of. The road to achieving the full potential of nanomaterials in sensing technologies will be shaped by addressing toxicity issues, advancing standardisation, and investigating creative uses (Kumar *et al.*, 2018; Albalawi *et al.*, 2021; Siontorou *et al.*, 2019; Kiani *et al.*, 2022; Gupta and Bhardwaj, 2020; Karpagavinayagam *et al.*, 2022).

5. Future Perspectives

The last section offers some perspectives on where nanomaterials in sensing and detection may go in the future. The integration of nanomaterial-based sensors with artificial intelligence and machine learning is one of the upcoming themes that are covered, along with the possibility of creating wearable and implantable sensing devices. The chapter ends with a picture of a future in which nanomaterials continue to be fundamental in transforming analytical capacities, resulting in revolutionary developments in environmental monitoring, personalised medicine, and diagnostics.

Nanomaterials in sensing and detection have a bright future ahead of them, as continued research indicates new uses and improved capabilities. Future sensors based on nanomaterials should grow into multifunctional nanocomposites by mergingdissimilar nanomaterials to recover sensor performance. Combining graphene with metal nanoparticles or quantum dots, for instance, will deliver a synergistic consequence that will permit for sensors with improved sensitivity, selectivity, and answer times. Sensors that are smoothly combined with biological systems will be developed as a result of advancements in biointegration. In order to progress biocompatible sensors for in-the-moment bodily monitoring, biofunctionalized nanomaterials will be vital to the growth of personalised healthcare and disease diagnoses.The Internet of Things (IoT) ecosystem will be dependent on nanomaterial-based sensors, which will assistancemake intelligent and networked sensing devices. Real-time data from these sensors will provision industrial automation, environmental monitoring, and smart cities. Future nanosensors could have energy-harvesting components that permit them to draw energy from their environments. The autonomy and sustainability of sensors will be better-quality by addition with energy-harvesting nanomaterials or nanogenerators, chiefly in distant or resource-constrained environments. One frontier area is the combination of quantum technologies into nanosensors. By using the ideas of quantum entanglement and superposition, quantum sensors may be capable to noticevariations in the physical and chemical world with unmatched accuracy, leading to growths in quantum metrology and quantum-enhanced sensing.Forthcoming research will focus on ecologically friendly nanosensors, with a stress on biodegradability, green synthesis methods, and negligible environmental consequence. The increasing focus on sustainable technology is in line with the development of environmentally friendly nanomaterials and sensor production techniques. These future paths have the potential to completely change industries, healthcare, and our perception of the world as long as nanomaterials continue to be the driving force behind sensing and detecting breakthroughs. They could also usher in a new age of highly well-organized, integrated, and sustainable sensing skills.

The aim of this chapter is to offer a thorough overview for scientists, researchers, and practitioners who are concerned in learning about the importanteffect that nanomaterials have on analytical chemistry and how they may alteration the field of sensing and detecting skills.

6. Conclusion

In summary, the combination of nanomaterials into sensing and detecting skills is a paradigm shift with broad ramifications for a diversity of fields. Progressive requests are now likely thanks to nanomaterials' unmatched capability to recover sensors' sensitivity, selectivity, and response times. The flexibility of nanomaterials from metal nanoparticles to graphene and quantum dots has led to the development of a wide range of sensors that meet definite analytical supplies.

Important development has been made in the expansion of bio-integrated sensors through the usage of nanomaterials in sensing technologies, permitting for real-time monitoring of biological processes. The possible of targeted medicines, personalised healthcare, and disease diagnostics has materialised. Additionally, the Internet of Things (IoT) and nanomaterials have come organised to deliver the framework for intelligent and networked sensing devices that provision the evolution of smart cities, environmental monitoring, and industrial automation.

Future directions for nanomaterials in sensing and detecting technologies include energy harvesting sensors, quantum sensing, multifunctional nanocomposites, and a dedication to environmental sustainability. It is anticipated that the synergistic combination of several nanomaterials would push sensor performance boundaries and offer answers to challenging problems in a range of industries.

It is crucial that we address issues with standardisation, morality, and environmental effect as we adopt new advancements. To fully use nanomaterials in sensing and detection, it will be imperative to strike a balance between scientific breakthroughs and safe deployment. Essentially, nanomaterials are the cornerstones of a future where sustainability, accuracy, and connection come together to completely change how we view, monitor, and engage with the environment through sensing technologies. They are not only catalysts for scientific breakthroughs.

References

Abid, N., Khan, A.M., Shujait, S., Chaudhary, K., Ikram, M., Imran, M., Haider, J., Khan, M., Khan, Q. and Maqbool, M., (2022). Synthesis of nanomaterials using various top-down and bottom-up approaches, influencing factors, advantages, and disadvantages: A review. Advances in Colloid and Interface Science, 300, 102597. https://doi.org/10.1016/j.cis.2021.102597

Ahmed, M., (2020). Nanomaterial synthesis. In Polymer science and nanotechnology, 361-399. Elsevier. https://doi.org/10.1016/B978-0-12-816806-6.00016-9

Albalawi, F., Hussein, M.Z., Fakurazi, S. and Masarudin, M.J., (2021). Engineered nanomaterials: The challenges and opportunities for nanomedicines. International journal of nanomedicine, 161-184. https://doi.org/10.2147/IJN.S288236

Andrews, D., Nann, T. and Lipson, R.H., (2019). Comprehensive nanoscience and nanotechnology. Academic press.

Asha, A.B. and Narain, R., (2020). Nanomaterials properties. In Polymer science and nanotechnology, 343-359). Elsevier. https://doi.org/10.1016/B978-0-12-816806-6.00015-7

Baig, N., Kammakakam, I. and Falath, W., (2021). Nanomaterials: A review of synthesis methods, properties, recent progress, and challenges. Materials Advances, 2(6), 1821-1871.10.1039/D0MA00807A

Bayda, S., Adeel, M., Tuccinardi, T., Cordani, M. and Rizzolio, F., (2019). The history of nanoscience and nanotechnology: from chemical–physical applications to nanomedicine. Molecules, 25(1), 112. https://doi.org/10.3390/molecules25010112

Bhagyaraj, S.M., Oluwafemi, O.S., Kalarikkal, N. and Thomas, S., (2018). Synthesis of inorganic nanomaterials. Woodhead publication, 1e18. https://link.springer.com/article/10.1186/s11671-024-04006-8

Binns, C., (2021). Introduction to nanoscience and nanotechnology. John Wiley & Sons.

Brahmkhatri, V., Pandit, P., Rananaware, P., D'Souza, A. and Kurkuri, M.D., (2021). Recent progress in detection of chemical and biological toxins in Water using plasmonicnanosensors. Trends in Environmental Analytical Chemistry, 30, e00117. https://doi.org/10.1016/j.teac.2021.e00117

Buckley, D.J., Black, N.C., Castanon, E.G., Melios, C., Hardman, M. and Kazakova, O., (2020). Frontiers of graphene and 2D material-based gas sensors for environmental monitoring. 2D Materials, 7(3),032002. https://iopscience.iop.org/article/10.1088/2053-1583/ab7bc5/meta

Buzea, C. and Pacheco, I., (2017). Nanomaterials and their classification. EMR/ESR/EPR spectroscopy for characterization of nanomaterials, 3-45. https://link.springer.com/chapter/10.1007/978-81-322-3655-9_1

Cheng, W., Ping, Y., Zhang, Y., Chuang, K.H. and Liu, Y., (2013). Magnetic resonance imaging (MRI) contrast agents for tumor diagnosis. Journal of healthcare engineering, 4, 23-46. https://doi.org/10.1260/2040-2295.4.1.23

Choi, S.J. and Kim, I.D., (2018). Recent developments in 2D nanomaterials for chemiresistive-type gas sensors. Electronic Materials Letters, 14, 221-260. https://link.springer.com/article/10.1007/s40820-020-00503-4

Čitaković, N.M., (2019). Physical properties of nanomaterials. Vojnotehničkiglasnik/Military Technical Courier, 67(1), 159-171. https://doi.org/10.5937/vojtehg67-18251

Dahiya, A. and Kumar, P.S., (2020), November. Plasmonic gold-copper alloy dimer as 'nanorulers'. In AIP Conference Proceedings (Vol. 2265, No. 1). AIP Publishing. https://doi.org/10.1063/5.0017257

Darweesh, H.H.M., (2018). Nanomaterials: classification & properties-Part I. Nanoscience, 1(1), 1-11.

Gubicza, J., (2017). Defect structure and properties of nanomaterials. Woodhead Publishing.

Gupta, C. and Bhardwaj, A., (2020). Summary and future perspectives of nanomaterials and technologies: Special emphasis on energy and environment. In Nanomaterials for Sustainable Energy and Environmental Remediation, 333-353. Elsevier. https://doi.org/10.1016/B978-0-12-819355-6.00011-X

Huston, M., DeBella, M., DiBella, M. and Gupta, A., (2021). Green synthesis of nanomaterials. Nanomaterials, 11(8), 2130. https://doi.org/10.3390/nano11082130

Jamkhande, P.G., Ghule, N.W., Bamer, A.H. and Kalaskar, M.G., (2019). Metal nanoparticles synthesis: An overview on methods of preparation, advantages and disadvantages, and applications. Journal of drug delivery science and technology, 53, 101174. https://doi.org/10.1016/j.jddst.2019.101174

Joudeh, N. and Linke, D., (2022). Nanoparticle classification, physicochemical properties, characterization, and applications: a comprehensive review for biologists. Journal of Nanobiotechnology, 20(1), 262. https://link.springer.com/article/10.1186/s12951-022-01477-8

Karpagavinayagam, P., Rajam, J.A., Suneetha, R.B. and Vedhi, C., (2022). Prospects of carbon nanomaterial-based sensors for sustainable future. In Carbon nanomaterials-based sensors, 417-428. Elsevier. https://doi.org/10.1016/B978-0-323-91174-0.00021-4

Khan,F.A.,(2020).Nanomaterials:types,classifications,andsources.Applicationsofnanomaterials in human health, 1-13. https://link.springer.com/chapter/10.1007/978-981-15-4802-4_1

Khanal, B.P. and Zubarev, E.R., (2020). Gold nanowires from nanorods. Langmuir, 36(49), 15030-15038. https://doi.org/10.1021/acs.langmuir.0c02571

Kiani, M., Rehman, M.U., Tian, X. and Yakobson, B., (2022). Two□Dimensional Nanomaterials for the Development of Efficient Gas Sensors: Recent Advances, Challenges, and Future Perspectives. Advanced Materials Technologies, 7(7), 2101252. https://doi.org/10.1002/admt.202101252

Kolahalam, L.A., Viswanath, I.K., Diwakar, B.S., Govindh, B., Reddy, V. and Murthy, Y.L.N., (2019). Review on nanomaterials: Synthesis and applications. Materials Today: Proceedings, 18, 2182-2190. https://doi.org/10.1016/j.matpr.2019.07.371

Kumar, R., Liu, X., Zhang, J. and Kumar, M., (2020). Room-temperature gas sensors under photoactivation: From metal oxides to 2D materials. Nano-Micro Letters, 12, 1-37.

Kumar, V., Kumar, P., Pournara, A., Vellingiri, K. and Kim, K.H., (2018). Nanomaterials for the sensing of narcotics: Challenges and opportunities. Trac Trends In Analytical Chemistry, 106, 84-115. https://doi.org/10.1016/j.trac.2018.07.003

Kusuma, S.A.F., Harmonis, J.A., Pratiwi, R. and Hasanah, A.N., (2023). Gold Nanoparticle-Based Colorimetric Sensors: Properties and Application in Detection of Heavy Metals and Biological Molecules. Sensors, 23(19), 8172. https://doi.org/10.3390/s23198172

Lue, J.T., (2007). Physical properties of nanomaterials. Encyclopedia of nanoscience and nanotechnology, 10(1), 1-46. https://citeseerx.ist.psu.edu/document?repid=rep1&type=pdf&doi=de03bec8400fe1b1f76c1c1caa6518a548fffc32

Ma, F., Li, C.C. and Zhang, C.Y., (2018). Development of quantum dot-based biosensors: principles and applications. Journal of Materials Chemistry B, 6(39),6173-6190. https://doi.org/10.1039/C8TB01869C

Masoomi, M.Y., Morsali, A., Dhakshinamoorthy, A. and Garcia, H., (2019). Mixed□ metal MOFs: unique opportunities in metal–organic framework (MOF) functionality and design. AngewandteChemie, 131(43), 15330-15347. https://doi.org/10.1002/ange.201902229

Nishat, S., Awan, F.R. and Bajwa, S.Z., (2019). Nanoparticle-based point of care immunoassays for in vitro biomedical diagnostics. Analytical Sciences, 35(2),123-131. https://doi.org/10.2116/analsci.18R001

Omanović-Mikličanin, E., Badnjević, A., Kazlagić, A. and Hajlovac, M., (2020). Nanocomposites: A brief review. Health and Technology, 10,51-59. https://link.springer.com/article/10.1007/s12553-019-00380-x

Rizwan, M., Shoukat, A., Ayub, A., Razzaq, B. and Tahir, M.B., (2021). Types and classification of nanomaterials. In Nanomaterials: synthesis, characterization, hazards and safety, 31-54. Elsevier.

Saha, S., Bansal, S. and Khanuja, M., (2022). Classification of nanomaterials and their physical and chemical nature. In Nano-enabled Agrochemicals in Agriculture, 7-34. Academic Press. https://doi.org/10.1016/B978-0-323-91009-5.00001-X

Salah, L.S., Ouslimani, N., Bousba, D., Huynen, I., Danlée, Y. and Aksas, H., (2021). Carbon nanotubes (CNTs) from synthesis to functionalized (CNTs) using conventional and new chemical approaches. Journal of Nanomaterials, 2021, 1-31. https://doi.org/10.1155/2021/4972770

Saleh, T.A., (2020). Nanomaterials: Classification, properties, and environmental toxicities. Environmental Technology & Innovation, 20, 101067. https://doi.org/10.1016/j.eti.2020.101067

Sannino, D., (2021). Types and classification of nanomaterials. Nanotechnology: Trends and Future Applications, 15-38. https://link.springer.com/chapter/10.1007/978-981-15-9437-3_2

Saxena, S.K., Nyodu, R., Kumar, S. and Maurya, V.K., (2020). Current advances in nanotechnology and medicine. NanoBioMedicine, 3-16. https://link.springer.com/chapter/10.1007/978-981-32-9898-9_1

Shahdeo, D., Roberts, A., Abbineni, N. and Gandhi, S., (2020). Graphene based sensors. In Comprehensive Analytical Chemistry, 91,175-199. Elsevier. https://doi.org/10.1016/bs.coac.2020.08.007

Siontorou, C.G., Nikoleli, G.P., Nikolelis, M.T. and Nikolelis, D.P., (2019). Challenges and future prospects of Nanoadvanced sensing technology. Advanced Biosensors for Health Care Applications, 375-396. https://doi.org/10.1016/B978-0-12-815743-5.00015-9

Siontorou, C.G., Nikoleli, G.P., Nikolelis, M.T. and Nikolelis, D.P., (2019). Challenges and future prospects of Nanoadvanced sensing technology. Advanced Biosensors for Health Care Applications, 375-396. https://doi.org/10.1016/B978-0-12-815743-5.00015-9

Sudha, P.N., Sangeetha, K., Vijayalakshmi, K. and Barhoum, A., (2018). Nanomaterials history, classification, unique properties, production and market. In Emerging applications of nanoparticles and architecture nanostructures, 341-384. Elsevier. https://doi.org/10.1016/B978-0-323-51254-1.00012-9

Tajik, S., Beitollahi, H., Nejad, F.G., Dourandish, Z., Khalilzadeh, M.A., Jang, H.W., Venditti, R.A., Varma, R.S. and Shokouhimehr, M., (2021). Recent developments in polymer nanocomposite-based electrochemical sensors for detecting environmental pollutants. Industrial & engineering chemistry research, 60(3),1112-1136. https://doi.org/10.1021/acs.iecr.0c04952

Toto, E., Laurenzi, S. and Santonicola, M.G., (2022). Recent trends in graphene/polymer nanocomposites for sensing devices: Synthesis and applications in environmental and human health monitoring. Polymers, 14(5), 1030. https://doi.org/10.3390/polym14051030

Trotta, F. and Mele, A., (2019). Nanomaterials: classification and properties. Nanosponges Synth. Appl, 1-26.

Tulinski, M. and Jurczyk, M., (2017). Nanomaterials synthesis methods. Metrology and Standardization of Nanotechnology: Protocols and Industrial Innovations, 75-98. https://doi.org/10.1002/9783527800308.ch4

Varghese, R.J., Parani, S., Thomas, S., Oluwafemi, O.S. and Wu, J., (2019). Introduction to nanomaterials: synthesis and applications. In Nanomaterials for solar cell applications, 75-95. Elsevier. https://doi.org/10.1016/B978-0-12-813337-8.00003-5

Verma, G. and Gupta, A., (2022). Recent development in carbon nanotubes based gas sensors. Journal of Materials NanoScience, 9(1),3-12. https://pubs.thesciencein.org/journal/index.php/jmns/article/view/282

Wang, Z., Zeng, W. and Li, Z. eds., 2019. Advanced Nanomaterials for Sensing Applications. Frontiers Media SA.

Wu, Q., Miao, W.S., Zhang, Y.D., Gao, H.J. and Hui, D., (2020). Mechanical properties of nanomaterials: A review. Nanotechnology Reviews, 9(1), 259-273. https://doi.org/10.1515/ntrev-2020-0021

Wu, X., Mu, F., Wang, Y. and Zhao, H., (2018). Graphene and graphene-based nanomaterials for DNA detection: a review. Molecules, 23(8), 2050. https://doi.org/10.3390/molecules23082050

Zhang, G., Zeng, H., Liu, J., Nagashima, K., Takahashi, T., Hosomi, T., Tanaka, W. and Yanagida, T., (2021). Nanowire-based sensor electronics for chemical and biological applications. Analyst, 146(22),6684-6725. https://pubs.rsc.org/en/content/articlelanding/2021/an/d1an01096d/unauth

Zhang, L., Khan, K., Zou, J., Zhang, H. and Li, Y., (2019). Recent advances in emerging 2D material□based gas sensors: potential in disease diagnosis. Advanced Materials Interfaces, 6(22), 1901329. https://doi.org/10.1002/admi.201901329

Zhang, Q., Zhang, X., Ma, F. and Zhang, C.Y., (2022). Advances in quantum dot-based biosensors for DNA-modifying enzymes assay. Coordination Chemistry Reviews, 469, p.214674. https://doi.org/10.1016/j.ccr.2022.214674

Zheng, Y., Sun, F.Z., Han, X., Xu, J. and Bu, X.H., (2020). Recent progress in 2D metal□organic frameworks for optical applications. Advanced Optical Materials, 8(13),2000110. https://doi.org/10.1002/adom.202000110

5

Nanoparticle Immune Interactions Insights from Diverse Living Species

Arpit Sharma[1*], Abhinav Saxena[2], Mansimran Kaur Randhaw[3] Gurlovleen Kaur[4], Robin Bansal[5], Amandeep Singh[6], Diksha Garg[7], Palki Sahib Kaur[8] and Rajinder Kaur[9]

[1&4]Department of Biotechnology and Food Technology, Punjabi University, Patiala Punjab 147002, India*

[2]PG Department of Agriculture, GSSDGS Khalsa College, Patiala, Punjab 147001 India

[3]Food Processing & Engineering, GSSDGD Khalsa College Patiala, Punjab 147001 India

[5]Civil Hospital, Nabha, District Patiala, Punjab 147201

[6]Department of Energy and Environment, Thapar Institute of Engineering and Technology, Patiala 147004, India

[7]Department of Food Technology, Mata Gujri College, Fatehgarh Sahib, Punjab 140407, India

[8]Department of Biotechnology, Chandigarh College of Technology, CGC Landran Mohali 140307, India

[9]PG Department of Zoology, Sri Guru Gobind Singh College, Sector-26, Chandigarh

Abstract

Nanomaterials, a cornerstone of nanotechnology, encompass a broad array of particles showcasing remarkable properties and diverse applications. This chapter outlines the classifications and applications of various nanomaterials, including organic, semiconductor, metal, metal oxide, and carbon-based nanoparticles. Each type offers unique characteristics, such as biodegradability, optical and electrical properties, making them valuable in fields like drug delivery, electronics, and catalysis. Furthermore, this study focuses on the interactions of nanoparticles with living organisms, emphasizing their impact on immune systems across plants, invertebrates, and humans. The intricate mechanisms underlying nanoparticle-immune interactions, including protein corona formation and innate and adaptive

* *Corresponding author e-mail: razor_arpit@yahoo.co.in*

immune responses, are elucidated. Notably, the chapter explores strategies for engineered immune evasion by nanoparticles, highlighting the dynamic interplay between nanomaterials and the immune system. Recent advancements in nanoparticle design, such as pH-dependent shell shedding and incorporation of "don't-eat-me" signals, offer promising avenues for targeted drug delivery and reduced immunogenicity. Additionally, the influence of nanoparticles on the complement system, a vital component of innate immunity, is discussed, underscoring the need for translational research to bridge murine models with human immunology. As researchers continue to unravel the complexities of nanoparticle-immune interactions, this review provides insights into fundamental mechanisms and innovative strategies at the forefront of nanomedicine development.

Keywords: *Nanomaterials, Protein corona, Innate immunity, Adaptive immunity, Immune evasion, Drug delivery, Complement system, Nanomedicine.*

1. Introduction

Diversity among living species has given rise to various interaction mechanisms, owing to variations in the types of immune systems present among them. These variations can either lead to the stimulation or repression of immune system and its components, resulting in diverse responses or effects. Nanoparticles (NPs) also interact with various species in specific ways, in accordance with the complexity of the immune system. Different species have one or multiple routes for uptake, processing, and interaction with these particles. These interactions are attributed to the unique cells and their types, distributed among various plant and animal species. Furthermore, interactions can shed light on the toxicity of engineered NPs at the cellular, tissue, or organ levels, as well as at the individual, population, or community levels, contributing to a better understanding of risk assessment and management associated with them.

In the present scenario, advancements in nanotechnology have acted as a transformative tool, revolutionizing the enhancement of targeted immune responses for both preventing and treating a myriad of infectious and non-infectious diseases. Through finely tuned nanostructures and precise engineering, it enables localized delivery of immunotherapeutic agents, significantly reducing systemic toxicity while amplifying the efficacy of immunostimulatory molecules. This sophisticated approach not only bolsters the body's defenses against pathogens but also holds promise for combating autoimmune disorders, cancer, and other conditions by harnessing the power of nanoscale interventions to modulate immune function with unprecedented precision and effectiveness (Viana, 2021).

This comprehensive chapter underscores the multifaceted potential of nanomaterials across various disciplines, emphasizing their pivotal role in understanding working of nanoparticles among various living species and giving and insight into human immunology.

2. Nanoparticles and their types

Nanomaterials exemplify an important part of nanotechnology that deals with innumerable studies which involves particles ranging from 1 to 100 nm as a minimum in one spatial dimension (Hasan, 2015). There are many types of NPs depending upon their composition, which are described as follows:

2.1 Organic nanoparticles

They are mostly environment friendly and non-toxic in nature. Some of these particles have hollow core and are called as nanocapsules for example liposomes, nano-emulsions, micelles etc. These are generally sensitive to light, heat, and electromagnetic radiations. They are frequently used for drug delivery applications owing to their exceptional characteristics (An *et al.*, 2002). Delivery systems, stability, efficacy, and drug holding capacity (either adsorbed on surface or entrapped within nanoparticle) determines their main range of applications. Their efficiency is also based upon on different characteristics for instance surface morphology, size, and composition (Fu and Yao, 2001). They are extensively employed in bio-medical field as drug delivery systems, and are injected on definite parts of the body which is called as targeted drug delivery (Figure 1).

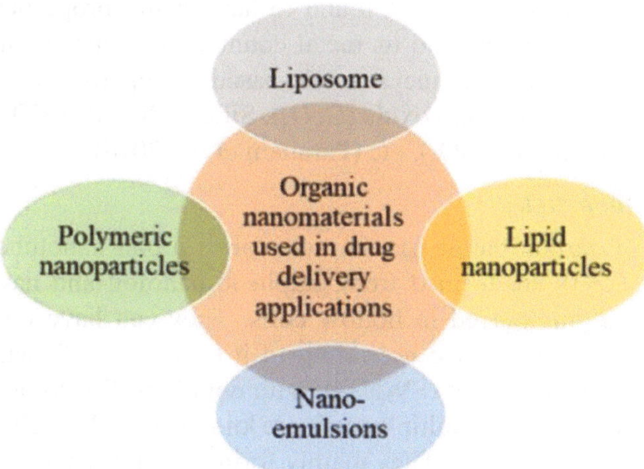

Fig. 1: Organic nanomaterials used in drug delivery applications

2.2. Semiconductor nanoparticles

They have characteristics of both metals, and nonmetals which enhances their wide range of applications such as Gallium Nitride (GaN), Gallium phosphide (GaP), Indium Phosphide (InP) etc. Moreover, researchers can alter their properties by bandgap tuning as they have huge bandgaps and band edge positions. They are mainly used in electronic devices, photocatalysis and photooptics (Chen *et al.*, 2004).

2.3. Metal nanoparticles

These are synthesized from metals by reducing their sizes to only few nanometers. They are mainly synthesized by two methodologies that is, top-down and bottom-up approaches which are also known as destructive or constructive methods (Kelly *et al.*, 2003). The most frequently used metals for fabrication of nanoparticle are zinc (Zn), iron (Fe), cadmium (Cd), copper (Cu), lead (Pb), gold (Au), aluminium (Al) and silver (Ag). These NPs have distinct features such as small size, high surface area to volume ratio, size dependent optical properties, size dependent electrical properties, high surface charge and charge density. They can be either crystalline or amorphous in nature and their shapes can vary from spherical, oval, cylindrical etc (Kelly *et al.*, 2003).

2.4. Metal oxide nanoparticles

These nanoparticles are produced by altering the characteristics of their relevant metal based nanoparticles such as Zinc oxide (ZnO) nanoparticles are synthesized from zinc (Zn) by oxidation reaction at room temperature in presence of oxygen. These Fe_2O_3 nanoparticles were found to have better properties, reactivity and efficiency in comparison to its metal counterpart (Chavali and Nikolova, 2019). Other examples of metal oxide-based nanoparticles are Aluminium oxide (Al_2O_3), ZnO, Cerium oxide (CeO_2), Silicon dioxide (SiO_2), Magnetite (Fe_3O_4), Titanium oxide (TiO_2), etc (Paraham *et al.*, 2016).

2.5. Carbon-based nanoparticles

The two main groups of carbon based nanoparticles are Carbon nanotubes (CNTs) and fullerenes. CNTs are formed from graphene nanofoil and have honeycomb structure of atoms placed in hollow coils. They can have one, two, or numerous walls and are named according to it such as CNTs with single wall are known as single-walled (SWNTs), with double wall is known as double-walled (DWNTs) and with multiple walls are known as multi-walled carbon nanotubes (MWNTs) and so on. They mainly form nanotubes and size of single-layered CNT is minute as 0.7 nm and size of multi-layered CNTs extends up to 100 nm. Also, their length ranges from few micrometers to several millimeters. Their ends are either hollow or are closed by half fullerene molecules (Salvetat *et al.*, 1999).

Fullerenes (C_{60}) are the other form of carbon-based nanoparticles which are spherical in shape and are made from sp^2hybridized carbon atoms. The structure mostly comprises 28 to 1500 carbon atoms. The diameter of single layered fullerenes ranges up to 10 nm and for multi-layered fullerenes size mostly ranges from 4-40 nm. They have excellent adaptability, electron affinity, and electrical conductivity (Pan *et al.*, 2020).

3. Nanoparticle interaction across various living species

Studying the role of nanoparticles in plants and invertebrates, we can compare their effects with humans and study much about broad and particular effects on immunity mechanisms of NPs as illustrated in Figure 2. The basic mechanisms of immunological defense system can be effectively studied by the interactions of NPs with cells and molecules of innate immune system.

In upcoming sections, we discuss about the current studies on the nanotoxicological evaluations and assessment of immune-related impacts in plants, invertebrates and humans, *in vivo* and *in vitro*.

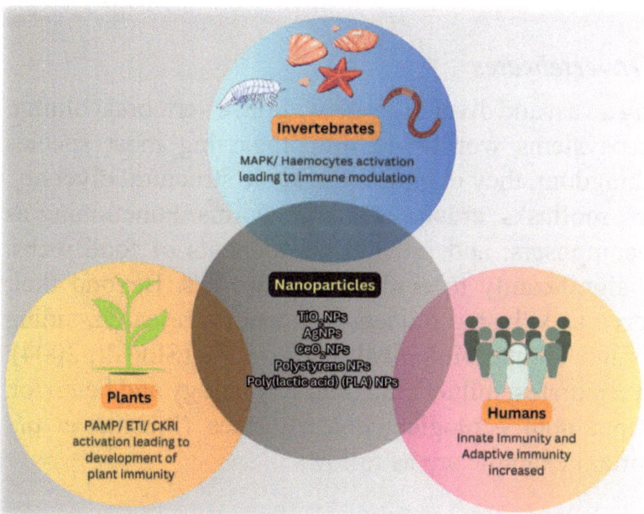

Fig. 2: Interaction of Nanoparticles with Plants, animals and Humans.

3.1. Effects of NPs on Plants

Plants are in continues contact with NPs present in the environment. The sensing capacity demonstrated by plant immunity serves as a remarkable illustration, relying on networks of immune receptors functioning as sensors of invading pathogens (Wu *et al.*, 2018). The plant's immune system consists of three tiers: the pathogen-associated molecular patterns (PAMP)-triggered immunity

(PTI) as the initial defense, effector-triggered immunity (ETI) as the secondary defense, and the exosome-mediated cross-kingdom RNA interference (CKRI) system serving as an additional defense mechanism (Zhang *et al.*, 2019).

The enhanced growth in spinach plant was observed as a result of treatment with TiO_2NPs which was due to accelerated function of Rubisco activase as well as increased light absorption efficiency. Conversely, they diminish photosynthetic rate, causing a slowdown in growth for *Ulmus elongate* (Gao *et al.*, 2013). Moreover, when leaf cells of plant *Arabidopsis thaliana*were exposed to substantial NPs of single-walled carbon nanotubes (SWCNTs), it resulted in formation of structures that were endocytosed in the plasma membrane. Though, these NPs were unable to cross the cell walls in other plant varieties (Shen *et al.*, 2010). *A. thaliana* was also treated with gold NPs that lead to increased level of antioxidant metabolic activity, growth as well as stimulated seed germination due to increased expression of microRNAs and particular regulatory molecules. *A. thaliana*, being a petite model plant with the brief life cycle, facilitates convenient manipulation and study (Kumar *et al.*, 2013).

3.2. Effects of NPs on Invertebrates

Invertebrates, constituting a vast and diverse group without a vertebral column, play pivotal roles in ecosystems worldwide. Encompassing most species richgroup in the animal kingdom, they exhibit remarkable structural diversity, including insects, spiders, mollusks, crustaceans, and worms. Functioning as essential pollinators, decomposers, and integral components of food webs, invertebrates contribute significantly to ecological processes. Beyond their ecological roles, they serve as valuable models in scientific research, aiding in the understanding of fundamental biological mechanisms (Siddall, 2004). Nanoparticles impact invertebrates, influencing their physiology and behavior, raising concerns about potential ecological consequences (Pinsino *et al.*, 2020). Some of the important example are as follows:

3.2.1. Earthworms

Earthworms, belonging to the Phylum Annelida, are the predominant invertebrates found in soils globally (Siddall, 2004). Their permeable skin and complex digestive tract keep them in continual contact with the soil particles, potentially exposing them to an undisclosed array of pathogens and pollutants (Bilej *et al.*, 2010). The intracellular signaling pathway responsible for the modulation of earthworm immunity induced by AgNPs appears to engage myeloid differentiation factor 88 (MyD88), which encodes a central adaptor protein for toll-like receptors, along with the mitogen-activated protein kinase (MAPK) cascades (Hayashi *et al.*, 2012; Hayashi *et al.*, 2013). The metabolomic analysis of *Eiseniafetida* earthworms exposed to TiO_2NPs in soil, as determined by ^{1}H NMR, revealed notable alterations in the metabolic profile indicative of oxidative stress as the primary mechanism underlying immune-related inflammation/toxicity (Whitfield *et al.*, 2012).

3.2.2. Mussels

Mussels (Molluscs), generally present in the continental shelves, are edible bivalves (Siddall, 2004). Mussels serve as a valuable model for exploring interplay among NPs and innate immunity. This is due to their well-documented biochemical and physiological reactions to diverse environmental stressors and pathogens (Canesi *et al.*, 2017). Additionally, their ease of experimental handling, including the non-invasive sampling of immune cells (haemocytes) and blood (haemolymph), enhances their suitability for research (OSPAR Commission, 2013). Examining the interaction between nanoparticles and mussel biological fluids, especially the haemolymph, facilitated the identification of specific proteins forming a stable bio-corona on NPs surface (Canesi and Pruzzo, 2016). This analysis also enabled the assessment of haemocyte interaction with the biocorona-coated NPs. Depending on the nature of the NPs and their biocorona, this interaction may trigger various responses (recognition, internalization, activation/inhibition of immune responses), leading to distinct immunomodulatory pathways (Canesi *et al.*, 2012).Investigating the interaction of titanium dioxide nanoparticles (TiO_2NPs) with marine invertebrate proteins reveals preferential binding of negatively charged proteins, particularly those involved in cellular adhesion and cytoskeletal organization, shedding light on immune cell recognition and interaction mechanisms in *Paracentrotuslividus*(Alijagic *et al.*, 2019).

3.2.3. Terrestrial isopoda

Isopods are crustaceans habited in the soil, are keystone species to study the effects of NPs on Arthropods (Siddall, 2004). Prolonged exposure to nanoparticles (AgNPs, CeO_2NPs) through food did not induce immune activation, despite causing alterations in proteins, lipids, nucleic acids, and carbohydrates at the tissue level (Novak *et al.*, 2019; Malev *et al.*, 2017).

3.2.4. Echinodermata

Echinoderms stand as the most advanced marine invertebrates, serving as a link to primitive chordates, including Urochordates and Cephalochordates. Echinoderms have a specialized water-based vascular system that is open to the aquatic environment. This feature is highly sensitive to swift changes in water chemistry, such as variations in ion content. This sensitivity has prompted the development of a complex, rapid, and sophisticated immune system in these animals (Zhu *et al.*, 2018). The research on the interaction between NPs and the immune system of the sea urchin reveals that, like humans, various extracellular proteins in sea urchins, including glycoproteins and cytoskeletal proteins, create a biocorona on the particle surface *in vitro* (Pinsino and Alijagic, 2019).

3.2.5. Humans

In Chordata, Humans, encompasses two primary focuses. Firstly, the investigation delves into the immunosafety of NPs, whether they are encountered as environmental pollutants or utilized in nanomedicines. Secondly, it examines the potential utilization of NPs for medically modulating immune responses (Wilding and Bodmer, 2014).

Many *in vitro* models employed to evaluate the immune effects of nanoparticles in humans utilize cancer cell lines or immortalized cell lines for their ease of handling and high reproducibility; however, their derivation from cancer cells imparts physiological and metabolic differences compared to normal primary immune cells, potentially limiting their predictability for human responses (Ninan *et al.*, 2018).

A highly effective and anticipatory method for evaluating the overall effect of NPs on blood of humans is by whole blood assay, where a portion of anticoagulated blood is exposed to NPs, involving NP coating with proteins present in serum and almost all circulating cells interacted with them. However, specific cell types that are affected by NPs cannot be identified by this assay. Recognizing phagocytes as primary defense against particulate matter, the most reliable models for innate immunity involve monocytes isolated from the blood of healthy donors (MAT, monocytes activation test), as well as macrophages and dendritic cells differentiated *in vitro* from fresh monocytes (Daneshian *et al.*, 2009; JRC ECVAM, 2002; Michelini *et al.*, 2018).

Following an extensive analysis of nanoparticle interactions across a range of living organisms, it is clear that these nanoparticles actively participate in direct engagements with components of the immune system. However, the implications of these findings go beyond direct interactions, as noteworthy observations reveal instances of indirect interactions facilitated by protein factors.

4. Insight into Human's Immune System and Nanoparticles

The complex processes present in various organisms not only enable evasion of immune responses but also instigate significant changes in various biochemical entities by interaction with nanoparticles. This intricate interplay between nanoparticles and biological systems, accentuates the multifaceted nature of their influence on immune responses and biochemical pathways.

4.1. Corona formation

Introduction of nanoparticles into biological system leads to coating of these particles with biomolecules, preferably proteins that are circulating in the system. These protein molecules when gets attached to NPs produce an effect known as "corona", forming protein clusters partially covering them. This phenomenon is of great importance to recognition by immune system, as conjugates of small particles such as NPs with proteins act as heptan-protein conjugates which are prominently able to produce immune response. Other complexes which are formed by interaction with NPs are lipid complexes, resulting from the binding of cholesterol, triglycerides and phospholipids forming high density lipoproteins (Fadeel, 2019).

4.1.1. Protein corona

Protein complexes are readily formed when NPs are entering the biological system, and are found to change over time. Their formation is also dependent upon the size of NPs. Interaction of these NPs with protein leads to change in the physical and chemical properties of NPs, leading to increased diameter and modifying charge present on their surface (Lima *et al.*, 2020).

4.1.2. Lipid complex recognition

Recognition of lipid complexes can be done using a technique known as immuno-epitope mapping, a complex between low density lipoprotein and immunoglobulin G can be simultaneously recognized by low density lipoprotein receptor and Fc- gamma receptor I. Ultimately leading body's system to consider this complex as an endogenous particle, like a lipoprotein and as exogenous, such as virus (Lara *et al.*, 2017).

4.1.3. Positively charged corona

Positively charged NPs adsorb proteins on their surface and produce corona, this leads them to their recognition and degradation by lysosomes. Furthermore, these changes lead to release of lysosomal proteases into cytosol leading to apoptosis (Fadeel, 2019).

4.1.4. Metal oxide corona

Metal nanoparticles are taken up by cells and due to their small size and high surface area, they start forming aggregates which can damage critical biological function. Certain metal oxide NPs like TiO_2, are able to bind to ribosomal and cytoskeletal proteins which found in macrophages which makes their availability less to another cell (Manuja *et al.*, 2021).

4.2. Innate immune response

Recognition and elimination of foreign particles using nonspecific defense mechanism comes under innate immunity. Underlying principle of innate defense includes recognition of antigen based upon pattern recognition receptors, physical barriers, generation of phagocytic and inflammatory response.

NPs depending on their size, shape and their composition can modulate immune response leading to oxidative stress, genotoxicity and cell division or function related changes. Many nanomaterials like carbon nanotubes, and metal oxides are responsible for generating reactive oxygen species (superoxide radicals, singlet oxygen, hydroxyl radicals) when they enter biological systems (Figure 3). This response generated by NPs or NMs is due to activation of inflammatory cells such as neutrophils and macrophages, leading to increased ROS production (Liu *et al.*, 2022). The response generated against NPs is variable and depends upon the following criteria as described by Aljabali *et al.*, (2023):

i) Redox Cycling – Surface property exhibited by transition metal-based NPs
ii) Functional Groups – Surface property of NPs due to chemical nature
iii) Cell specific interactions

Uptake of NPs by phagocytic cells is dependent upon its particle size, as 30-100nm sized particles are found appropriate for internalization, as compared to lower sized NPs. Also, after entering the biological system, proteins get attached to the surface of NPs, this effect of increased surface protein density of NPs leads to increased uptake by phagocytosis (Swartzwelter *et al.*, 2020). However, this effect becomes independent of the protein density beyond a threshold level.

Complement as a part of innate immune system, contains protein which when gets in contact with foreign particles or pathogens gets cleaved in a sequential manner to produce enzymatic functions, leading to clearance of antigen. It was observed that complement qualitatively affects distribution of NPs, when studies compared distribution of iron oxide NPs among complement deactivated models (C3 KO mice) as compared to wild type (Moghimi *et al.*, 2020).

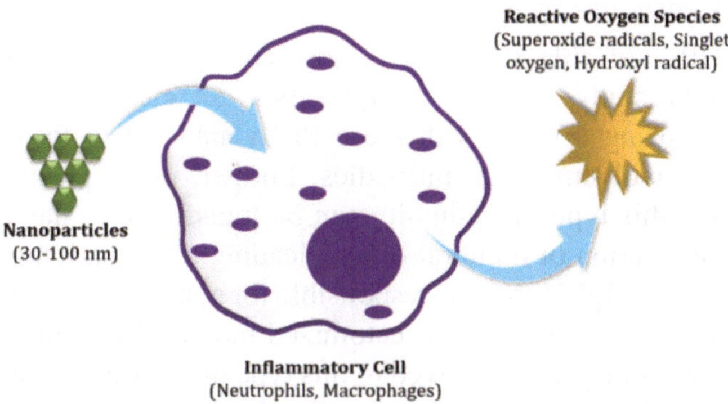

Fig. 3: Effect of nanoparticles on inflammatory cells for the production of reactive oxide species to provide immunity against pathogens.

4.3. Adaptive immune response

B-cells and T-cells activation constitutes adaptive immunity, leading to generation of humoral and cell mediated immune response, respectively. B-cells when activated produce antibodies and T cells by taking help from antigen presenting cells (APCs) perform effector functions, adapting to fight antigens (Figure 4). Together these two branches work interdependently to achieve specific immune response against incoming pathogens.

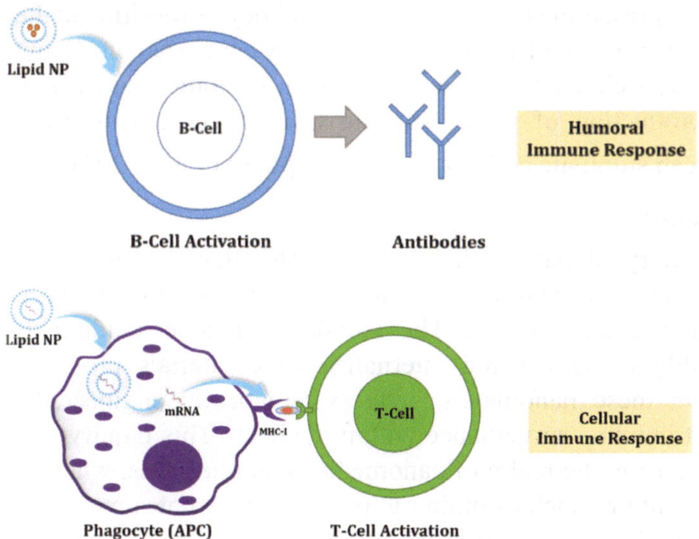

Fig. 4: Activation of humoral and cellular immune response via nanoparticles interactions

4.3.1. Humoral immune response by B-cell

Surface properties of NPs are able to promote interactions with immune system, as in a particular study, PEGylated liposomes get cleared from biological system. Prior dosing a week with these PEGylated led to faster response, suggesting the involvement of antibodies. Further, using plasma transfusions indicated that this type of immunity can be transferred to naïve animals, confirming the activation of humoral branch leading to formation of antibodies. Anti-PEG antibody, IgM type was responsible for generation of this type of response, which was not seen in splenectomized models. To remove the misconception that T cells might be involved in this type of response, mice without T cells were reported to develop antibodies (Pondman *et al.*, 2023).

4.3.2. Cellular immune response by T cell

NPs with size of 80nm, constituting protein derived from papaya mosaic virus, were able to initiate Interferon alpha in tumor induced animals. This phenomenon led to further induction of interferon gamma inducible proteins, monocyte chemoattractant protein and inflammatory cytokines, making $CD8^+$ T lymphocytes to perform their cytotoxic functions. Another type of technology using lipid NPs is their application as vehicles for transport of mRNA to phagocytes. These phagocytes act as APCs, processing these mRNAs in their cytoplasm and expressing them on their surface with major histocompatibility complex I, ultimately presenting them to $CD8^+$ T lymphocytes resulting in their clonal expansion. Transfection of T cells using antibody coated nanomedicines containing mRNAs (specialized for chimeric antigen receptor against tumor cell), stimulated production of tumor specific CAR-T cells. This clearly indicates that NPs can stimulate cellular immune response (Viana, 2021).

4.4. Immune Evasion

In the intricate tapestry of nanomedicine, our exploration delves into the nuanced interplay between nanomaterials and the immune system, with a particular focus on immune evasion. The foundational understanding that immune cells readily recognize and internalize nanomaterials prompts a crucial inquiry: can these nanomaterials be systematically engineered to circumvent vigilant immune surveillance within the host? This inquiry gains heightened significance in the realm of nanomedicine applications, where the precise targeting of entities such as brain tumor cells necessitates meticulous avoidance of unscheduled immune clearance.

4.4.1 Active and Passive Immune Evasion Strategies in Nanomedicines

Within the orchestration of nanomedicine, the strategic modulation of nanomaterials for immune evasion takes centre stage. If the objective is to target the immune system itself, as exemplified in antigen delivery for vaccination, the design of NPs must be purposefully aligned with this immunological objective. The conventional approach to evade nonspecific clearance by the reticulo-endothelial system (RES) and prolong circulation times involves the surface grafting of polyethylene glycol (PEG) onto particles. While effective in reducing protein adsorption, this method poses challenges due to the non-biodegradable nature of PEG polymers, potentially limiting their utility, especially in scenarios requiring repeated or chronic administration. The emergence of anti-PEG antibodies further compounds concerns, potentially accelerating the clearance of subsequent doses. Schöttler *et al.*, (2016) investigation sheds light on the influence of specific proteins, notably clusterin, on the cellular uptake of polymer-modified NPs. Employing polystyrene NPs modified with PEG or poly(ethyl ethylene phosphate) (PEEP), the study elucidates the role of clustering in reducing nonspecific cellular uptake. However, the dynamic nature of bio-corona formation and its impact on cellular internalization, especially under *in vivo* conditions, warrant meticulous exploration.

4.4.2. Exploring the Dynamic Interplay of Bio-Corona Formation and Cellular Internalization

Bertrand *et al.*, (2017) recent findings challenge prevailing assumptions about the alteration of blood circulation time with the enrichment of clusterin/ ApoJ on polymer-based NPs possessing high PEG densities. The intriguing possibility of protein desorption upon contact with blood introduces a layer of complexity to bio-corona dynamics, necessitating further studies to decipher the biological significance of this phenomenon. Striking a delicate balance between cellular uptake by specific target cells and evasion of phagocytic cells within the immune system or RES emerges as a critical imperative. The pursuit of absolute stealth proves of limited academic interest, as most nanomedicine applications necessitate cellular uptake for subsequent payload delivery during the remarkable journey of NPs through the intricate landscapes of the human body. Song *et al.*, (2017) recent elucidation of the cellular "tropism" of poly(lactic acid) (PLA) NPs modified with diverse polymers enriches our understanding. Drawing inspiration from virology, where tropism denotes the targeted preference of viruses, the study exemplifies the tunability of NP surfaces to modulate cell uptake in healthy and tumor cells. This emphasis on a nuanced approach underscores the intricate considerations required in the design of nanomedicine formulations. Prompted by the question of whether NPs can be engineered to evade unwanted immune clearance in the bloodstream, remaining available for on-site activation in specific locales, recent work by Qiao *et al.*, (2017) introduces a reconfigurable nanotherapeutic. This innovative design, capable of shedding its PEG shell in the tumor microenvironment in a pH-dependent manner, offers the potential for on-demand cytotoxicity towards tumor cells without observed short-term systemic toxicity.

4.4.3. Active Stealth Strategies: Exploiting "Don't Eat ME" Signals

Beyond passive means of immune evasion, researchers are exploring active stealth properties using natural "don't-eat-me" signals, such as CD47, which is a marker of self-expressed on cell membranes, associates with phagocytes, inhibiting macrophage uptake. Rodriguez *et al.*'s (2013) pioneering work, attaching minimal self-peptides derived from CD47 onto NPs, demonstrates delayed macrophage-mediated clearance and enhanced drug delivery to tumor sites. This innovative approach offers a promising alternative to traditional PEGylation, potentially mitigating concerns related to immunogenicity.

4.4.4. Complement System Dynamics in Nanomedicine

The complement system, an integral component of the innate immune system, introduces an additional layer of complexity. Nanomaterials have been shown to activate the complement system through various pathways, leading to opsonization and clearance. Chen *et al.*,(2017) study on superparamagnetic iron oxide "nanoworms" highlights the opsonization of these nanomaterials with C3, a key protein in the complement cascade. The intricate dynamics of complement binding, influenced by NP coatings like PEG, prompt questions regarding the translatability of "stealth" approaches from murine models to the unique immunological landscape of humans.

5. Application of Nanoparticles in Treatment of Diseases

As we traverse the intricate landscape of immunological strategies in nanomedicine (Table 1), each chapter unfolds new layers of complexity, beckoning researchers to decode the intricate language spoken between nanomaterials and the immune system. This journey promises not only insights into the fundamental mechanisms but also innovative solutions at the forefront of scientific exploration. NPs forming nanoclusters such as silver, gold and platinum have been widely used for in the field of nanomedicine due to their bactericidal, anti-cancerous and chemotherapeutic potential respectively. These NPs also act as delivery vehicles for moving drugs to particular target site e.g. liposomes, polymeric nanoparticles (utilizing both active and passive transport) (Ray *et al.*, 2021).Treatment of spherical fullerenes as NPs, with human mast cells and peripheral blood basophils, result in reduced IgE receptor-mediated signalling, leading to lower production of ROS and lesser degranulation (Smith *et al.*, 2013).NPs can be used to administered cancer immunotherapy in a controlled manner, PEGylated gold NPs has been successful in treatment of solid tumors by delivering recombinant TNF-α into tumor micro-environment. NP vaccines can deliver a gene to block programmed cell death-ligand 1 (immunosuppressive protein element expressed in tumor micro-environment) in cancer cells (Pondman *et al.*, 2023). Using iron oxide NP and loading ferumoxytol, TAM (Tumor associated macrophages) can be repolarised to produce ROS leading to killing of tumor cells (Zanganeh *et al.*, 2016).

Table 1: Medical Applications of Nanoparticles (Smith *et al.*, 2013)

Types of NPs	Medical application
Spherical fullerenes	Allergy
Carbon nanotubes Single-walled carbon nanotubules Multi-walled carbon nanotubules	Induction of apoptosis, Inhalation exposure
Nanocrystals	Transplant rejection
Polymeric nanoparticles Polyamidoamine dendrimers Poly(lactide- co-glycolide)	Cerebral palsy, scar formation and gastroenteritis Arthritis and autoimmune disease
Lipid nanoparticles Liposomes Liposomes with DC-targeting ligands Sterically stabilized phospholipid Micelles Nanoemulsions	Transplant rejection Arthritis, coronary artery stenosis and acute lung injury Autoimmune disease Rheumatoid arthritis Autoimmune thyroiditis

6. Conclusion

The exploration of nanomaterials and their interactions with the immune system represents a pivotal frontier in both scientific research and practical applications. Nanotechnology, particularly through the lens of nanomaterials, offers a plethora of opportunities across various domains, from medicine to environmental science. Understanding the diverse types of nanoparticles and their effects on living organisms, ranging from plants to humans, is crucial for harnessing their potential benefits while mitigating potential risks.

Nanomaterials, defined as particles ranging from 1 to 100 nanometers in size, exhibit unique properties that make them suitable for a wide array of applications. The types of nanoparticles include organic nanoparticles, semiconductor nanoparticles, metal nanoparticles, metal oxide nanoparticles, and carbon-based nanoparticles. Each type possesses distinct characteristics and applications, ranging from drug delivery systems to electronic devices and environmental remediation.

One of the key aspects highlighted in this chapter is the interaction of nanoparticles with various living organisms, including plants, invertebrates, and humans. These interactions encompass a broad spectrum of effects, from enhancing growth and physiological functions to inducing immune responses and toxicity. For example, in plants, nanoparticles such as TiO_2NPs can influence photosynthesis and growth, while invertebrates like earthworms and mussels exhibit immune responses and physiological alterations upon exposure to nanoparticles.

Moreover, the intricate mechanisms underlying nanoparticle-immune system interactions provides better understating for their role in biological processes. Nanoparticles can elicit responses from both the innate and adaptive immune systems, leading to inflammation, oxidative stress, and modulation of immune cell functions. Understanding these mechanisms is crucial for assessing the immunological safety of nanoparticles and designing tailored approaches for biomedical applications.

Furthermore, strategies for modulating nanoparticle interactions with the immune system, including the formation of protein coronas and the engineering of nanoparticles for immune evasion.By manipulating surface properties and coating compositions, researchers aim to enhance nanoparticle biocompatibility and reduce immune recognition, thereby improving their efficacy in targeted drug delivery and other applications.

Overall, the comprehensive examination of nanoparticle-immune system interactions underscores the intricate interplay between nanomaterials and biological systems. This knowledge is essential for advancing nanotechnology and leveraging its potential for diverse applications while ensuring safety and efficacy in various contexts. As research in this field continues to evolve, it holds promise for addressing pressing challenges in healthcare, environmental sustainability, and beyond, shaping the future of science and technology.

References

Alijagic, A., Benada, O., Kofroňová, O., Cigna, D. & Pinsino, A. (2019). Sea Urchin Extracellular Proteins Design a Complex Protein Corona on Titanium Dioxide Nanoparticle Surface Influencing Immune Cell Behavior. Frontiers in Immunology,10,2261.https://doi.org/10.3389/fimmu.2019.02261

Aljabali, A.A., Obeid, M.A., Bashatwah, R.M., Serrano-Aroca, Á., Mishra, V., Mishra, Y., El-Tanani, M., Hromić-Jahjefendić, A., Kapoor, D.N., Goyal, R., Naikoo, G.A. & Tambuwala, M.M. (2023).Nanomaterials and their impact on the immune system. International Journal of Molecular Sciences, 24,2008. https://doi.org/10.3390/ijms24032008

An, B.K., Kwon, S.K., Jung, S.D. & Park, S.Y. (2002). Enhanced emission and its switching in fluorescent organic nanoparticles. Journal of the American Chemical Society, 124,14410–14415. https://doi.org/10.1021/ja0269082

Bertrand, N., Grenier, P., Mahmoudi, M., Lima, E.M., Appel, E.A., Dormont, F., Lim, J.M., Karnik, R., Langer, R. & Farokhzad, O.C. (2017). Mechanistic understanding of in vivo protein corona formation on polymeric nanoparticles and impact on pharmacokinetics. NatureCommunications,8,777. https://doi.org/10.1038/s41467-017-00600-w

Bilej, M., Procházková, P., Silerová, M. & Josková, R. (2010). Earthworm immunity. Advances in experimental medicine and biology, 708,66–79. https://doi.org/10.1007/978-1-4419-8059-5_4

Canesi, L.&Pruzzo, C.(2016). Specificity of Innate Immunity in Bivalves: A Lesson from Bacteria. Lessons in Immunity,6,9–91.https://doi.org/10.1016/B978-0-12-803252-7.00006-0

Canesi, L., Ciacci, C.& Balbi, T. (2017). Invertebrate models for investigating the impact of nanomaterials on innate immunity: The example of the marine mussel Mytilus spp. Current Bionanotechnology,2,77–83. https://dx.doi.org/10.2174/2213529402666160601102529

Canesi, L., Ciacci, C., Fabbri, R., Marcomini, A., Pojana, G. & Gallo, G.(2012). Bivalve molluscs as a unique target group for nanoparticle toxicity. Marine Environmental Research,76,16–21. https://doi.org/10.1016/j.marenvres.2011.06.005

Chavali, M.S. & Nikolova, M.P. (2019). Metal oxide nanoparticles and their applications in nanotechnology. SN Applied Sciences, 1,607. https://doi.org/10.1007/s42452-019-0592-3

Chen, F., Wang, G., Griffin, J.I., Brenneman, B., Banda, N.K., Holers, V.M., Backos, D.S., Wu, L., Moghimi, S.M. & Simberg, D. (2017). Complement proteins bind to nanoparticle protein corona and undergo dynamic exchange in vivo. Nature Nanotechnology,12,387–393. https://doi.org/10.1038/nnano.2016.269

Chen, W., Zhang, J.Z. & Joly, A.G. (2004). Optical properties and potential applications of doped semiconductor nanoparticles. Journal of Nanoscience and Nanotechnology, 4,919–947.https://doi.org/10.1166/jnn.2004.142

Daneshian, M., von Aulock, S. & Hartung, T. (2009). Assessment of pyrogenic contaminations with validated human whole-blood assay. Nature Protocols, 4,1709–1721. https://doi.org/10.1038/nprot.2009.159

Fadeel, B. (2019). Hide and seek: Nanomaterial interactions with the immune system. Frontiers in Immunology, 10,133. https://doi.org/10.3389/fimmu.2019.00133

Fu, H.B. & Yao, J.N.(2001).Size effects on the optical properties of organic nanoparticles. Journal of the American Chemical Society, 123,1434–1439. https://doi.org/10.1021/ja0026298

Gao, J., Xu, G., Qian, H., Liu, P., Zhao, P. & Hu, Y. (2013). Effects of Nano-TiO2 on photosynthetic characteristics of Ulmuselongata seedlings. Environmental Pollution, 176,63–70. https://doi.org/10.1016/j.envpol.2013.01.027

Hasan, S. (2015).A review on nanoparticles: Their synthesis and types. Research Journal of recent Sciences, 2277,2502.

Hayashi, Y., Engelmann, P., Foldbjerg, R., Szabó, M., Somogyi, I., Pollák, E., Molnár, L., Autrup, H., Sutherland, D.S., Scott-Fordsmand, J. & Heckmann, L.H. (2012). Earthworms and humans in vitro: Characterizing evolutionarily conserved stress and immune responses to silver nanoparticles. Environmental Science and Technology, 46,4166–4173. https://doi.org/10.1021/es3000905

Hayashi, Y., Heckmann, L.H., Simonsen, V. & Scott-Fordsmand, J.J. (2013). Time-course profiling of molecular stress responses to silver nanoparticles in the earthworm Eiseniafetida. Ecotoxicology and Environmental Safety, 98,219–226. https://doi.org/10.1016/j.ecoenv.2013.08.017

JRCECVAM (2002). In Vitro Pyrogen Test using Human Whole Blood/IL-6. https://tsar.jrc.ec.europa.eu/test-method/tm2002-05.

Kelly, K.L., Coronado, E., Zhao, L.L. & Schatz, G.C. (2003). The optical properties of metal nanoparticles: The influence of size, shape, and dielectric environment. Journal of Physical Chemistry B,107,668–677. https://doi.org/10.1021/jp026731y

Kumar, V., Guleria, P., Kumar, V. & Yadav, S.K. (2013). Gold nanoparticle exposure induces growth and yield enhancement in Arabidopsis thaliana. Science of the Total Environment, 461–462, 462–468. https://doi.org/10.1016/j.scitotenv.2013.05.018

Lara, S., Alnasser, F., Polo, E., Garry, D., Lo Giudice, M.C., Hristov, D.R., Rocks, L., Salvati, A., Yan, Y. & Dawson, K.A. (2017). Identification of receptor binding to the biomolecular corona of nanoparticles. ACS Nano,11, 1884–1893. https://doi.org/10.1021/acsnano.6b07933

Lima, T., Bernfur, K., Vilanova, M. & Cedervall, T. (2020). Understanding the lipid and protein corona formation on different sized polymeric nanoparticles. Scientific Reports,10,1129. https://doi.org/10.1038/s41598-020-57943-6

Liu, J., Liu, Z., Pang, Y.&Zhou, H. (2022). The interaction between nanoparticles and immune system: Application in the treatment of inflammatory diseases. Journal of Nanobiotechnology, 20, 127. https://doi.org/10.1186/s12951-022-01343-7

Malev, O., Trebše, P., Piecha, M., Novak, S., Budič, B., Dramićanin, M. D., & Drobne, D. (2017). Effects of CeO2 Nanoparticles on Terrestrial Isopod Porcellioscaber: Comparison of CeO2 Biological Potential with Other Nanoparticles. Archives of environmental contamination and toxicology, 72(2), 303-311. https://doi.org/10.1007/s00244-017-0363-3

Manuja, A., Kumar, B., Kumar, R., Chhabra, D., Ghosh, M., Manuja, M., Brar, B., Pal, Y., Tripathi, B.N. & Prasad, M. (2021). Metal/metal oxide nanoparticles: Toxicity concerns associated with their physical state and remediation for biomedical applications. Toxicology Reports, 8,1970–1978. https://doi.org/10.1016/j.toxrep.2021.11.020

Michelini, S., Sarajlic, M., Duschl, A.&Horejs-Hoeck, J. (2018). IL-1β induces expression of costimulatory molecules and cytokines but not immune feedback regulators in dendritic cells. Human Immunology, 79, 610–615. https://doi.org/10.1016/j.humimm.2018.06.002

Moghimi, S.M., Simberg, D., Papini, E.&Farhangrazi, Z.S. (2020). Complement activation by drug carriers and particulate pharmaceuticals: Principles, challenges and opportunities. Advanced Drug Delivery Reviews, 157,83–95. https://doi.org/10.1016/j.addr.2020.04.012

Ninan, N., Albrecht, H.&Blencowe, A. (2018). Mammalian cell-basedassays for studyingbio-Nanointeractions. InS. M. Bhagyaraj, O. S. Oluwafemi, N. Kalarikkal & S. Thomas (Eds.), Micro and Nano Technologies, Characterization of Nanomaterials (pp. 129–166). Woodhead Publishing. https://doi.org/10.1016/B978-0-08-101973-3.00005-5

Novak, S., Romih, T., Drašler, B., Birarda, G., Vaccari, L., Ferraris, P., Sorieul, S., Zieba, M., Sebastian, V., Arruebo, M., Hočevar, S.B., JemecKokalj, A.J. & Drobne, D. (2019). Thein vivoeffects of silver nanoparticles on terrestrial isopods, Porcellioscaber, depend on a dynamic interplay between shape, size and nanoparticle dissolution properties. Analyst, 144,488–497. https://doi.org/10.1039/C8AN01387J

OSPAR Commission (2013). Background Documents and TechnicalAnnexes for Biological Effects Monitoring. https://www.ospar.org/documents?v=7316.

Pan, Y., Liu, X., Zhang, W., Liu, Z., Zeng, G., Shao, B., Liang, Q., He, Q., Yuan, X., Huang, D.

& Chen, M. (2020). Advances in photocatalysis based on fullerene C60 and its derivatives: Properties, mechanism, synthesis, and applications. Applied Catalysis B, 265,118579. https://doi.org/10.1016/j.apcatb.2019.118579

Parham, S., Wicaksono, D.H.B., Bagherbaigi, S., Lee, S.L. & Nur, H. (2016). Antimicrobial treatment of different metal oxide nanoparticles: A critical review. Journal of the Chinese Chemical Society, 63,385–393. https://doi.org/10.1002/jccs.201500446

Pinsino, A.&Alijagic, A. (2019). Sea urchin Paracentrotus lividus immune cells in culture: Formulation of the appropriate harvesting and culture media and maintenance conditions. Biology Open, 8,bio039289. https://doi.org/10.1242/bio.039289

Pinsino, A., Bastús, N.G., Busquets-Fité, M., Canesi, L., Cesaroni, P., Drobne, D., Duschl, A., Ewart, M.A. & Gispert, I. (2020). Probing the immune responses to nanoparticles across environmental species. A perspective of the EUHorizon 2020 project PANDORA. EnvironmentalScience: Nano, 7,3216–3232.https://doi.org/10.1039/D0EN00732C

Pondman, K., Le Gac, S. & Kishore, U. (2023). Nanoparticle-induced immune response: Health risk versus treatment opportunity? Immunobiology, 228, 152317. https://doi.org/10.1016/j.imbio.2022.152317

Qiao, Z.Y., Zhao, W.J., Gao, Y.J., Cong, Y., Zhao, L., Hu, Z. & Wang, H. (2017). Reconfigurable peptide nanotherapeutics at tumor microenvironmental pH.ACS AppliedMaterials and Interfaces, 9,30426–30436. https://doi.org/10.1021/acsami.7b09033

Ray, P., Haideri, N., Haque, I., Mohammed, O., Chakraborty, S., Banerjee, S., Quadir, M., Brinker, A. & Banerjee, S. (2021).The impact of nanoparticles on the immune system: A gray zone of nanomedicine. Journal of Immunological Sciences, 5,19–33. https://doi.org/10.29245/2578-3009/2021/1.1206

Rodriguez, P.L., Harada, T., Christian, D.A., Pantano, D.A., Tsai, R.K. & Discher, D.E. (2013). Minimal 'self' peptides that inhibit phagocytic clearance and enhance delivery of nanoparticles. Science, 339, 971-975. https://www.ncbi.nlm.nih.gov/pmc/articles/PMC3966479/

Salvetat, J.-P., Bonard, J.-M., Thomson, N.H., Kulik, A.J., Forró, L., Benoit, W. & Zuppiroli, L. (1999). Mechanical properties of carbon nanotubes. Applied Physics A, 69,255–260. https://doi.org/10.1007/s003390050999

Schöttler, S., Becker, G., Winzen, S., Steinbach, T., Mohr, K., Landfester, K., Mailänder, V.&Wurm, F.R. (2016). Protein adsorption is required for stealth effect of poly(ethylene glycol)- and poly(phosphoester)-coated nanocarriers. NatureNanotechnology, 11,372–377. https://doi.org/10.1038/nnano.2015.330

Shen, C.X., Zhang, Q.F., Li, J., Bi, F.C. & Yao, N. (2010). Induction of programmed cell death in Arabidopsis and rice by single-wall carbon nanotubes. American Journal of Botany, 97,1602–1609. https://doi.org/10.3732/ajb.1000073

Siddall, M.E. (2004). Invertebrates. R.C. Brusca and G.J. Brusca, Sinauer Associates, Sunderland, Massachusetts. Systematic Biology, 53, 664-666.https://doi.org/10.1080/10635150490472968

Smith, D.M., Simon, J.K. & Baker, Jr, J.R. (2013). Applications of nanotechnology for immunology. Nature Reviews. Immunology, 13,592–605. https://doi.org/10.1038/nri3488

Song, E., Gaudin, A., King, A.R., Seo, Y.E., Suh, H.W., Deng, Y., Cui, J., Tietjen, G.T., Huttner, A. & Saltzman, W.M. (2017). Surface chemistry governs cellular tropism of nanoparticles in the brain. NatureCommunications,8,15322. https://doi.org/10.1038/ncomms15322

Swartzwelter, B. J., Fux, A. C., Johnson, L., Swart, E., Hofer, S., Hofstätter, N., Geppert, M.,

Italiani, P., Boraschi, D., Duschl, A., &Himly, M. (2020). The impact of nanoparticles on innate immune activation by live bacteria. International Journal of Molecular Sciences, 21, 9695. https://doi.org/10.3390/ijms21249695

Viana, I.M.O., Roussel, S., Defrêne, J., Lima, E.M., Barabé, F.&Bertrand, N. (2021). Innate and adaptive immune responses toward nanomedicines. Acta Pharmaceutica Sinica B,11,852–870. https://doi.org/10.1016/j.apsb.2021.02.022

Whitfield Åslund, M.L., McShane, H., Simpson, M.J., Simpson, A.J., Whalen, J.K., Hendershot, W.H.&Sunahara, G.I. (2012). Earthworm sublethal responses to titanium dioxide nanomaterial in soil detected by 1H NMR metabolomics. Environmental Science and Technology, 46,1111–1118. https://doi.org/10.1021/es202327k

Wilding, J.L.& Bodmer, W.F. (2014). Cancer cell lines for drug discovery and development. Cancer Research, 74,2377–2384. https://doi.org/10.1158/0008-5472.CAN-13-2971

Wu, C.H., Derevnina, L. & Kamoun, S. (2018). Receptor networks underpin plant immunity. Science, 360,1300–1301. https://doi.org/10.1126/science.aat2623

Zanganeh, S., Hutter, G., Spitler, R., Lenkov, O., Mahmoudi, M., Shaw, A., Pajarinen, J.S., Nejadnik, H., Goodman, S., Moseley, M., Coussens, L.M. & Daldrup-Link, H.E. (2016). Iron oxide nanoparticles inhibit tumour growth by inducing pro-inflammatory macrophage polarization in tumour tissues. Nature Nanotechnology, 11,986–994. https://doi.org/10.1038/nnano.2016.168

Zhang, R., Zheng, F., Wei, S., Zhang, S., Li, G., Cao, P. & Zhao, S. (2019). Evolution of disease defense genes and their regulators in plants. InternationalJournal of MolecularSciences, 20,335. https://doi.org/10.3390/ijms20020335

Zhu, B.K., Fang, Y.M., Zhu, D., Christie, P., Ke, X. & Zhu, Y.G. (2018). Exposure to nanoplastics disturbs the gut microbiome in the soil oligochaete Enchytraeus crypticus. Environmental Pollution, 239,408–415. https://doi.org/10.1016/j.envpol.2018.04.017

6

Nanoparticulate Metal Composites Particles: Future Prospects and Therapeutic Implications in Acne Treatment

Ajay Kumar Shukla[1], Vimal Kumar Yadav[1], Manish Kumar[2] Sunita Panchawat[3], and Joohee Pradhan[3*]

[1]Institute of Pharmacy, Dr Rammanohar Lohia Avadh University Ayodhya 224001

[2]Department of Pharmacy, Madhav University, 307026, Pindwara, India

[3]Department of Pharmaceutical Sciences, Mohanlal Sukhadia University Udaipur Rajasthan 313001

Abstract

Synthesis of nanoparticulate metal composites (NMCP) presents a multitude of advantages compared to conventional dosage forms. Acne, a multifactorial skin condition, predominantly involves the pilosebaceous unit and is primarily caused by bacterial species such as Propionibacterium acnes and Acne vulgaris. A staggering 95% of the global population is afflicted by acne. Leveragitng NMCP strategies holds immense promise for innovating novel, efficacious, and low-dose treatments for managing acne. This review delves into various nanotechnological methodologies tailored for effectively addressing acne. ***Methods:*** *This review encompasses recent advancements pertaining to NMCP carriers aimed at treating acne. A variety of nanocarriers are used in modern practice, such as liposomes, niosomes, microsponges, microemulsions, nanoparticles, microspheres, and NMCP greatly improves anti-acne therapy. Encapsulation of anti-acne agents within diverse NMCP carriers enhances their therapeutic efficacy while mitigating adverse effects. These carriers exhibit better drug penetration and regulated release, including access to the skin's pilosebaceous unit.* ***Conclusion:*** *NMCP-based interventions have the potential to improve several aspects of healthcare delivery, from diagnostic to therapeutic intervention, by effectively treating and managing skin diseases. Such interventions facilitate tailored approaches based on specific dysregulated biomarkers. Despite challenges*

**Corresponding author e-mail: juhi.pradhan@mlsu.ac.in*

inherent in drug delivery for dermatological conditions like atopic dermatitis and acne.

Keywords: *Nanoparticulate metal composites (NMCP), bioactive compounds, carriers, advantages, toxicity, acne disease*

1. Introduction

The properties of Nanoparticulate Metal Composite Particles (NMCP) are superior to those of standard formulations that rely on the creation and manufacturing of nanostructures (Rahman *et al.*, 2017). Numerous metallic-nanoparticulate carriers are now being studied as efficient drug delivery methods. These include lipidic micro/nano-emulsion systems, lipidic vesicles, metallic nanoparticles, lipid nanoparticles, polymeric nanoparticles, lipid-polymer hybrid nanoparticle systems, etc. The topical application of silver nanoparticles as nanocarriers has garnered significant attention among these formulations and holds promise as a model for developing effective treatments against various pathogenic microorganisms. Among the many advantages of silver nanomedicine were also targeted drug delivery, eliminating the medication's solubility or stability issues, low dosage, and lowering drug-induced side effects (Patel *et al.*, 2019). Furthermore, metallic nanoparticle production is an inexpensive, simple, and ecologically safe process. Herbal plants are an effective way to synthesis a wide range of greener nanoparticles, such as cobalt, copper, silver, gold, palladium, platinum, zinc oxide, and magnetite. In addition, a number of diseases, such as hepatitis, HIV, cancer, malaria, and other acute disorders, may be treated by plant-mediated nanoparticles (Benner *et al.*, 2013). One of the greatest common skin conditions affecting teenagers is acne. Additionally, it is sporadically persisting into maturity (Garg *et al.*, 2014).

Comedones, or black and white pimples, scaly red skin (seborrhea), pinhead-sized pimples, papules, and pimples on the face, neck, and back are characteristics that set it apart (Garg *et al.*, 2015). Bacteria are the usual cause of acne. Secondary factors contributing to the development of acne include hormone fluctuations, anxiety, and genetics, all of which can be caused by *Propionibacterium acnes*, also known as Acne vulgaris (Knutsen-Larson *et al.*, 2012). Acne is caused by increased sebum production and its accumulation in skin pores as a result of hormonal stimulation of the sebaceous gland (Brajac *et al.*, 2004). Acne affects almost 90% of the population. Teenagers are the primary population affected because of this, as their bodies contain higher levels of hormones like testosterone (Tanghetti *et al.*, 2013, Thakur *et al.*, 2015). Therefore, the current review paper covered metallic nanoparticles in drug delivery systems, including their types, roles, toxicities, and potential applications for treating acne.

Topical infections, which include a variety of conditions like impetigo, eczema, acne, psoriasis, and infected seborrheic dermatitis, provide serious health risks to people. Traditional treatments have struggled due to ineffective anti-infectives, prolonged treatment durations, and incomplete recovery. Moreover, the rise of antibiotic-resistant bacteria like Staphylococcus aureus and Methicillin-Resistant Staphylococcus aureus (MRSA) underscores the necessity for novel delivery systems to enhance therapeutic effectiveness. Nanotechnology, particularly in the realm of medicine (nanomedicine), is revolutionizing the approach to combating topical infections by engaging at the sub-atomic level with skin tissue. Vesicular carriers such as liposomes, lipidic nanoparticles, and silver nanoparticles show promise as delivery solutions, facilitating interactions at barrier membranes. Recent efforts in leveraging nanomedicine-based delivery systems against topical infections are summarized in this review, highlighting their potential in addressing current treatment challenges. This includes insights into formulation considerations, penetration mechanisms, and relevant patents. However, despite advancements in technology and delivery systems, concerted efforts are still required to effectively tackle the global challenge of topical infections (Thakur *et al.*, 2018).

2. NMCP in Drug Delivery Systems

A more effective drug delivery system is desperately needed in order to solve the problems of low selectivity, water insolubility, poor stability, and toxicity related to a wide range of drugs. To reduce the negative effects of free medication on organs that are not intended targets, nanoparticle-mediated drug delivery aims to enhance target cell absorption or delivery. Furthermore, it has been debated whether or not these metallic nanoparticles are highly biocompatible, which means that when employed in in vitro and in vivo experiments, they have very little to no negative impacts. Although not much has been done in this area, a study of the literature indicates that further research could be done on green nanoparticle-mediated medicine administration and bioactive molecule delivery. Metallic NMCP has a number of uses outside of cosmetics, such as in skin care products, hydration, anti-aging, and acne treatment. Examples of inorganic NMCPs that can be used in topical sunscreen compositions are ZnO and TiO2. Not only are they clear, but they also have a lovely, fitting appearance. ZnO and TiO2 NMCP work together to protect the skin from broad-band UV rays. Additionally, the microsized ZnO and TiO_2 (NPs) (<100 nm) in these opaque sunscreens address the cosmetic fault and give good cosmetic formulation properties when combined with TiO_2 NMCP. Good UV-A and UV-P protection is provided by the combination of micro- and nanosized ZnO dispersions with nanosized TiO2 particles (Smijs *et al.*, 2011). Silver has antibacterial qualities, as do other NMCPs. Treatments for cutaneous leishmaniasis (Crissey *et al.*, 1998), trichophyton (Elkeeb *et al.*, 2010), and onychomycosis (Pinheiro *et al.*, 2007) can all be effectively completed using silver and methicillin-resistant Aureus Staphylococcus.

3. Advantages of NMCP

The following are some benefits that plant-derived NMCPs have over traditional drug delivery dosage forms:

1. Plant-derived non-steroidal cofactor polymers (NMCPs) are utilized to deliver different bioactive chemicals to specific areas without damaging healthy tissues;
2. Plant-derived NMCPs exhibit high levels of encapsulation efficiency, skin penetration ability, and sustained drug release capacity;
3. Plant-derived NMCP carriers can encapsulate hydrophilic and lipophilic medicines while reducing their toxicity;
4. Plant-derived NMCP carriers shield the medication from chemical and physical deterioration and lengthen its half-life in the target organ;
5. An enhancement in the horny layer qualities shows that the encapsulated medication molecules' physicochemical properties are maintained throughout time (Garg *et al.*, 2014, Vyas *et al.*, 2014).

3.1 Several cited benefits of MNCP include

NMCP syntheses mediated by plants are widely used because they are inexpensive, readily accessible, and simple to make. NMCP prepared environmentally by utilizing plant extracts. It turns out to be a significant topic of study for bio-nanotechnology in this discipline. The aqueous herbal extracts are primarily used in the biogenic synthesis process to fabricate noble NMCP due to the extract's greater availability of reducing agent than the entire plant (Huang *et al.*, 2007).

Comparing plant-mediated synthesis of NMCP to conventional physical and chemical procedures, it is also much simpler and easier to perform without the need for specialized equipment or special operating conditions. This process is also more environmentally friendly because the synthesized NMCP of products includes non-corrosive solvent. However, strong and weak chemical reducing agents and capping agents, such as sodium citrate, sodium borohydride, and alcohols—which are typically toxic, flammable, and difficult to decompose—are needed for the physical and chemical processes (Lalitha *et al.*, 2013).

It has been demonstrated that the rate of bio-reduction of metal ions employing biological agents is significantly faster, even under ambient temperature and pressure conditions, independent of the features of reproducibility and stability (Pasupuleti *et al.*, 2013). Conversely, past research findings indicated that the herbal extracts' capacity for bio-reduction. In comparison to the microbial culture, it is higher (Khalil *et al.*, 2014). Furthermore, the microbial-based process produced the waste products. Depending on the kind of bacteria utilized in the synthesis, it may be more detrimental to the ecosystem (Moghaddam *et al.*, 2010). Therefore, there is little to no contamination associated with plant-mediated NMCP production, which lessens its negative environmental effects. The biosynthetic method using herbal extracts has now emerged as a straightforward, practical, and efficient technique with all the benefits and exceptional qualities over other methods mentioned above. It is also a good alternative to traditional chemical and physical NMCP preparation methods, as well as microbial methods (Huang *et al.*, 2007).

Herbal extracts are used in the development of the green NMCP. These carriers improve the bioactive compounds' topical administration, and they work wonders for a variety of skin conditions like burns, fungus, and acne. Through these metallic carriers, the bioactive compounds are easily transported and delivered to dermal location, with a decrease in their adverse effects. Effective anti-acne therapy depends on the controlled release of medications from a nanotechnological carrier and their biodegradation. A variety of NMCP carriers, such as liposomes, niosomes, microemulsions, microsponge, microspheres, and NMCP, are used in topical formulations to treat various topical illnesses, such as anti-acne therapy (Castro *et al.*, 2008).

4. Types of Nano-Carriers

Numerous carriers of nanoparticles that can be utilized to convey bioactive compounds, as well as their metallic counterparts (Papakostas *et al.*, 2011):

1. Liposomes: hydrophilic and lipophilic compounds that, when combined with medicine, exhibit high hair follicle penetration (Jung *et al.*, 2006).
2. Immunostimulatory complexes, or ISCOMs, are vaccine-use matrices made of phospholipids, cholesterol, saponifiers, and antigens (Peek *et al.*, 2008).
3. Virosomes: Protein-liposome hybrids produced by viruses. It can be a part of the HBV and HPV vaccines (Ludwig *et al.*, 2007).
4. Polymerized particles (PLA, PLGA, and CL): Due to their cytotoxicity and lack of biodegradability, these particles are primarily used for the delivery of anticancer medications, which restricts their usage in human medicine (Mahe *et al.*, 2009).

5. Fullerenes: These supramolecular structures have the ability to absorb UV light and neutralize free radicals, making them useful as NMCPs in topical formulations for UV protection; (Chirico *et al.*, 2007, Kato *et al.*, 2010)
6. Dendrimers: These help move active molecules through the epidermis (Villalonga-Barber *et al.*, 2008).
7. PAMAM (Polyamidoamine): This has been utilized to improve active molecule penetration through the skin (Venuganti *et al.*, 2009).
8. When delivering active substances to the skin, Solid Lipid Nanoparticles (SLN) and Nanostructured Lipid Carriers (NLC) are incredibly safe since they maintain their stability in both lipophilic and hydrophilic environments while releasing the liposome micelle structure (Cho *et al.*, 2023, Krist *et al.*, 2010, Saupe *et al.*, 2005, Schäfer-Korting *et al.*, 2007). Several types of nanostructured porous materials (NMCPs), such as nanopores, nanotubes, nanorods, nanoclusters, and nanostars, have been created using a variety of elements, including gold, silver, platinum, zirconium, copper, palladium, iron, selenium, and strontium (Walther *et al.*, 2008).

5. Green NMCP'S Roles

i. **Reducing and stabilizing:** Natural materials like gold and silver are commonly employed in the synthesis of NMCP. Using gold chloride, a reducing agent, and a stabilizer, plants like neem (Azadirachta indica), Aloe vera, Camellia sinensis, Catharanthus roseus, geranium leaf, Datura metel, Cinnamomum camphora, and lemongrass (Cymbopogon sp.) are used in this procedure to reduce the plant material (Shankar *et al.*, 2016; Mittal *et al.*, 2013).

ii. **Potential anti-inflammatory and antibacterial substances:** As a result of their remarkable properties particularly their small size NMCP has been reported by several researchers to exhibit strong antimicrobial and anti-inflammatory properties. These properties can be combined with natural or Synthetic anti-inflammatory agents to manage skin diseases. Currently, research is being done on the topical treatment of skin conditions with gold, silver, and platinum NMCP. The applications of NMCP-polyphenols are especially in topical drug delivery. Because polyphenolic compounds are reactive oxygen species (ROS)-producing enzymes are inhibited, these are naturally occurring compounds, found in many diet products. These compounds have been shown to have anti-inflammatory and antioxidant properties that can be beneficial in the treatment of skin infections (Miquel-Jeanjean *et al.*, 2012).

iii. **Pass through the skin with ease:** The NMCP, such as silver and gold NMCP, were synthesized using an environmentally safe method and then put to an ointment. Gold nanoparticle carrier systems (NPCs) in particular are gaining popularity due to their advantageous properties, which make them ideal for topical drug delivery. These properties include their ease of functionalization, large surface area relative to volume, and small particle size, as well as their inherent anti-inflammatory effect, which produces a synergistic effect when loaded with anti-inflammatory agents (Wang *et al.*, 2018). In addition to improving solubility and stability, targeting, and reducing the necessary dosage of the medicine, the silver NMCP also has the benefit of increasing efficacy and preventing side effects (Kalasin *et al.*, 2009).

   According to Miquel-Jeanjean and colleagues, there was no difference between the two types of skin since TiO2 NMCP was contained in the epidermis and did not penetrate the dermis layer in both cases when tested for penetration in normal skin (David *et al.*, 2014). Through trans-appendegeal, intracellular, and intercellular pathways, NMCP cross the skin barrier. However, the altered condition of psoriatic skin should be considered because of the broken tight connection between cells. Even though NMCP is used to treat skin irritation and has antimitotic effects. Greenly produced gold nanoparticle composites exhibit low cytotoxicity (Perde-Schrepler *et al.*, 2016, Zolotarenko *et al.*, 2017), which calls for more research.

iv. **The regulator known as nuclear factor kappa light chain enhancer (KFKB):** According to Zolotarenko *et al.* (Niska *et al.*, 2017), the NFKB (nuclear factor kappa light chain enhancer of activated B cells) family of transcription factors, which controls a variety of crucial cellular behaviors like inflammatory responses, cellular growth, and the apoptosis signaling pathway, is one of the primary transcriptional regulators involved in skin inflammation. Platinum NMCP has been used as an anti-inflammatory medication by several researchers, who have found that it has a down-regulating effect on this transcription factor family.

v. **Good spreading agent:** Particle size, shape, and functional groups were found to be the three primary determinants impacting skin penetration and retention. According to Niska *et al.*, NMCPs, or very small particles (<4 nm), can pass through the skin barrier, whereas small particles (4–20 nm) can only pass through damaged skin. Particles are greater than 45 nm cannot penetrate skin that is either intact or damaged (Mehrotra *et al.*, 2016).

vi. **NMCP as nano-biosensor:** MNP-based biosensors' distinct physico-chemical characteristics present new opportunities for clinical research and diagnostics. These biosensors may help with imaging, therapy, cell tracking, marker analysis, disease diagnosis, and disease pathogenesis monitoring, according to the reports (Malaekeh-Nikouei *et al.*, 2020).

vii. **UV protective agents:** Furthermore, this method is also known as green synthesis or biosynthesis, which makes it both economical and safe for the environment. Furthermore, because of these plants' anti-inflammatory properties, NMCP generated from them has been demonstrated to be promising in the treatment of dermatological conditions (Crisan *et al.*, 2016). Due to their effectiveness as antibacterial, antifungal, anti-skin cancer, UV radiation protecting, targeting, and good skin penetrating agents, they are currently being studied extensively as skincare products to be used in both dermatology and cosmetology fields (Haider *et al.*, 2004).

viii. **Synergistic property:** NMCP evaluated and validated their synergistic properties using an in vitro experiment. With the help of copper and zinc sulfates, they produced two gels and a cream that have antibacterial qualities. It was shown that zinc sulfate and copper sulfate had a strong synergistic antibacterial effect in the creams and gels. It was shown that 3% w/w was the lowest effective concentration for both active ingredients against Escherichia coli and Staphylococcus aureus, the two bacteria that cause skin infections. This indicates that it can be used in a variety of topical formulations (Lolis *et al.*, 2009).

ix. **Extremely stable and reproducible:** Because the green-NMCP synthesis process exhibits greater stability and reproducibility, it is an excellent choice for large-scale manufacturing at a lower cost per unit, as well as being safe for use in human therapy (Lolis *et al.*, 2009).

6. Current Approaches to Acne Treatment

Many factors affect how a patient is treated for acne. The clinical type of acne, the number and size of lesions, the level of inflammation, the oil content of the skin, the presence of hyperandrogenism in females, and the extent of scarring are some of these variables (49). The type or clinical class of acne ultimately determines the course of treatment. The main objectives of the treatment plan are to reduce the amount of sebum, bacterial count, irritation, and hyperkeratinization (Sykes *et al.*, 1994). The patient's skin type, whether oily or dry, affects the anti-acne drug delivery system selection (Kumasaka *et al.*, 1992). Because gels have the ability to dry out skin, those with oily skin can use them. Creams and lotions can help people with dry skin since they hydrate. Large areas of the body can be easily covered with solutions, despite the possibility of dryingness. There is potential for using foams to apply anti-acne drugs to body parts where there is a lot of hair growth (Mekonnen *et al.*, 2013). These days, acne can be treated with topical, systemic, surgical, and photodynamic treatments.

A topical skin care cream including cucumber plant was made using a conventional technique; this cream strengthens the antiacne benefits by increasing transepidermal water loss (Zahoor *et al.*, 2021). In addition to *Ocimum gratissimum* (leaves) of Silver NMCP, it is stable, spherical, and polydisperse, with an average size of 207 nm. It forms a herbal gel that inhibits Propionibacterium acne, S. aureus, and E. coli (Ghyczy *et al.*, 1996). Acne treatment involves the efficient distribution of bioactive chemicals through the application of current nanotechnological approaches. In a clinical study, it was demonstrated that flexible liposomes might help reduce wrinkles and provide other advantages including improved skin smoothness and decreased efflorescence when treating acne (Guterres *et al.*, 2007). Consequently, topical formulations based on nanotechnology, such as nanospheres, are becoming more and more common in the cosmetics industry, particularly in skin care products like anti-wrinkle, anti-acne, and moisturizing creams (Yapar *et al.*, 2012). Shampoos, gels for hairstyles, sunscreen, and anti-acne treatments are just a few cosmetic products that benefit from the application of dendrimers (Novi *et al.*, 2020). Aqueous extract of *Artemisia scoparia* (AS) has been used as a reducing and stabilizing agent to produce Au, Ag, Cu, AuAg, AuCu, and AgCu, among other forms of NMCP. Additionally, gelatin/NMCP composite films were made. When combined, clindamycin and silver NMCP have demonstrated some limited synergistic benefits. In contrast to the combination dose of silver NPCP and clindamycin, the Minimum Bacteriocidal Concentration (MBC) value drops from the previous version. It has been confirmed that clindamycin and silver NMCP combinations can reduce side effects and conquer resistance (Paralikar *et al.*, 2014). It is clear that adding antibiotics to silver non-metal zinc chloride particle (NMCP) increased its efficacy against test pathogens. Strong antibacterial activity is exhibited by silver nanoparticle cyanide (NMCP) made from *E. oxypetalum* leaf extract, which makes it a valuable tool for a variety of biological applications (Palanivel *et al.*, 2016). Green synthesised AgNPs produced with *C. sativum* are used in biomedicine for anti-acne, anti-dandruff, and anti-breast cancer therapy (Nel *et al.*, 2006). A gel for the treatment of acne vulgaris was created and assessed. The combination of it with artificial silver nanocluster polymer (NMCP) derived from *Ocimum gratissimum* showed promising results against acne-causing bacteria. Because of this, the silver NMCP of *Ocimum gratissiumum* in an aqueous gel base is one formulation that works well for treating acne vulgaris (Nasir *et al.*, 2010). There is great promise for the use of *C. sativum*-based green synthesised AgNPs in biomedical applications such as anti-acne, anti-dandruff, and the treatment and management of breast cancer.

7. Silver NMCP Toxicity

Overall, silver nanoparticle capture polymers (NPCPs) is a great carrier for a number of uses in a variety of industries, particularly the biomedical one, where it is used for cell imaging, drug delivery, diagnosis, and implantation. However, a number of studies have indicated that nanoparticles can have negative effects on both humans and the environment. Due to their production of reactive oxygen species (ROS), the NMCP were accused of cytotoxic effects on epidermal keratinocytes and stem cells in hair follicles. Metal ions are also believed to irritate skin. TiO2 and ZnO NPs are incorporated into the stratum corneum during skin covering from NP-containing sunscreens. This might modify certain NP attenuation qualities because of physicochemical interactions between particles, between particles and skin, and between skin particles and light. Both sunscreen nanoparticles (NPs) cause (photo) cytotoxicity and genotoxicity; inviable skin layers have been reported sometimes, particularly in the case of prolonged exposure to ZnO. According to studies, carbon nanotubes cause granulomas in the lungs of lab animals and are cytotoxic. The safety issues of the nano-engineered materials being discharged into our environment and into consumer and health care items have not been well studied (Jo *et al.*, 2015, Ferreira *et al.*, 2022).

Hence, it is imperative that we, as a community that is intimately linked to individuals' health, understand this new technology, impart knowledge about it to our own members, and take an active role in evaluating it and formulating guidelines for its responsible and advantageous application. It will take time to determine if nanotechnology turns out to be advantageous or damaging despite all the hype surrounding it.

As a pharmaceutical and metal ion carrier in addition to being an active component, chitosan has two roles. Reactive group and acetylation degree changes can be made to change its characteristics. It is used in films, scaffolds and nanoparticles, among other forms. When paired with other polymers or bioactive substances, chitosan efficiently elicits the intended biological reactions. It is a good choice for skincare because of its biocompatibility, water entrapment capacity, antioxidant, antibacterial, and anti-inflammatory qualities, which guard against germs and oxidative damage. Chitosan-based nanoparticles can be used topically or systemically to deliver medications for the treatment of wounds, cancers, acne, and inflammation. The method of producing chitosan, its properties, alterations, and interactions with bioactive substances are covered, in addition to its molecular mechanisms for skin healing and protection (Ferreira *et al.*, 2022, Cho *et al.*, 2023).

8. Future Perspectives

Since NMCP are essential to nanomedicine and other biological applications, they seem to have a significant future in the twenty-first century. To make these NMCP, one should choose stable, reasonably priced, environmentally friendly, and chemical-free herbal resources. The production of monodispersed NMCP is crucial for next studies. However, the exact method by which NMCP is synthesized is yet unknown. New approaches are emerging with the potential for advancement in nanoscience to use noble NMCP to overcome the obstacles. It has to be thoroughly researched in the future. Because of their distinctive atomic and supramolecular characteristics, noble NMCPs can be effective active agents in a variety of medicinal and diagnostic applications. There is currently a great need for a range of metallic nanoparticles and their composites in the pharmaceutical, biological, and health care sectors. Thus, in order to safeguard human health, it is imperative that safety measures come first. It is vital to investigate the NMCP interactions at the laboratory scale and alter the NMCP for better biocompatibility with cells (cancerous) in order to prevent damage to normal cells. Highly controlled processes are used to modify the herbal plant-mediated synthesis such that it is appropriate for the creation of NMCP. In addition, biological NMCP synthesis is an inexpensive, one-step, ecologically safe method. Moreover, these plant-mediated nanoparticles may be used as pharmaceutical agents to treat a variety of diseases, such as malaria, HIV, cancer, and hepatitis. Future research should therefore look into the mechanism that controls the size and form of nanoparticles. Reducing the toxicity of NMCP and increasing its use in medical applications are two more major challenges. The NMCP formulation and a few current reported applications are displayed in tabular form.

Table 1: List of NMCP formulations and their applications

S.N.	Types of nanoparticulate metal composites (NMCP)	Applications in acne	References
1	Gold nanoparticles	Gold photothermal therapy for acne management	Suh *et al.*, 2021
2	Curcumin-containing liposomal gold nanoparticles (Au Lipos Cur NPs)	Au Lipos Cur NPs shows promise as an effective treatment for recurrent acne	Alvi *et al.*, 2021
3	Zinc porphyrin-based metal-organic framework	Acne treatment by utilizing ultrasound-responsive interface engineering	Alvi *et al.*, 2023
4	Gold and silver nanoparticles	Gold and silver nanoparticles hold promise for acne treatment	Artounian *et al.*, 2021
5	Silver nanoparticles (AgN-CA) in a gel matrix	AgN-CA gel proved to be a promising carrier system for topical acne treatment with Citrus sinensis extract	Alam *et al.*, 2023
6	Gold nanoshell-mediated photothermal therapy	Gold nanoshell-mediated photothermal therapy as a safe and efficient option for treating recurrent acne vulgaris	Park *et al.*, 2020
7	Carboxymethyl chitosan/sodium alginate (CMCS/SA) hydrogel loaded with HHC36 and silver nanoparticles (AgNPs) conjugates (Ag-H2/CMCS/SA hydrogel)	Ag-H2/CMCS/SA hydrogel, composed of natural polysaccharides, ensures cell compatibility and suggests its potential in acne therapy	Liu *et al.*, 2023
8	Controlled Synthesis of Zein-Functionalized TiO(2) Nanoparticles	Hybrid nanocuboids could serve as effective environmentally friendly materials with promising antimicrobial properties, suitable for cosmeceutical applications	Badgujar *et al.*, 2021
9	Silver nanoparticles (AgNPs) of *Coriandrum sativum* leaf extract	AgNPs synthesized using *C. sativum* leaf extract hold promise for biomedical applications such as treating acne, dandruff, and breast cancer	Sathishkumar *et al.*, 2016
10	Zinc oxide nanoparticles of *Dendrobium anosum* for anti-acne nanogels	Zinc oxide nanoparticles nanogels as green solutions for treating skin infections	Tan *et al.*, 2023
11	Silver Nanoparticle Gel	Silver nanoparticle gel appears as effective and safe as clindamycin gel for treating moderate acne vulgaris	Jurairattanaporn *et al.*, 2017

S.N.	Types of nanoparticulate metal composites (NMCP)	Applications in acne	References
12	ZnO Nanoparticle	ZnO-isotretinoin (ISO) exhibits significantly enhanced anticancer properties	Zhao *et al.*, 2017
13	Carboxyl-modified gold nanoparticles to phospholipid liposomes	Acid-responsive liposomes offer potential for treating skin diseases like acne and staph infections	Pornpattananangkul *et al.*, 2010
14	Silica-gold nanoshells	Ultrasonically-delivered silica-gold nanoshells in inducing photothermal disruption of sebaceous glands, while preserving the integrity of surrounding skin tissue	Paithankar *et al.*, 2015

9. Final Thoughts and Future Views

As a promising technique, nanotechnology has raised hopes for several scientific domains, most notably topical medicine. Even after years of intensive study, green NMCP which uses plant tissue remains a compelling field for investigation. NMCP synthesis is induced by the bioactive plant extract. It is easily accepted, non-toxic, easy to use, cheap, and simple. Treatments for a number of skin infections and serious illnesses, including cancer, HIV, hepatitis, and malaria, have been tried with this NMCP. Nevertheless, there hasn't been a lot of pilot and commercial work done in this area. Consequently, next study should focus on creating simple procedures by comprehending the kinetics and mechanism of NMCP at large-scale synthesis from plant sources.

References

Alam, A., Foudah, A. I., Alqarni, M. H., & Yusufoglu, H. S. (2023). Microwave-assisted and chemically tailored chlorogenic acid-functionalized silver nanoparticles of Citrus sinensis in gel matrix aiding QbD design for the treatment of acne. Journal of Cosmetic Dermatology, 22(5), 1613-1627. https://doi.org/10.1111/jocd.15611

Alvi, S.B., Rajalakshmi, P.S., Jogdand, A., Sanjay, A.Y., B V., John, R., & Rengan, A.K. (2021). Iontophoresis mediated localized delivery of liposomal gold nanoparticles for photothermal and photodynamic therapy of acne. Biomater Sci, 9(4), 1421-1430. doi: 10.1039/d0bm01712d

Artounian, K., Bundogji, N., Hoss, E., & Boen, M. (2021). Applications of gold and silver nanoparticles in the treatment of acne vulgaris: A systematic review. Journal of Drugs in Dermatology, 20(6), 666-670. https://doi.org/10.36849/JDD.2021.5762

Badgujar, H. F., & Kumar, U. (2021). Green approach towards morphology-controlled synthesis of zein-functionalized TiO(2) nanoparticles for cosmeceutical application. European Journal of Pharmaceutical Sciences, 167, 106010. https://doi.org/10.1016/j.ejps.2021.106010

Benner, N., & Sammons, D. (2013). Overview of the treatment of acne vulgaris. Osteopathic Family Physician, 5, 185-190.

Brajac, I., Bili-Zulle, L., Tkalcić, M., Lončarek, K., & Gruber, F. (2004). Acne vulgaris: Myths and misconceptions among patients and family physicians. Patient Education and Counseling, 54(1), 21-25.

Castro, G. A., & Ferreira, L. A. (2008). Novel vesicular and particulate drug delivery systems for topical treatment of acne. Expert Opinion of Drug Delivery, 5(6), 665-679.

Chirico, F., Fumelli, C., Marconi, A., Tinari, A., Straface, E., Malorni, W., Pellicciari, R., & Pincelli, C. (2007). Carboxyfullerenes Localize Within Mitochondria and Prevent the UVB-Induced Intrinsic Apoptotic Pathway. Experimental Dermatology, 16, 429-436.

Cho, Y. H., Oh, S. M., Choi, J. Y., & Jeong, K. H. (2023). Acne treatment based on selective photothermolysis with topically delivered light-absorbing platinum nanoparticles. Lasers in Medical Science, 38(1), 125. https://doi.org/10.1007/s10103-023-03787-1

Crissey, J. T. (1998). Common dermatophyte infections. A simple diagnostic test and current management. Postgraduate Medicine, 103(2), 191-192, 197-200, 205. https://doi.org/10.3810/pgm.1998.02.359

David, L., Moldovan, B., Vulcu, A., Olenic, L., Perde-Schrepler, M., Fischer-Fodor, E., Florea, A., Crisan, M., Chiorean, I., Clichici, S., *et al.* (2014). Green synthesis, characterization and anti-inflammatory activity of silver nanoparticles using European black elderberry fruits extract. Colloids and Surfaces B: Biointerfaces, 122, 767–777.

Elkeeb, R., AliKhan, A., Elkeeb, L., Hui, X., & Maibach, H. I. (2010). Transungual drug delivery: Current status. International Journal of Pharmaceutics, 384, 1–8.

Ferreira, P.G., Ferreira, V.F., da Silva, F.C., Freitas, C.S., Pereira, P.R., & Paschoalin, V.M.F. (2022). Chitosans and Nanochitosans: Recent Advances in Skin Protection, Regeneration, and Repair. Pharmaceutics, 14(6), 1307. doi: 10.3390/pharmaceutics14061307

Garg, T. (2014). Artificial cells. Nanomedicine, and Biotechnology, 44(1), 98-105.

Ghyczy, M., Nissen, H. P., & Biltz, H. (1996). The treatment of acne vulgaris by phosphatidylcholine from soybeans. Journal of Applied Cosmetology, 14(4), 137-145.

Guterres, S. S., Alves, M. P., & Pohlmann, A. R. (2007). Polymeric nanoparticles, nanospheres and nanocapsules, for cutaneous applications. Drug Target Insights, 2, 147-157.

Haider, A., & Shaw, J. C. (2004). Treatment of acne vulgaris. Journal of the American Medical Association, 292(6), 726-735

Huang, J., Qingbiao, L., Sun, Daohua., Yinghua, Lu., Yuanbo, Su., Yang, Xin., Wang, Huixuan., Wang, Yuaneng., Shao, Wenyao., He, Ning., Hong, Jinqing., & Chen, C. (2007). Nanotechnology, 18(10), 1-12.

Jo, H., Her, J., & Ban, C. (2015). Dual aptamer-functionalized silica nanoparticles for the highly sensitive detection of breast cancer. Biosensors and Bioelectronics, 71, 129-136.

Jung, S., Otberg, N., Thiede, G., Richter, H., Sterry, W., Panzner, S., & Lademann. (2006). Nanoparticles and their interactions with the dermal barrier. Journal of Investigative Dermatology, 126, 1728-1732.

Jurairattanaporn, N., Chalermchai, T., Ophaswongse, S., & Udompataikul, M. (2017). Comparative Trial of Silver Nanoparticle Gel and 1% Clindamycin Gel when Used in Combination with 2.5% Benzoyl Peroxide in Patients with Moderate Acne Vulgaris. Journal of the Medical Association of Thailand, 100(1), 78-85.

Kalasin, S., & Santore, M. M. (2009). Non-specific adhesion on biomaterial surfaces driven by small amounts of protein adsorption. Colloids and Surfaces B: Biointerfaces, 73, 229-236

Kato, S., Taira, H., Aoshima, H., Saitoh, Y., & Miwa, N. (2010). Clinical evaluation of fullerene-C60 dissolved in squalane for anti-wrinkle cosmetics. Journal of Nanoscience Nanotechnology, 10, 6769-6774.

Khalil, M. M. H., Ismail, E. H., El-Baghdady, K. Z., & Mohamed, D. (2014). Arab Journal of Chemistry, 7(6), 1131-1139.

Knutsen-Larson, S., Dawson, A. L., Dunnick, C. A., & Dellavalle, R. P. (2012). Acne vulgaris: Pathogenesis, treatment, and needs assessment. Dermatologic Clinics, 30(1), 99-106.

Krist, J., Teskac, K., & Grabnar, P. A. (2010). Current View on Nanosized Solid Lipid Carriers for Drug Delivery to the Skin. Journal of Biomedical Nanotechnology, 6, 529-542.

Kumasaka, B. H., & Odland, P. B. (1992). Acne vulgaris. Topical and systemic therapies. Postgrad Med, 92(5), 181-183.

Kuppusamy, P., Yusoff, M. M., Maniam, G. P., & Govindan, N. (2016). Biosynthesis of metallic nanoparticles using plant derivatives and their new avenues in pharmacological applications - An updated report. Saudi Pharmaceutical Journal, 24(4), 473-484.

Lalitha, A., Subbaiya, R., & Ponmurugan, P. (2013). Green synthesis of silver nanoparticles from leaf extract Azhadirachta indica and to study its anti-bacterial and antioxidant property. International Current Microbiology and Applied Sciences, 2(6), 228-235

Liu, H., Zhan, J., Lin, R., Yin, Y., & Ren, L. (2023). Strong infiltrative HHC36 antimicrobial peptide/silver nanoparticles-loaded carboxymethyl chitosan/sodium alginate hydrogel for acne vulgaris therapy. Nanotechnology, 34(49). https://doi.org/10.1088/1361-6528/acf5f5

Lolis, M. S., Bowe, W. P., & Shalita, A. R. (2009). Acne and systemic disease. Medicine of Clinical North American, 93, 1161-1181.

Ludwig, C., & Wagner, R. (2007). Virus-like particles-universal molecular toolboxes. Current Opinion in Biotechnology, 18, 537-545.

Mahe, B., Vogt, A., Liard, C., Duffy, D., Abadie, V., Bonduelle, O., Boissonnas, A., Sterry, W., Verrier, B., Blume-Peytavi, U., & Combadiere, B. (2009). Applications of Nanotechnology in Dermatology. Journal of Investigative Dermatology, 129, 1156-1164.

Malaekeh-Nikouei, B., Bazzaz, B. S. F., Mirhadi, E., Tajani, A. S., & Khameneh, B. (2020). The role of nanotechnology in combating biofilm-based antibiotic resistance. Journal of Drug Delivery Science and Technology, 60, 101880.

Mekonnen, A., Sidamo, T., Asres, K., & Engidawork, E. (2013). In vivo wound healing activity and phytochemical screening of the crude extract and various fractions of Kalanchoe petitiana A. Rich (Crassulaceae) leaves in mice. Journal of Ethnopharmacology, 145(2), 638-646.

Miquel-Jeanjean, C., Crépel, F., Raufast, V., Payre, B., Datas, L., Bessou-Touya, S., & Duplan, H. (2012). Penetration study of formulated nanosized titanium dioxide in models of damaged and sun-irradiated skins. Photochemistry and Photobiology, 88(6), 1513-1521.

Mittal, A. K., Chisti, Y., & Banerjee, U. C. (2013). Synthesis of metallic nanoparticles using plant extracts. Biotechnology Advances, 31(3), 346-356.

Moghaddam, K. M. (2010). An Introduction to Microbial Metal Nanoparticle Preparation Method. Journal of Young Investigation, 19(19), 1-6.

Nasir, A. (2010). Nanodermatology: a glimpse of caution just beyond the horizon - part II. Skin Therapy Letters, 15, 4-7.

Nel, A., Xia, T., Mädler, L., & Li. N. (2006). Toxic potential of materials at the nanolevel. Science, 311, 622-627.

Niska, K., Zielinska, E., & Radomski, M. (2017). Metal nanoparticles in dermatology and cosmetology: Interactions with human skin cells. Chemico-Biological Interactions, 1, 38-51.

Novi, Haryanti., Iskandarsyah., & Yeva Rosana. (2020). Sinergicity test of silver nanoparticles and clindamycin against Staphylococcus aureus. International Journal of Research in Pharmaceutical Sciences, 11(1), 1192-1198.

Paithankar, D., Hwang, B. H., Munavalli, G., Kauvar, A., Lloyd, J., Blomgren, R., Faupel, L., Meyer, T., & Mitragotri, S. (2015). Ultrasonic delivery of silica-gold nanoshells for photothermolysis of sebaceous glands in humans: Nanotechnology from the bench to clinic. Journal of Controlled Release, 206, 30-36. https://doi.org/10.1016/j.jconrel.2015.03.004

Palanivel, S.K., Johnson, P., Raji, V., Abdull, R., Yusoff, M., Fuad A., Sadhasivam, S., Ramasamy, B., & Thayumanavan P. (2016). Biogenic Synthesis of Silver Nanoparticles Using Leaves Extract of Epiphyllum Oxypetalum and its Antibacterial Activity. Journal of Photochemistry and Photobiology, 163, 69-76.

Papakostas, D., Rancan, F., Sterry, W., Blume-Peytavi, U., & Vogt, A. (2011). Nanoparticles in dermatology. Archives of Dermatology Research, 303, 533-550.

Paralikar, P. (2014). Biogenic Synthesis of Silver Nanoparticles Using Leaves Extract of Epiphyllum Oxypetalum and its Antibacterial Activity. Austin Journal of Biotechnology & Bioengineeringis, 1(7), 1-5.

Parikha, Mehrotra. (2016). Biosensors and their applications - A review. Journal of Oral Biology and Craniofacial Research, 6(2), 153-159.

Park, K. Y., Han, H. S., Hong, J. Y., Seo, S. J., & Lee, S. J. (2020). Gold nanoshell-mediated photothermal therapy for acne vulgaris. Dermatologic Therapy, 33(1), e13189. https://doi.org/10.1111/dth.13189

Pasupuleti, V. R., Prasad, T., Shiekh, R. A., Balam, S. K., Narasimhulu, G., Reddy, C. S., Rahman, I. A., & Gan, S. H. (2013). International Journal of Nanomedicine, 8, 3355-3364.

Patel, K., Kumar, V., Verma, A., Rahman, M., & Patel, D. K. (2019). Amarogentin as topical anticancer and anti-infective potential: Scope of lipid-based vesicular in its effective delivery. Recent Patents on Anti-Infective Drug Discovery, 14(1), 7-15.

Peek, L. J., Middaugh, C. R., & Berkland, C. (2008). Nanotechnology in vaccine delivery. Advanced Drug Delivery Reviews, 60, 915-928

Perde-Schrepler, M. (2016). Plant-based green synthesis of metallic nanoparticles: scientific curiosity or a realistic alternative to chemical synthesis. Journal of Nanomaterials, 2016, 1-13.

Pinheiro, T., Pallon, J., Alves, L. C., Veríssimo, A., Filipe, P., Silva, J. N., & Silva, R. (2007). The influence of corneocyte structure on the interpretation of permeation profiles of nanoparticles across skin. Nuclear Instruments and Methods in Physics Research Section B, 260, 119–123.

Pornpattananangkul, D., Olson, S., Aryal, S., Sartor, M., Huang, C. M., Vecchio, K., & Zhang, L. (2010). Stimuli-responsive liposome fusion mediated by gold nanoparticles. ACS Nano, 4(4), 1935-1942. https://doi.org/10.1021/nn9018587

Rahman, M., Beg, S., Verma, A., Kazmi, I., Patel, D. K., Anwar, F., & Kumar, V. (2017). Therapeutic applications of liposomal based drug delivery and drug targeting for immune linked inflammatory maladies: A contemporary viewpoint. Current Drug Targets, 18(13), 1558-1571.

Sathishkumar, P., Preethi, J., Vijayan, R., Mohd Yusoff, A. R., Ameen, F., Suresh, S., ... Palvannan, T. (2016). Anti-acne, anti-dandruff and anti-breast cancer efficacy of green synthesized silver nanoparticles using Coriandrum sativum leaf extract. Journal of Photochemistry and Photobiology B: Biology, 163, 69-76. https://doi.org/10.1016/j.jphotobiol.2016.08.005

Saupe, A., Wissing, S. A., Lenk, A., Schmidt, C., & Müller, R. H. (2005). Solid lipid nanoparticles (SLN) and nanostructured lipid carriers (NLC) -- structural investigations on two different carrier systems. Bio-medical Material Engineering, 15, 393-402.

Schäfer-Korting, M., Mehnert, W., & Korting, H. C. (2007). Lipid nanoparticles for improved topical application of drugs for skin diseases. Advanced Drug Delivery Reviews, 59, 427-443.

Shankar, P. D., Shobana, S., Karuppusamy, I., Pugazhendhi, A., Ramkumar, V. S., Arvindnarayan, S., & Kumar, G. A. (2016). A review on the biosynthesis of metallic nanoparticles (gold and silver) using bio-components of microalgae: Formation mechanism and applications. Enzyme and Microbial Technology, 95, 1-58.

Smijs, T. G., & Pavel, S. (2011). Titanium dioxide and zinc oxide nanoparticles in sunscreens: Focus on their safety and effectiveness. Nanotechnology, Science and Applications, 4, 95-112.

Suh, D.H., Park, T.J., Jeong, J.Y., Kim, K.E., Park, J.H., Lee, S.J., Kim, H.J., & Ryu, H.J. (2021). Photothermal therapy using gold nanoparticles for acne in Asian patients: A preliminary study. Dermatol Ther, 34(3), e14918. doi: 10.1111/dth.14918

Sykes, N. L. Jr., & Webster, G. F. (1994). Saving face: A treatment update for acne. Drugs, 48(1), 59-70.

Tan, Y. Y., Wong, L. S., Nyam, K. L., Wittayanarakul, K., Zawawi, N. A., Rajendran, K., ... Dhanapal, A. C. T. (2023). Development and evaluation of topical zinc oxide nanogels formulation using Dendrobium anosmum and its effect on acne vulgaris. Molecules, 28(19), 6749. https://doi.org/10.3390/molecules28196749

Tanghetti, E. A. (2013). The role of inflammation in the pathology of acne. Journal of Clinical and Aesthetic Dermatology, 6, 27-35

Thakur, K., Sharma, G., Singh, B., Chhibber, S., & Katare, O. P. (2018). Current state of nanomedicines in the treatment of topical infectious disorders. Recent Patents on Anti-Infective Drug Discovery, 13(2), 127-150. https://doi.org/10.2174/1574891X13666180529103804

Venuganti, V. V., & Perumal, O. P. (2009). Poly (amidoamine) Dendrimers as Skin Penetration Enhancers: Influence of Charge, Generation, and Concentration. Journal of Pharmaceutical Science, 98, 2345-2356.

Villalonga-Barber, C., Micha-Screttas, M., Steele, B. R., Georgopoulos, A., & Demetzos, C. (2008). Dendrimers as Biopharmaceuticals: Synthesis and Properties. Current Topics of Medicinal Chemistry, 8, 1294-1309.

Vyas, A., Kumar, S. A., & Gidwani, B. (2014). Carrier-based drug delivery system for treatment of acne. The Scientific World Journal, 2014, 276260.

Walther, C., Meyer, K., Rennert, R., & Neundorf, I. (2008). Bioconjugate Chemistry, 19, 2346-2356.

Wang, M., Lai, X., Shao, L., & Li, L. (2018). Evaluation of immunoresponses and cytotoxicity from skin exposure to metallic nanoparticles. International Journal of Nanomedicine, 13, 4445-4459.

Xiang, Y., Lu, J., Mao, C., Zhu, Y., Wang, C., Wu, J., ... Yeung, K. W. K. (2023). Ultrasound-triggered interfacial engineering-based microneedle for bacterial infection acne treatment. Science Advances, 9(10), eadf0854. https://doi.org/10.1126/sciadv.adf0854

Yapar, E. A., & Inal, O. (2012). Nanomaterials and cosmetics. Journal of Pharmacy of Istanbul University, 42(1), 43-70

Zahoor, M., Nazir, N., Iftikhar, M., Naz, S., Zekker, I., Burlakovs, J., Uddin, F., Kamran, A. W., Kallistova, A., Pimenov, N., *et al.* (2021). A Review on Silver Nanoparticles: Classification, Various Methods of Synthesis, and Their Potential Roles in Biomedical Applications and Water Treatment. Water, 13(16), 2216. https://doi.org/10.3390/w13162216

Zhao, W., Wei, J. S., Zhang, P., Chen, J., Kong, J. L., Sun, L. H., Xiong, H. M., & Möhwald, H. (2017). Self-Assembled ZnO Nanoparticle Capsules for Carrying and Delivering Isotretinoin to Cancer Cells. ACS Applied Materials & Interfaces, 9(22), 18474-18481. https://doi.org/10.1021/acsami.7b02542

Zolotarenko, A., Chekalin, E., & Mehta, R. (2017). Tackling the various classes of nano-therapeutics employed in topical therapy of psoriasis. In R. Mehta (Ed.), Nanotechnology in Diagnosis, Treatment, and Prophylaxis of Infectious Diseases (pp. 355-370). Humana Press.

7

Nanomedicine and Healthy Gut

Habiba Bano*

Department of Zoology, University of Rajasthan, Jaipur

Abstract

Nanomedicine, an interdisciplinary field at the intersection of nanotechnology and medicine, has emerged as a promising avenue for addressing complex health challenges. One area of significant interest is the application of nanotechnology in maintaining and improving gut health. This chapter provides a concise overview of the current state of research in nanomedicine's role in promoting gastrointestinal well-being.

Nanoparticles, with their unique physicochemical properties, offer novel approaches for targeted drug delivery, imaging, and diagnostics in the gastrointestinal tract. Engineered nanoparticles can navigate the harsh conditions of the digestive system, delivering therapeutic agents precisely to the intended sites within the gut. This targeted drug delivery minimizes systemic side effects and enhances the therapeutic efficacy of drugs.

Furthermore, nanoscale materials are employed in developing innovative formulations for enhancing nutrient absorption, managing inflammatory bowel diseases, and promoting the growth of beneficial gut microbiota. Nanoencapsulation of bioactive compounds and probiotics provides a protective environment, ensuring their stability through the acidic environment of the stomach and facilitating their release in the intestines.

In addition to drug delivery and nutrient enhancement, nanomedicine contributes to diagnostic imaging of gastrointestinal disorders. Nanoparticles can be functionalized to target specific biomarkers associated with gastrointestinal diseases, enabling early detection and precise monitoring of pathological conditions.

This chapter concludes by highlighting the potential of nanomedicine in revolutionizing gastrointestinal healthcare. However, it also acknowledges the need for continued research to address challenges such as biocompatibility, long-term safety, and scalability of nanomedicine approaches for widespread clinical applications. The evolving landscape of nanomedicine in gut health holds promise for personalized and effective interventions, paving the way for enhanced therapeutic outcomes and improved overall well-being.

Keywords: Gut Microbiota, Nanomedicine, Nanotheranostics, Healthy gut

* *Corresponding author e-mail: banohabiba@gmail.com*

1. Introduction: Unlocking the Potential of Nano Medicine

The human gastrointestinal tract, or gut, is a marvel of biological engineering, responsible for digestion, nutrient absorption, and immune defense. It houses trillions of microorganisms collectively known as the gut microbiota, which play crucial roles in maintaining gut homeostasis and overall health. However, disturbances in the gut microbiota, known as dysbiosis, can lead to a myriad of health issues, including inflammatory bowel diseases (IBD), irritable bowel syndrome (IBS), and metabolic disorders like obesity and diabetes.

Recent advancements in nano medicine, the application of nanotechnology in medicine, have opened up new frontiers in understanding and treating gut-related disorders. At the heart of nano medicine lies the manipulation of materials at the nanoscale, typically ranging from 1 to 100 nanometers. At this scale, materials exhibit unique properties that can be harnessed for diagnostic, therapeutic, and preventive purposes.

One of the most significant contributions of nanomedicine to gut health lies in targeted drug delivery. Traditional oral medications often suffer from poor bioavailability and off-target effects, leading to suboptimal therapeutic outcomes and systemic side effects. Nano carriers, such as liposomes, polymeric nanoparticles, and dendrimers, offer a solution by encapsulating therapeutic agents and delivering them specifically to the gastrointestinal tract. This targeted approach not only enhances the efficacy of treatments but also minimizes side effects by reducing exposure to healthy tissues (Qiu *et al.*, 2023).

Moreover, nano medicine has revolutionized diagnostic imaging in gastroenterology. Nanoparticles can serve as contrast agents for various imaging modalities, including magnetic resonance imaging (MRI), computed tomography (CT), and fluorescence imaging. By selectively accumulating in diseased tissues or microbial populations, nano probes enable early detection and precise localization of gut pathologies, facilitating timely intervention and personalized treatment strategies.

In addition to targeted drug delivery and diagnostic imaging, nano medicine offers innovative approaches for modulating the gut microbiota. Engineered nano formulations can deliver probiotics, prebiotics, or antimicrobial agents directly to the gut, promoting the growth of beneficial bacteria while suppressing harmful pathogens. Furthermore, nanoparticles can be designed to selectively target microbial populations associated with specific diseases, offering a novel strategy for precision medicine in gut health.

Despite the tremendous potential of nano medicine in reshaping the landscape of gastrointestinal medicine, several challenges remain. Ensuring the safety and biocompatibility of nano materials, optimizing drug loading and release kinetics, and navigating regulatory hurdles are among the key obstacles that must be addressed. Nonetheless, ongoing research efforts continue to push the boundaries of nano medicine, offering hope for improved outcomes and enhanced quality of life for patients with gut-related disorders.

In conclusion, the convergence of nano medicine and gut health represents a paradigm shift in our approach to gastrointestinal disorders. By harnessing the power of nanotechnology to target therapies, diagnose diseases, and modulate the gut microbiota, researchers and clinicians are poised to unlock new avenues for precision medicine and personalized healthcare in gastroenterology (Malik *et al.*, 2023).

1.1 Understanding Gut Health: A Microscopic Universe

The gut microbiota, often referred to as the "forgotten organ," represents a diverse and dynamic community of microorganisms that inhabit the gastrointestinal tract. Comprising bacteria, fungi, viruses, and other microorganisms, the gut microbiota forms a complex ecosystem that interacts intricately with the host organism. This symbiotic relationship is essential for maintaining gut homeostasis and overall health (Thursby and Juge, 2017)

The gut microbiota plays a multitude of critical roles in the body, with perhaps its most well-known function being digestion (AboNahas *et al.*, 2022). These microorganisms aid in the breakdown of complex carbohydrates, proteins, and fats, facilitating the absorption of nutrients and energy from food. In addition to digestion, the gut microbiota also contributes to the synthesis of vitamins, such as vitamin K and certain B vitamins, which are essential for various metabolic processes.

Furthermore, the gut microbiota plays a pivotal role in regulating immune function. It serves as a barrier against invading pathogens, competing for resources and space within the gut and producing antimicrobial peptides that help fend off harmful bacteria. Moreover, the gut microbiota interacts with the immune system, influencing the development and maturation of immune cells and modulating immune responses to maintain tolerance to harmless antigens while mounting appropriate defenses against pathogens (Ahlawat and Sharma, 2021).

Metabolism is another area profoundly influenced by the gut microbiota. These microorganisms participate in the metabolism of dietary components and host-derived compounds, producing metabolites that can have far-reaching effects on host physiology. Short-chain fatty acids (SCFAs), for example, are byproducts of bacterial fermentation in the gut and serve as an energy source for intestinal epithelial cells, regulate immune responses, and exert anti-inflammatory effects.

However, disruptions to the delicate balance of the gut microbiota, known as dysbiosis, can have detrimental effects on health. Imbalances in microbial composition or function have been linked to a range of gastrointestinal disorders, including inflammatory bowel diseases (IBD) such as Crohn's disease and ulcerative colitis, irritable bowel syndrome (IBS), characterized by abdominal pain, bloating, and altered bowel habits, and metabolic disorders such as obesity, insulin resistance, and type 2 diabetes.

Inflammatory responses triggered by dysbiosis can lead to tissue damage and chronic inflammation in the gut, contributing to the pathogenesis of IBD. Similarly, alterations in the gut microbiota composition and metabolic activity have been implicated in the development of IBS symptoms, including visceral hypersensitivity and abnormal gut motility. Moreover, dysbiosis-induced changes in energy metabolism and adipose tissue function can contribute to obesity and metabolic syndrome, highlighting the intricate interplay between the gut microbiota and host metabolism.

In summary, the gut microbiota represents a vital component of human physiology, influencing digestion, immune function, and metabolism. Disruptions to this complex ecosystem can have profound implications for health, underscoring the importance of understanding and maintaining gut microbiota balance in the prevention and treatment of gastrointestinal and metabolic disorders.

2. Nano Medicine: The Power of Small

Nano medicine represents a groundbreaking approach to healthcare that capitalizes on the unique properties of materials at the nanoscale (typically 1 to 100 nanometers) to revolutionize diagnostics, therapeutics, and preventive strategies. At this scale, materials exhibit novel physical, chemical, and biological properties that differ from their bulk counterparts, making them ideally suited for a wide range of biomedical applications.

Furthermore, nano medicine holds great promise for tissue engineering and regenerative medicine. Nano materials can be engineered to mimic the native extracellular matrix, providing a scaffold for cell attachment, proliferation, and differentiation. By controlling the physical and biochemical properties of nano scaffolds, researchers can create tissue-specific microenvironments that promote tissue regeneration and repair (Ventola, 2012).

Additionally, nano particles can be functionalized with bioactive molecules growth factors, or stem cells to enhance tissue regeneration and promote functional recovery in various organs and tissues, including bone, cartilage skin, and nerve tissue. Despite the immense potential of nano medicine several challenges must be addressed to realize its full clinical impact. These include ensuring the safety and biocompatibility of nano materials, optimizing their pharmacokinetics and pharmacodynamics, and scaling up production methods for large-scale manufacturing. Moreover, regulatory considerations and ethical concerns surrounding the use of nano materials in healthcare must be carefully addressed to ensure patient safety and public trust. In conclusion, nano medicine represents a paradigm shift in modern healthcare, offering innovative solutions to longstanding challenges in diagnostics, therapeutics, and regenerative medicine (Mir *et al.*, 2023). By harnessing the power of nanotechnology, researchers and clinicians are poised to revolutionize patient care and improve outcomes across a wide range of medical conditions, from cancer and cardiovascular disease to neurological disorders and regenerative medicine (Matsuoka and Yashiro., 2023 and Afzal *et al.,* 2022).

3. Applications of Nano Medicine in Gut Health

Nano medicine offers a transformative approach to drug delivery by harnessing the unique properties of nano carriers to enhance targeted delivery to the gastrointestinal tract. Liposomes, polymeric nanoparticles, and dendrimers can be engineered to encapsulate therapeutic agents, protecting them from degradation and facilitating their transport through the harsh acidic environment of the stomach to the target site in the intestines. This targeted approach minimizes systemic side effects by reducing drug exposure to healthy tissues and organs, thereby improving patient tolerance and compliance. Furthermore, nano carriers can be functionalized with ligands or peptides that selectively bind to receptors or biomarkers overexpressed in diseased tissues, enabling precise localization and enhanced therapeutic efficacy (Siddiqui *et al.*, 2022). By overcoming barriers such as poor drug solubility and absorption, nano medicine holds great promise for optimizing treatments for gut-related disorders, including inflammatory bowel diseases, colorectal cancer, and gastrointestinal infections (Badrunnisa and Menghani, 2023).

3.1 Diagnostic Imaging

Nano medicine has revolutionized diagnostic imaging in gastroenterology by leveraging nanoparticles as contrast agents for non-invasive imaging modalities. These nanoparticles can be engineered to exhibit specific physicochemical properties that enable them to selectively accumulate in diseased tissues or microbial populations within the gastrointestinal tract. For example, magnetic nanoparticles can be targeted to tumor cells using specific ligands, allowing for precise localization and characterization of gastrointestinal tumors using magnetic resonance imaging (MRI). Similarly, fluorescent nanoparticles can be functionalized with targeting moieties that bind to specific bacteria or inflammatory markers, enabling real-time imaging of gut pathologies using fluorescence imaging techniques. By providing high sensitivity and resolution, nano probes enable early detection of gastrointestinal disorders, accurate staging of diseases, and monitoring of treatment responses, thereby facilitating timely intervention and personalized patient care.

3.2 Therapeutic Modulation of the Gut Microbiota

Nano medicine offers innovative strategies for modulating the composition and function of the gut microbiota, which plays a crucial role in gut health and disease. Conventional drug formulations often suffer from limitations such as poor solubility, rapid clearance from the body, and non-specific distribution, leading to suboptimal therapeutic outcomes and systemic side effects while nano formulations can be designed to deliver probiotics, prebiotics, or antimicrobial agents directly to the gastrointestinal tract, promoting the growth of beneficial bacteria while suppressing harmful pathogens. For example, probiotic-loaded nanoparticles protect probiotic bacteria from gastric acidity and bile salts, ensuring their survival and targeted delivery to the intestines, where they can exert beneficial effects on gut microbiota composition and function (Qiu *et al.*, 2023). Similarly, prebiotic nanofibers can serve as substrates for the growth of beneficial bacteria, enhancing their colonization and activity in the gut. This targeted approach enhances drug efficacy while minimizing off-target effects and reducing the required dose, thereby improving patient compliance and safet (Abed *et al.*, 2022).

Additionally, microbial-targeting nanomaterials can selectively bind to and neutralize pathogenic bacteria or toxins, offering a targeted approach for treating dysbiosis-related conditions such as *Clostridioides difficile* infection or inflammatory bowel diseases. By restoring gut microbiota homeostasis, nano medicine holds promise for alleviating symptoms and improving outcomes in patients with gastrointestinal disorders.

3.3 Nanotheranostics and Precision Medicine

Theranostic medicine represents a groundbreaking approach that integrates therapeutic and diagnostic capabilities within a single formulation, known as nanotheranostics. This innovation has emerged from the parallel development of nanomedicine for both therapeutics and diagnostics, offering a multifaceted approach to patient care (National Research Council, 2011).

Nanotheranostics serve various purposes, including monitoring the distribution and accumulation of nanomedicines within the body, visualizing drug release in response to specific triggers, and assessing therapeutic effectiveness over time (Xue *et al.*, 2021). By combining diagnostic and therapeutic functionalities nanotheranostics enable personalized treatment interventions by identifying patients who are likely to respond positively to treatment and allowing for the longitudinal monitoring of treatment responses.

However, the significance of nanotheranostics extends beyond simply combining diagnosis and treatment. Its core objective lies in patient preselection for therapy suitability, predicting therapeutic responses, and monitoring treatment effectiveness over time, rather than solely focusing on disease diagnosis. This approach not only helps in identifying patients who will benefit most from nanomedicine-based treatments but also aids in identifying those at risk of adverse effects due to off-target drug localization. In essence, nanotheranostics represent a transformative approach to patient care, offering personalized and proactive strategies for managing diseases and optimizing treatment outcomes.

Personalized nano medicine approaches leverage advances in genomic sequencing, metabolomics, and bioinformatics to tailor interventions to individual patients based on their unique gut microbiome profiles (Schork, 2015). By integrating patient-specific data with nanotechnology-enabled therapies, clinicians can optimize treatment outcomes and minimize adverse effects. For example, nanoparticle-based drug delivery systems can be customized to encapsulate drugs with known efficacy against specific gut microbiota profiles or disease phenotypes, ensuring targeted delivery to the intestines while minimizing systemic exposure and off-target effects. Furthermore, nano sensors and imaging probes can be employed to monitor changes in gut microbiota composition and function over time, enabling clinicians to adjust treatment regimens and personalize interventions based on dynamic microbiome profiles. By embracing a precision medicine approach, nano medicine holds the potential to revolutionize the management of gastrointestinal disorders, improving patient outcomes and quality of life (Alghamdi *et al.*, 2022).

4. Challenges

Safety and Biocompatibility: Ensuring the safety and biocompatibility of nano materials is paramount to their clinical translation. While nano carriers offer unique advantages for targeted drug delivery and imaging, concerns regarding their potential toxicity, immunogenicity, and long-term effects on the host organism must be carefully addressed through rigorous preclinical testing and regulatory scrutiny.

Optimizing Drug Loading and Release Kinetics: Achieving optimal drug loading and release kinetics is essential for maximizing therapeutic efficacy while minimizing off-target effects. Nano formulations must be engineered to efficiently encapsulate therapeutic agents and release them in a controlled manner at the target site within the gastrointestinal tract. Fine-tuning the physicochemical properties of nano carriers, such as size, shape, surface charge, and composition, can help optimize drug loading and release profiles for specific applications.

Navigating Regulatory Hurdles: Navigating regulatory hurdles is a significant challenge in the development and commercialization of nano medicine products for gut health. Regulatory agencies, such as the FDA and EMA, require comprehensive safety and efficacy data to support the approval of nano formulations for clinical use. Meeting regulatory requirements entails conducting rigorous preclinical studies, optimizing manufacturing processes, and adhering to stringent quality control standards to ensure product consistency and safety.

5. Future Directions

Multifunctional Nano Platforms: Future research efforts will focus on the development of multifunctional nano platforms capable of integrating diagnostic imaging and therapeutic functionalities into a single nanomaterial. These next-generation nano systems will enable simultaneous imaging and therapy, providing clinicians with real-time feedback on treatment responses and disease progression while delivering targeted therapies to the gastrointestinal tract.

Novel Biomarkers for Early Disease Detection: The identification and validation of novel biomarkers for early disease detection represent a promising avenue for advancing nano medicine in gut health. By leveraging advances in omics technologies, such as genomics, proteomics, and metabolomics, researchers can identify biomolecular signatures associated with specific gut pathologies, enabling early diagnosis and intervention before the onset of clinical symptoms.

Integration with Emerging Fields: The integration of nanotechnology with other emerging fields, such as microbiome engineering and tissue regeneration, holds great promise for advancing the field of gut health. Nano materials can be engineered to interact with the gut microbiota, modulating microbial composition and function to promote gut homeostasis and treat dysbiosis-related conditions. Furthermore, nano scaffolds and bioactive nanoparticles can be utilized in tissue engineering approaches to repair damaged intestinal tissue and restore organ function in patients with gastrointestinal disorders.

By addressing these challenges and embracing future research directions, nano medicine has the potential to revolutionize the diagnosis, treatment, and prevention of gut-related disorders, paving the way for personalized and precision medicine approaches that improve patient outcomes and quality of life.

6. Conclusion

Nano medicine indeed marks a significant paradigm shift in the management of gut-related disorders, offering unprecedented opportunities for targeted therapies with enhanced efficacy and reduced side effects. By leveraging the unique properties of nanomaterials, researchers and clinicians are revolutionizing the diagnosis, treatment, and prevention of gastrointestinal conditions, ultimately paving the way towards a new era of precision medicine for gut health.

One of the key advantages of nano medicine lies in its ability to precisely target diseased tissues or microbial populations within the gastrointestinal tract. Nano carriers, such as liposomes, polymeric nanoparticles, and dendrimers, can be engineered to encapsulate therapeutic agents and deliver them specifically to the intended site of action. This targeted approach minimizes systemic exposure to drugs, reducing off-target effects and improving therapeutic outcomes. For example, in the treatment of inflammatory bowel diseases (IBD), nano formulations can deliver anti-inflammatory drugs directly to inflamed regions of the intestine, mitigating inflammation and promoting tissue healing while minimizing systemic side effects such as immunosuppression.

Moreover, nano medicine offers innovative strategies for modulating the gut microbiota, which plays a crucial role in gut health and disease. Nano formulations can be designed to selectively target and manipulate microbial populations, promoting the growth of beneficial bacteria while suppressing harmful pathogens. For instance, probiotic-loaded nanoparticles can deliver probiotic bacteria to the intestines, where they can colonize and exert beneficial effects on gut microbiota composition and function. Similarly, prebiotic nanofibers can serve as substrates for the growth of beneficial bacteria, enhancing their colonization and activity in the gut. By restoring gut microbiota homeostasis, nano medicine holds promise for treating dysbiosis-related conditions and improving gastrointestinal health.

Furthermore, nano medicine enables precise imaging and monitoring of gut pathologies, facilitating early detection and intervention. Nanoparticles can serve as contrast agents for various imaging modalities, allowing for non-invasive visualization of diseased tissues or microbial populations within the gastrointestinal tract. This enables clinicians to diagnose gastrointestinal disorders at earlier stages, when interventions are most effective, and to monitor treatment responses in real-time, optimizing therapeutic strategies for individual patients.

In summary, nano medicine represents a transformative approach to gastrointestinal health, offering targeted therapies with enhanced efficacy and reduced side effects. By harnessing the power of nanotechnology to manipulate the intricate interplay between the gut microbiota and host physiology, researchers and clinicians are ushering in a new era of precision medicine for gut-related disorders. Through continued innovation and collaboration, nano medicine holds the potential to improve patient outcomes and quality of life for individuals affected by gastrointestinal conditions.

References

Abed, O. A., Attlassy, Y., Xu, J., Han, K., & Moon, J. J. (2022). Emerging Nanotechnologies and Microbiome Engineering for the Treatment of Inflammatory Bowel Disease. Molecular Pharmaceutics, 19(12), 4393–4410. https://doi.org/10.1021/acs.molpharmaceut.2c00222.

AboNahas, H. H., Darwish, A. M., Abd El-kareem, H. F., AboNahas, Y. H., Mansour, S. A., Korra, Y. H., *et al.* (2022). "Trust Your Gut: The Human Gut Microbiome in Health and Disease," in Microbiome-Gut-Brain Axis, eds R. Z. Sayyed and M. Khan (Singapore: Springer), 53–96.

Afzal, O., Altamimi, A. S. A., Nadeem, M. S., Alzarea, S. I., Almalki, W. H., Tariq, A., Mubeen, B., Murtaza, B. N., Iftikhar, S., Riaz, N., & Kazmi, I. (2022). Nanoparticles in Drug Delivery: From History to Therapeutic Applications. Nanomaterials (Basel, Switzerland), 12(24), 4494. https://doi.org/10.3390/nano12244494.

Ahlawat, S., and Sharma, K. K. (2021). Gut–organ axis: A microbial outreach and networking. Lett. Appl. Microbiol. 72, 636–668. doi: 10.1111/lam.13333.

Alghamdi, M. A., Fallica, A. N., Virzì, N., Kesharwani, P., Pittalà, V., & Greish, K. (2022). The Promise of Nanotechnology in Personalized Medicine. Journal of Personalized Medicine, 12(5), 673. https://doi.org/10.3390/jpm12050673.

Badrunnisa S., Menghani S.S. Review on Health Sciences Applications of Nanotech. Advances in Gastroenterology, 6(4), 295–308. https://doi.org/10.1177/1756283X13482996.

Malik, S., Muhammad, K., & Waheed, Y. (2023). Emerging Applications of Nanotechnology in Healthcare and Medicine. Molecules (Basel, Switzerland), 28(18), 6624. https://doi.org/10.3390/molecules28186624.

Matsuoka, T., & Yashiro, M. (2023). Novel biomarkers for early detection of gastric cancer. World Journal of Gastroenterology, 29(17), 2515–2533. https://doi.org/10.3748/wjg.v29.i17.2515.

Mir Najib Ullah, S. N., Afzal, O., Altamimi, A. S. A., Ather, H., Sultana, S., Almalki, W. H., Bharti, P., Sahoo, A., Dwivedi, K., Khan, G., Sultana, S., Alzahrani, A., & Rahman, M. (2023). Nanomedicine in the Management of Alzheimer's Disease: State-of-the-Art. Biomedicines, 11(6), 1752. https://doi.org/10.3390/biomedicines11061752.

National Research Council (2011) Toward precision medicine: building a knowledge network for biomedical research and a new taxonomy of disease. National Academies Press. Cambridge, MA, USA, 2011; ISBN 0309222222.

Qiu, S., Cai, Y., Yao, H., Lin, C., Xie, Y., Tang, S., & Zhang, A. (2023). Small molecule metabolites: discovery of biomarkers and therapeutic targets. Signal Transduction and Targeted Therapy, 8(1), 132. https://doi.org/10.1038/s41392-023-01399-3.

Schork NJ (2015) Personalized medicine: time for one-person trials. Nature 520(7549): 609–611.

Siddiqui, R., Mungroo, M. R., Alharbi, A. M., Alfahemi, H., & Khan, N. A. (2022). The Use of Gut Microbial Modulation Strategies as Interventional Strategies for Ageing. Microorganisms, 10(9), 1869. https://doi.org/10.3390/microorganisms10091869.

Thursby, E., & Juge, N. (2017). Introduction to the human gut microbiota. The Biochemical Journal, 474(11), 1823–1836. https://doi.org/10.1042/BCJ20160510

Ventola C. L. (2012). The nanomedicine revolution: part 1: emerging concepts. P & T : a peer-reviewed journal for formulary management, 37(9), 512–525.

Xue, Y., Gao, Y., Meng, F., & Luo, L. (2021). Recent progress of nanotechnology-based theranostic systems in cancer treatments. Cancer biology & medicine, 18(2), 336–351. Advance online publication. https://doi.org/10.20892/j.issn.2095-3941.2020.0510

8

Nanotechnology Used for Detection of Parasitic Diseases

Sanweer Khatoon[1*], Naresh Kumar[2] and Nirmala[1]

[1] *Veterinary Parasitology, College of Veterinary and Animal Science, Navania, Udaipur*
[2] *Vetlab Diagnostics, Jaipur*

Abstract

Despite substantial advances in the diagnostics and treatment of diseases, parasitic infections are a major global health issue causing significant mortality and morbidity. Currently, chemotherapeutic antiprotozoal drugs, anthelmintic and anti-ectoparasitic agents are used to treat, but resistance to these drugs has developed due to overuse. In this scenario, nanoparticles are proving a major breakthrough in the treatment and control of parasitic diseases. In the last decade, there has been enormous development in the field of nano medicine for parasitic control. Gold and silver nanoparticles have shown promising results in the treatments of various types of parasitic infections. These nanoparticles are synthesized through the use of various conventional and molecular technologies and have shown great efficacy. They work in different ways that include damaging the parasite membrane, DNA (Deoxyribonucleic acid) disruption, protein synthesis inhibition and free-radical formation. These agents are effective against intracellular parasites as well. Other nanoparticles, such as iron, nickel, zinc and platinum, have also shown good results in the treatment and control of parasitic infections. No doubt in future it will become the future of modern drug development.

Keywords: *Parasitic, Nanoparticle toxicity, Safety*

1. Introduction

Parasites are organisms that live on or within other organisms and utilize the resources of their hosts in terms of food, shelter and safety (Hotez *et al.*, 2007). Parasitic infections are a major global health issue causing significant mortality and morbidity. Despite substantial advances in the diagnostics and treatment of these diseases, the currently available options fall far short of expectations.

* *Corresponding author e-mail: drsanvir@gmail.com*

From diagnosis and treatment to prevention and control, nanotechnology-based techniques show promise as an alternative approach. Nanoparticles can be designed with specific properties to target parasites and deliver antiparasitic medications and vaccines. Nanoparticles such as liposomes, nanosuspensions polymer-based nanoparticles, and solid lipid nanoparticles have shown to overcome limitations such as limited bioavailability, poor cellular permeability nonspecific distribution, and rapid drug elimination from the body. These nanoparticles also serve as nanobiosensors for the early detection and treatment of these diseases.

Professor Norio Taniguchi of the Tokyo Science University, introduced the term "nanotechnology", in an article published in 1974. He described nanotechnology as the processing of, separation, consolidation, and deformation of materials by one atom or by one molecule. Nanotechnology involves imaging, measuring modeling, and manipulating matter at this length scale.

Parasitic infections are highly prevalent, because of a lack of suitable treatment options and a lack of research in pharmaceutical sciences. Most of these diseases are considered to be neglected tropical diseases, and the death toll from such parasitic infections is high. These parasitic infections have been treated customarily by chemotherapeutic agents and ethnobotanicals. The chemotherapeutics are antiprotozoal, antihelminthic, insecticide and acaricide agents. Among antiprotozoal drugs, antimalarials, antiamoebic, antigiardial, trypanocidal, antileishmanial and anti-toxoplasmic agents are used to treat various protozoal diseases. These agents act in a variety of ways: they disrupt the parasite DNA, inhibit protein synthesis, damage the parasite membrane and kill the protozoa (Figure 1). However, due to the development of resistance by these protozoa, some of these treatments are no longer effective (Capela *et al.*, 2019).

Treating parasitic diseases presents challenges, including drug resistance, side effects, limited efficacy, and high costs (Figure 1).

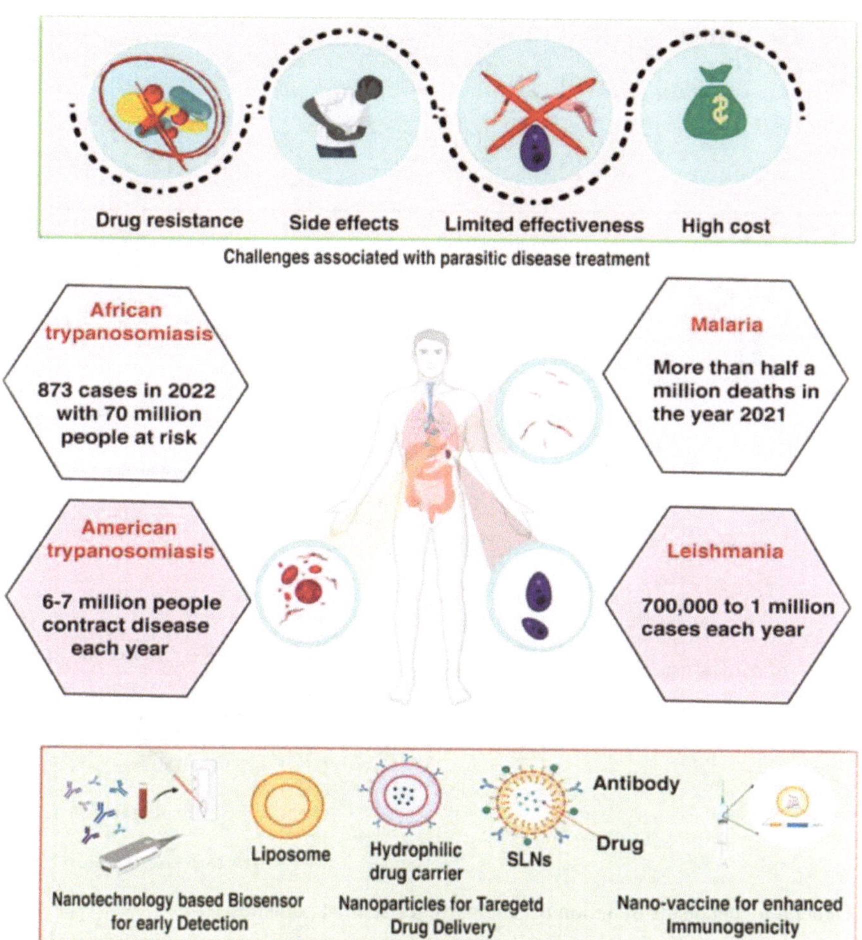

Fig. 1. Challenges associated with parasitic disease treatment and nanoparticle based potential solutions

Nanotechnology-based treatments, such as liposomes and SLNs, enhance drug effectiveness and bioavailability. Nanotechnology aids disease diagnosis and prevention through highly sensitive and specific diagnostic tools. Nano vaccines offer a superior alternative to conventional vaccination methods. General mechanism and mode of action of nanoparticles against parasitic diseases is shown in Figure 2.

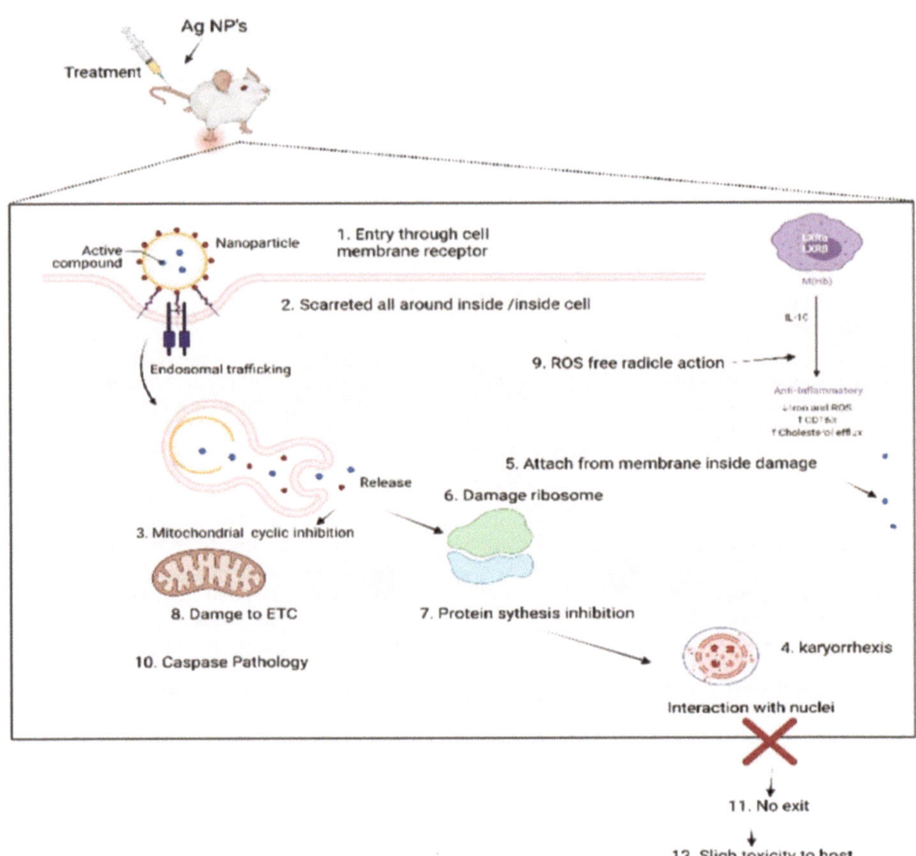

Fig. 2. General mechanisms of action of nanoparticles against parasites.

Nanotechnology-based biosensors also play a role in the fight against parasites. Most parasitic disease detection methods currently rely on microscopy, culture, and molecular techniques, requiring specialized expertise, being time-intensive, and incurring high costs, which hinder swift diagnosis. This position places us on the brink of developing more affordable technologies suitable for deployment in resource-constrained settings. Nanomaterials possess unique physical, chemical, and fluorescence properties that can be combined with conventional biosensors to increase the sensitivity and enable early disease diagnosis. This noninvasive approach provides an alternative to invasive techniques and holds promise for improved detection. While the application of nanotechnology in the treatment of parasitic diseases is still in its early stages, it shows great potential. Further research is needed to fully understand its capabilities and to develop it into a common clinical tool.

Nanoparticles are defined as particulate dispersions or solid particles with a size in the range of 10-1000 nm. The current state of various nanotechnology-based therapeutic and diagnostic interventions and prospects they bring in the fight against parasitic diseases is discussed. There are various types of nanoparticles.

1.1 Types of nanoparticles

1. *Solid Lipid Nanoparticles:* New type of colloidal drug carrier system for i.v. Consists of spherical solid lipid particles in the nm range, dispersed in water or in aqueous surfactant solution.
2. *Polymeric Nanoparticles:* are defined as particulate dispersions or solid particles with size in the range of 10-1000nm.
3. *Ceramic Nanoparticles:* nanoparticles made up of inorganic(ceramic) compounds silica. Exist in size less than 50 nm, which helps them in evading deeper parts of the body.
4. *Hydrogel Nanoparticles:* Polymeric system involving the self assembly and self aggregation of natural polymer cholesterol dextran and agarose cholesterol groups provide cross linking points.
5. *Copolymerized Peptide Nanoparticles:* Drug moiety is covalently bound to the carrier instead of being physically entrapped.
6. *Nanocrystals and Nanosuspensions:* Pure drug coated with surfactant, aggregation of these particles in crystalline form.Drug powder dispersed in aqueous surfactant solution.

1.2 Synthesis of nanoparticles

NPs are synthesized or prepared through the use of in-vitro or conventional techniques. Some of these methods are explained below:

a. *Photochemical and Thermal Disintegration:* Photochemical and thermal disintegration involves the decomposition of NP precursors at high temperatures in boiling solvents

b. *Bimetallic Nanoparticle Synthesis:* Bimetallic NPs are combinations of NPs with different architectures. These NPs optimize the energy of plasmon adsorption, a band of metallic solution which is a multipurpose tool for bio-sensing

c. *Chemical and Electrochemical Reduction:* This production method involves two processes: reduction and interaction. The interaction takes place between the polymeric and metallic species. Reducing agents that are used in this process include sodium borohydride, Tollen's reagent, hydrogen and ascorbate. This method is most commonly used in the synthesis of silver NPs (AgNPs), which have shown promising results in parasite control

d. *Sol-Gel Technique:* The sol in this technique is a colloid-based solid particle suspension in a liquid.

e. *Green Synthesis:* Synthesis of NPs through the use of various plant extracts offers a method that is cheaper, less toxic and simpler to set up than the methods described above. Extracts are prepared from leaves, roots, shoots, stems or bark. The active ingredients in these parts reduce the metals in NPs by acting as reducing agents.

2. Need of Nanotechnology in Parasitology

- Parasitic diseases are hardly diagnosed timely and thus bringing great economic losses to animal husbandry. (Roberts *et al.*1994).
- Parasites can reside in host cells and establish reservoirs - results in long term and repeated infections.
- Lack of functional vaccine against many of parasitic infections
- Most antiparasitic drugs have low bioavailability due to their insolubility and their short half-life.
- Treatment needs frequent dosage for long-time - long-life cycles of parasites. (Vercruysse *et al.*, 2007).
- Some antiparasitic drugs, e.g. ivermectin and praziquantel, are susceptible to enzymatic degradation or inactivation in Animals and thus exhibiting strong first-pass effects.
- They also show poor penetration across the biological membrane barriers of tissues and cells, which also result in reduced bio-availability and hardly achieve the expected therapeutic effect (Lu *et al.*, 2017; Babita *et al.*, 2018).
- Resistances of antiparasitic drugs are becoming more and more serious due to irrational usages (Balana-Fouce *et al.*, 1998).

With the people's increasing requirements for the treatment of parasitic diseases, nano-particles, especially organic nanoparticles, have attracted people's attention for antiparasitic drug delivery.

The general applications of nanotechnology are

1. As drug delivery system
2. Diagnosis purposes
3. For treatment
4. For immunization and vaccination

3. Metal-Based Nanoparticles and their Antiparasitic Significance:

Aurum (Gold)-Based Nanoparticles: The effectiveness of AuNPs against various parasites of public health importance, like *Toxoplasma gondii, Trypanosoma* spp., *Leishmania* spp. *(L. tropica, L. donovani*) and *Cryptosporidium* spp. *(C. parvum*); helminths including trematodes (*Schistosoma* spp.) and cestodes *(Raillietina* spp.*);* and vectors including mosquito, genus *Aedes (*spp. *A. aegypti L.), Anopheles (*spp. *A. stephensi),* and *Culex (*spp. *C. quinquefasciatus*); and dipteran flies.

The development of a new drug delivery system favoring longer dose intervals and the increase of drug activity against parasites are some of the major issues studied in nano-parasitology. Active and nonactive targeting has been explored using liposomes, solid lipid NP and hydrophobic polymeric NP. NP carriers have also improved the bioavailability and drug selectivity, even in very complicated treatments like cerebral malaria by *Plasmodium falciparum* infection. One of the most important problems in parasitology research as well as, to the Veterinary counterpart is the development of drug resistance (Manning, *et al*., 2011). NP may offer a potent tool to reduce the risk of resistance to traditional drugs, avoiding some of the resistance mechanisms increasing drug bioavailability and potentiating the target of the treatment. New antiparasitic drugs can also be developed with nanotechnology like the case of Tio2@Ag NP, being an eco-friendly and inexpensive approach for use in veterinary medicine (Chaaban., *et al*., 2017; Grando, *et al*., 2016). The use of NP has overcome the major challenge for the antiparasitic drug albendazole (ABZ), related to its poor water solubility and subsequently low bioavailability. Solid NP composed of Fe3O4 increase the in vitro activity of ABZ (Abidi *et al*., 2018). In mice, solid lipid NP of ABZ could be a promising formulation for the treatment of *Toxocara canis* infection (Gershkovich, *et al*.,2009).

Nanomaterials are currently used for many biological, biomedical and environmental purposes, because of outstanding physical and chemical properties. The synthesis of gold nanoparticles (AuNPs) is of high interest for research in parasitology and entomology, since these nanomaterials showed promising applications, ranging from detection techniques to drug development, against wide range of parasites of public health relevance, as well as insect vectors.

It has been pointed out that nano-drug delivery systems result in high increases target delivery effect, decreases the toxicity with effect on the metabolism of drug including absorption, distribution, and excretion, enhance drug absorption by facilitating diffusion through the epithelium, releasing drugs with controlled system and prolonged systemic circulation lifetime. Typically delivery using nano- and micro-carriers can increase bioavailability and protect drugs combined from metabolisms with prolonging drug residence in the body organs of mammalians, such as liver, kidneys, and spleen, accordingly prolonging time between prescriptions (Kudtarkar *et al*., 2017). The second approach to drug formulation is the nano-nization of the drug, essential to the penetration of the cuticle of parasites and to intracellular parasites treatment. Nano-encapsulation of antileishmanial agents can improve the bioavailability of drugs, as it increases the intracellular penetration and distribution, mainly into macrophages. As a consequence, primaquine was 21 times more efficient than its free form in reducing the parasite burden in these cells (Siddiqi, *et al.* 2018).

3.1 Nanoparticle Toxicity and Safety

The safety and efficacy of NPs greatly rely on their physicochemical properties Different nanocarriers, in the form of nanocrystals, solid lipid nanoparticles (SLNs) and liposomes are introduced that can be administered through oral intestinal, intravenous, and pulmonary routes. These nanocarriers provide physical stability and targeted and controlled release of drugs. Furthermore they protect the loaded drug from degradation by enzymes and it has been observed that the bioavailability and mean residence time of praziquantel was increased by 5.67 and 4.94 folds with hydrogenated castor oil SLNs. NPs are metabolized easily into the liver and change physicochemical properties through enzymatic degradation. These nanoparticles can be metabolized by endocytosis facilitated by the reticuloendothelial system. It has been observed that NPs with smaller sizes (80–120 nm) can be excreted by the kidneys easily, while larger particles undergo hepatobiliary excretion. Furthermore the excretion of NPs also depends on the charge. It was observed that the NPs with higher charge (+34.4 mV) were excreted through faeces, bypassing the gastrointestinal tract; however, the particles with lesser charge (−17.6 mV) remained isolated in the liver.

4. New Developments for A Brighter Future

Polymers like (poly-alkyl cyanoacrylate, poly lactic-co-glycolic acid, dimercaptosuccinic acid) allow production of NP because of long circulation time and rapid clearance period by the phagocyte system, reducing their requirement for a repeated dose (Dmour and Tahamo, 2018). Pegylated or polyplex NP may also offer fantastic options to increase the biodistribution of a compound. NP can also be of great assistance to veterinary medicine in mass treatments against ectoparasites, facilitating the drug parasite interaction. The effectiveness of phytotherapeutic products can be improved, like the recent treatment of antiparasitic silver NP using *Cissus quadrangularis* against

Hippobosca aculata and *Rhipicephalus (Boophilus) microplus* (Santhosh., *et al.*,2018). It could also allow the potential recombinant proteins to act at different drug binding sites, as nanovectors (Devalapally *et al.*, 2007). Our research groups in Cuba and Brazil are currently running preliminary test using nanopeptides against the ruminant helminths *Haemonchus contortus* and *Fasciola hepatica* with the objective of developing a new therapeutic possibility to control these two devastating infections for humans and animals. Metal oxide NP (MNP), such as copper (Cu), titanium (Ti), silver (Ag), gold (Au), and zinc (Zn) can be an alternative for the current antibiotics and anthelmintics, potentiating the treatments of bacteria and parasite infections. MNP can have more effective mechanisms of action by developing multi-complexes against these diseases (Mahoochehri., *et al.*,2013). The active gold coated glass particles commonly known as nanoshells can help veterinary medicine to improve the treatment of parasites facilitating the target of the organism and drug delivery, physically destroying the infection (Dkhil., *et al.*, 2016). Mice receiving silver NP showed a statistically significant decrease in the oocyst burden and the number of viable oocysts in stool, and a statistically significant increase in serum.

IFN-γ. Silver NP proved their effectiveness against Cyclospora sp. infection with no evidence of host toxicity (Gaafar MR, *et al., 2019).* The antiparasitic activity mechanism of the metal and metal oxide NP are based on the generation of reactive oxygen species (ROS) (Khezerlou., *et al.*,2018). ROS leads to severe oxidative stress and damage to the cell's macromolecules, which overall causes lipid, peroxidation, alteration of proteins, inhibition of enzymes, and RNA/DNA damage. MNP can also impair the lipidic bilayer and the cytosolic proteins affecting energy production (Gold., *et al.*, 2018). Among the examples, metallic compounds and MNP inhibits the enzyme of trypanothione metabolism that is vital in the survival of *Leishmania* parasites (Navarro., *et al.*, 2010). ZnO and FeO NP exerted their anthelmintic effects against *Toxocara vitulorum* via the induction of oxidative/nitrosative stress. Ticks from cattle and pets are one of the major targets in the use of MN in Veterinary Medicine, mainly by their transmission of zoonotic pathogens and poor welfare condition to the animals. The NP showed acaricidal toxicity from both MNP synthesized through chemical and green-based fabrication routes. The most common studies assessed the toxicity of silver NP against ticks followed the use of titanium dioxide NP, and to a minor extent zinc oxide, nickel, and copper NP.(Benelli ., *et al.*, 2017).In particular, the so-called "green synthesis" include plant extracts and microbial filtrates for the synthesis of NP. Green synthesis reduce and capp/stabilize agents during metal and metal oxide NP production. Several eco-friendly preparations have been proposed to control *R. microplus* (Benelli ., *et al.,* 2017). However, the lack of in vivo tests is a major challenge.(Gaafar., *et al.*, 2019)

5. Conclusion

Although various benefits were demonstrated in this chapter, but a more active research is necessary to apply this technology to the real needs of animal health, taking into account the economic constraints of this sector and the enormous potential in the area. Nanotechnology, as any other technology, is getting good worldwide exposure but researchers must take into account the importance of ethical and regulatory issues. There is a need to strength our scientific monitoring, risk assessment, target species, aberrant results and safety to other organisms, before its widespread use. Despite the potentialities of the use of nanotechnology in veterinary parasitology research, only a few successful applications have been accomplished so far. A small number of parasites have been the target for nanotechnology, focusing primarily in *Leishmania sp.* and *Plasmodium sp.* research. So need is there with the use of nanotechnology to develop new therapies to control *Haemonchus contortus* and other important zoonotic parasites bringing in broader term a more holistic approach in the field.

References

Abidi H, Ghaedi M, Rafiei A, Jelowdar A, Salimi A, Asfaram A, Ostovan A (2018). Magnetic solid lipid nanoparticles co-loaded with albendazole as an antiparasitic drug: Sonochemical preparation, characterization, and in vitro drug release. Journal of Molecular Liquids.;268:11-18.

Babita S, Utpal J, Jyotirmaya S, *et al.* (2018). Systematic approach for the formulation and optimization of atorvastatin loaded solid lipid nanoparticles using response surface methodology. Biomed Micodevices, 20:53.

Balana-Fouce R, Reguera RM, Cubr~ ιA JC, *et al.* (1998). The pharmacology of leishmaniasis. Gen Pharmacol-Vasc S 30:435–43.

Benelli G, Maggi F, Romano D, Stefanini C, Vaseeharan B, Kumar S, Higuchi A, Alarfaj AA, Mehlhorn H, Canale A.(2017), Nanoparticles as effective acaricides against ticks: A review. Ticks and Tickborne Diseases.;8:821-826. 7.

Capela R.; Moreira R.; Lopes F. An overview of drug resistance in protozoal diseases. International journal of molecular sciences 2019, 20 (22), 5748. 10.3390/ijms20225748.

Chaaban A, Gomes EN, Santos VMCS, Deschamps C, Molento MB.(2017), Essential oils for myiasis control: Potentialities for ecofriendly insecticides. European Journal for Medicinal Plants. ; 21(4):125.

Devalapally H, Chakilam A, Amiji MM. (2007) Role of nanotechnology in pharmaceutical product development. J. Pharm. Sci.;96:2547-2565.

Dkhil MA, Khalil MF, Bauomy AA, Diab MSM, Al-Qura S. (2016) Efficacy of gold nanoparticles against nephrotoxicity induced by schistosoma mansoni infection in mice. Biomedical and Environmental Sciences.;29:773-781.

Dmour I, Taha MO. (2018) Natural and semisynthetic polymers in pharmaceutical nanotechnology, In: Grumezescu, A.M. (Ed.) Organic Materials as Smart Nanocarriers for Drug Delivery. William Andrew Publishing.;35-100.

Gaafar MR, El-Zawawy LA, El-Temsahy MM, Shalaby TI, Hassan AY.(2019) Silver nanoparticles as a therapeutic agent in experimental cyclosporiasis. Experimental Parasitology.;207:107772.

Gold K, Slay B, Knackstedt M, Gaharwar AK.(2018) Antimicrobial activity of metal and metal-oxide based nanoparticles. Advanced Therapeutics.;1:1700033.

Grando TH, De Sá MF, Baldissera MD, Oliveira CB, De Souza ME, Raffin RP, Monteiro SG. (2016)In vitro activity of essential oils of free and nanostructured Melaleuca alternifolia and of terpinen-4-ol on eggs and larvae of Haemonchus contortus. J Helminthol.;90(3):377-382.

Gershkovich P, Wasan EK, Lin M, Sivak O, Leon CG, Clement JG, Wasan KM.(2009) Pharmacokinetics and biodistribution of amphotericin B in rats following oral administration in a novel lipid-based formulation. Journal of Antimicrobial Chemotherapy.;64(1):101-108.

Hotez P. J.; Molyneux D. H.; Fenwick A.; Kumaresan J.; Sachs S. E.; Sachs J. D.; Savioli L.(2007) Control of neglected tropical diseases. New England journal of medicine, 357 (10), 1018–1027. 10.1056/NEJMra064142.

Khezerlou A, Alizadeh-Sani M, Azizi-Lalabadi M, Ehsani A. (2018) Nanoparticles and their antimicrobial properties against pathogens including bacteria, fungi, parasites and viruses. Microbial Pathogenesis.;123:505-526.

Kudtarkar A., Shinde U.P., Bharkad G., Singh K. (2017) Solid lipid nanoparticles of albendazole for treatment of Toxocara canis infection: in vivo efficacy studies. Nanoscience & Nanotechnology-Asia.;7:80-91.

Kumar S., Rahuman T., Bagavan A., Marimuthu S, Jayaseelan C, Kirthi AV. (2012) Evaluation of stem aqueous extract and synthesized silver nanoparticles using Cissus quadrangularis against Hippobosca maculata and Rhipicephalus (Boophilus) microplus. Exp. Parasitol.; 132:156-165.

Lu M, Dan X, Sun W, *et al.* (2017). Sustained release ivermectin-loaded solid lipid dispersion for subcutaneous delivery: in vitro and in vivo evaluation. Drug Deliv 24:622–31.

Manning L., Laman M., Law I., Bona C., Aipit S., Teine D., Warrell J., Rosanas-Urgell A., Lin E., Kiniboro B. (2011) Features and prognosis of severe malaria caused by Plasmodium falciparum, Plasmodium vivax and mixed Plasmodium species in Papua New Guinean children. PloS one, 6 (12), e29203.

Navarro M., Gabbiani C., Messori L., Gambino D. (2010) Metal-based drugs for malaria, trypanosomiasis and leishmaniasis: recent achievements and perspectives. Drug Discovery Today;15(23-24):10701078.

Roberts T., Murrell K.D., Marks S. (1994). Economic losses caused by foodborne parasitic disease. Parasitol Today, 10:419–23.

Siddiqi KS, Husen A, Rao RA. (2018) A review on biosynthesis of silver nanoparticles and their biocidal properties. Journal of Nanobiotechnology.;16:14.

Vercruysse J., Schetters T.P., Knox D.P., *et al.* (2007). Control of parasitic disease using vaccines: an answer to drug resistance? Rev Sci Tech Oie.

9

Applications of Nanotechnology in Medical Science

Palak Chawla*

Department of Chemistry, Ramniranjan Jhunjhunwala College, Ghatkopar (W) Mumbai - 400086

Abstract

The main objective of nanomedicine broadly defined as the comprehensive monitoring, repairing and improvement of all human biological systems, workingfrom the molecular level using engineered devices and nanostructures to achieve medical benefit. Nanomedicine is the nanomachine of diagnosing, treating, and preventing disease and traumatic injury, relieving pain, and of preserving and improving human health, using molecular tools and molecular knowledge of the human body. Nanotechnology applications in medical science can be essential in achieving this goal. These applications involves extensive surveillance, control, creation, repair, and defence of all biological human systems, employing designed nanodevices and nanostructures which works in molecular level. It is useful for covering several fields, such as preparation of vaccines, medication delivery, wearable equipment, diagnostic and imaging equipment, and antimicrobial products.

Keywords: *Nanomedicine, Nanodevices, Nanomachine, Health Care*

1. Introduction

Nanomedicines are one of the best applications of nanotechnology to medical science. Nanomedicine is defined as the use of nanomaterial (dimension ranging from 1-100 nm) are target specific, drug delivery and diagnostic imaging at molecular level. Nanomedicine is a cutting-edge field at the intersection of nanotechnology and medicine, leveraging tiny particles and devices at the nanoscale to revolutionize healthcare. At this scale, materials exhibit unique properties that can be harnessed for a myriad of medical applications, from diagnostics to therapeutics and beyond (Saha, 2009).

* *Correspondence author e-mail: palakchawla@rjcollege.edu.in*

Nanomedicine encompasses a broad spectrum of research, development and applications aimed at improving diagnosis, treatment, and prevention of diseases. By manipulating materials at the nanoscale, scientists and medical researchers are able to design highly targeted and efficient systems that can interact with biological processes at a cellular specific biomarkers associated with diseases, enabling early detection and precise monitoring of treatment responses. Furthermore, nanotechnology is paving the way for innovative therapies such as photothermal and photodynamic therapy, where nanoparticles are used to selectively destroy cancer cells with minimal damage to surrounding healthy tissues. Additionally, nanomaterials are being explored for regenerative medicine applications, including tissue engineering and wound healing (Haleem *et al.*, 2023).

The potential of nanomedicine extends far beyond these examples, with ongoing research exploring its applications in infectious disease management, neurological disorders, cardiovascular health, personalized medicine, and more. As the field continues to evolve, nanomedicine holds immense promise for transforming healthcare by providing safer, more effective, and personalized approaches to diagnosing, treating, and preventing diseases.

1.1. Size

The size of nanomedicine particles can vary greatly depending on their intended application and composition. Generally, nanomedicine refers to medical interventions at the nanoscale, typically ranging from 1 to 100 nanometers. These nanoscale dimensions are advantageous because they allow for interactions at the molecular and cellular level, enabling targeted delivery of drugs, imaging agents, or therapeutic agents to specific tissues or cells in the body. The precise size and composition of nanomedicine particles are often tailored to optimize their functionality, such as enhancing circulation time, cellular uptake, or tissue penetration. [Lovestam *et al.*, 2010; Bleeker *et al.*, 2013; Boverhof *et al.*, 2015]. Pharmaceutical manufacturing of nanomaterials follows two different approaches: top-down and bottom-up. In a top-down process, the breakdown of bulk material into a smaller one or smaller pieces by mechanical or chemical energy takes place. Conversely, the bottom-up process starts with atomic or molecular species allowing the precursor particles to increase in size through a chemical reaction (Boverhof *et al.*, 2015; Luther, 2004; Oberdorster, 2015). Nanoparticles have a great surface-area-to-volume ratio due to their nano size, which makes them a good candidate to absorb significant amounts of medications and move quickly throughout the bloodstream. Greater surface area increases their mechanical, magnetic, optical, and catalytic qualities and makes them suitable for pharmaceutical applications.

2. Key Areas of Nanomedicines

2.1. Drug Delivery

Nanoparticles are modified to encapsulate drugs, saving them from degradation and helping them for targeted delivery to specific sites within the body. This targeted drug delivery has fewer side effects and increases the efficiency of the drugs.

2.2. Diagnostic imaging

Nanoparticles can be engineered to interact with specific molecules or structures in the body, which makes them useful for imaging techniques like magnetic resonance imaging (MRI), computed tomography (CT) and fluorescence imaging. This helps in early detection and the exact site of diseases.

2.3. Therapeutics

Nanoparticles can be utilized in therapeutic agents such as drugs, genes, or proteins, allowing for the modulation of biological processes. Apart from that, nanomaterials themselves may act as unique therapeutics, such as photothermal or photodynamic therapy for cancer.

2.4. Regenerative Medicine

Nanotechnology can be applied to tissue engineering and regenerative medicine to design scaffolds and delivery systems that promote tissue repair and regeneration at the nanoscale level.

2.5. Diagnostic

Nanoparticles can be used to detect biomarkers associated with diseases, offering sensitive and specific diagnostic tools for early disease detection and monitoring of treatment responses.

Nanomedicine holds great promise for revolutionizing healthcare by providing more effective treatments with fewer side effects, enabling earlier disease detection, and offering new approaches to regenerative medicine. However, challenges such as biocompatibility, scalability, and regulatory issues must be addressed to fully realize the potential of nanomedicine in clinical practice.

3. Medical Devices Developed For Nanomedicine

Nanotechnology applications in medicine can be essential in achieving miraculous remedies to relieve sickness and injury pain for hundreds of years. Nanotechnology can improve medical research radically and create a new sector of human improvement [Moshed *et al.*, 2017; Fakruddin *et al.*, 2012].

Nanomedicine holds tremendous promise for revolutionizing healthcare, and with it comes a variety of innovative medical devices designed to operate at the nanoscale level. The examples are as follows:

3.1. Nano Drug Delivery System

Nanoparticles can be engineered to carry drugs to specific targets within the body, improving drug efficacy and reducing side effects. These nanoparticles can be functionalized with ligands that target specific cells or tissues, allowing for precise drug delivery.

3.2. Nano-robots

Also known as nanobots or nanorobots, these tiny machines can perform specific tasks within the body, such as targeted drug delivery, tissue repair, or even surgery on a microscopic scale.

3.3. Nanosensors

Nanoscale sensors can detect biomarkers or other molecules indicative of disease. They can be used for early disease detection, monitoring of chronic conditions, or even as part of implantable devices to provide real-time health monitoring.

3.4. Nanofiber Scaffolds

These scaffolds can mimic the extracellular matrix of tissues and organs and are used in tissue engineering and regenerative medicine. They provide a structural framework for cells to grow and differentiate into functional tissue.

3.5. Nanopore Sequencing Devices

Nanopore sequencing technology allows for rapid and portable DNA sequencing by passing DNA strands through a nanoscale pore. This technology has the potential to revolutionize genomics and personalized medicine.

3.6. Nanoparticle Contrast Agent

These are used in imaging techniques such as MRI, CT scans, and ultrasound to enhance contrast and improve the visualization of tissues and organs, aiding in the diagnosis and monitoring of diseases.

3.7. Implantable Nanoscale Devices

These devices can monitor physiological parameters, deliver drugs, or even stimulate nerves or tissues for therapeutic purposes. They offer the potential for minimally invasive treatments and long-term health monitoring.

3.8. Nanoparticle-Based Therapeutics

These include a wide range of treatments such as cancer therapies (e.g., photothermal therapy, chemotherapy), antimicrobial agents, and gene therapies, all designed to take advantage of the unique properties of nanoparticles for targeted and effective treatment.

These are just a few examples of the diverse range of medical devices being developed for nanomedicine. As technology advances, we can expect even more sophisticated and precise nanoscale tools to revolutionize the diagnosis, treatment, and prevention of diseases.

Apart from the above-listed devices various nanotechnology-based diagnostic kits such as nanosensors, nanoparticle-based imaging agents, nanoparticle-based PCR Assays, Lab-on-a-Chip devices, along with modern drugs and medicines such as nanoparticle-based drug delivery vehicles, liposomal formulations and polymeric nanoparticles, Nanomedicines (such as Abraxane (nanoparticle albumin-bound paclitaxel) and Doxil (pegylated liposomal doxorubicin)), nanotechnology in gene therapy, nanoparticle-based vaccines, and antimicrobial agents, etc. are being commercialized for research and clinical usage [Dessale *et al.*, 2022].

Nanoscaling of medical technologies provides better efficiency, a quick response rate, and functional effectiveness in most biological and chemical processes used to manufacture medical materials. Thus, research provides constant hope for the upcoming new applications of nanomedicine (Pramanik *et al.*, 2020; Fox *et al.*, 2019). Traditional stents are small mesh tubes inserted into narrowed or weakened arteries to improve blood flow. However, they have a risk of restenosis, which is the re-narrowing of the artery due to excessive tissue growth. To address this issue, drug-eluting stents were developed. One example of a medical device developed for nanomedicine is a drug-eluting stent (DES) used in the treatment of cardiovascular diseases.

In a drug-eluting stent, nanoparticles loaded with therapeutic drugs are embedded in a polymer coating covering the stent. These nanoparticles slowly release the drug over time, inhibiting tissue growth and reducing the risk of restenosis. The use of nanotechnology enables controlled and sustained drug release, improving the efficacy of the treatment while minimizing side effects.

4. Application of Nanomedicine in Different Area of Health Care

The healthcare revolution is driven by nanotechnology that emphasises preventive population health management. Nanotechnology helps solve the problem of focused treatment administration, minimizes the danger of side effects and improves the therapeutic effectiveness. This technology is very helpful in the diagnosis, treatment, and gene therapy of cancers. Nanomedicine is the most promising product of nanorobotics. Its applications serve various fields, such as the creation of vaccines, medication delivery, wearable equipment, diagnostic and imaging equipments, and antimicrobial products. The development of more target-oriented drugs, improved gadgets, and early diagnosis of various diseases is predicted to lead to nanomedicine. Combining current anti-cancer drugs with nanotechnology can be crossed intact and circulated inside the brain. This technology offers several potential markets and benefits, whole classes of present drugs. It is possible to develop tailored mechanisms for medication administration, new diagnostic methods, and nanoscale medical devices.

4.1. Nanoparticle Drug Delivery

Nanoparticles serve as carriers for therapeutic drugs, allowing for precise control over drug release kinetics and dosage.

4.2. Targeted Drug Discovery

The drug is delivered directly to the site of arterial injury, reducing systemic exposure and minimizing adverse effects on healthy tissues.

4.3. Biocompatible Materials

The polymer coatings and nanoparticles are designed to be biocompatible and non-toxic to the body, minimizing the risk of inflammation or allergic reactions.

4.4. Long-Term Efficacy

The sustained release of drugs from the nanoparticles helps maintain therapeutic levels in the arterial wall over an extended period, reducing the risk of restenosis.

4.5. Customization

Nanotechnology enables the customization of drug-eluting stents with specific drug combinations and release profiles tailored to individual patient needs.

The application of nano techniques in health care has been termed nanomedicine and it utilizes nanomaterials for detection, monitoring, control, prevention and treatment of disease. However, the definition of nanomaterial is not clear among the various scientific and international regulatory corporations. Some efforts have been made in order to find a consensual definition due to the fact that nanomaterials possess novel physiochemical properties, different from those of their conventional bulk material equivalents due to their small size. These properties greatly increase a set of opportunities in drug development, however, some questions about safety issues are raised. Some concerns such as alteration of pharmacokinetics namely the absorption, distribution, elimination, more easily cross biological barriers, and toxic properties over the applications of the nanomaterials (Bleeker *et al.*, 2013; Tinkle *et al.*, 2014)

5. Future Perspectives of Advanced Nanomedicine

Nanomedicine holds incredible promise for the future of healthcare. Here are some perspectives on its potential:

- ***Precision medicine:*** Nanotechnology allows for the precise targeting of drugs to specific cells or tissues in the body, minimizing side effects and improving treatment efficacy. This could revolutionize the way we treat diseases like cancer, where targeted drug delivery to tumors can enhance treatment outcomes while reducing harm to healthy cells.
- ***Early disease detection:*** Nanoscale sensors and imaging agents can detect biomarkers of diseases at an early stage, enabling proactive intervention before symptoms appear. This could lead to more effective management and even prevention of diseases like cancer, cardiovascular disorders, and neurodegenerative conditions.
- ***Regenerative disease:*** Nanomaterials can be engineered to mimic the extracellular matrix and promote tissue regeneration. This opens up possibilities for regenerative therapies to repair damaged organs and tissues, offering hope for patients with conditions ranging from spinal cord injuries to degenerative joint diseases.
- ***Personalized drug delivery:*** Nanomedicine enables the customization of drug delivery systems based on individual patient characteristics such as genetics, metabolism, and disease state. This personalized approach can optimize treatment regimens, enhance drug efficacy, and minimize adverse effects.
- ***Non-invasive treatment:*** Nanotechnology facilitates the development of minimally invasive or non-invasive treatment modalities, such as targeted drug delivery using nanoparticles, focused ultrasound therapy, or magnetic hyperthermia for cancer treatment. These approaches offer patients alternatives to traditional surgical procedures with reduced risk and recovery time.
- ***Combination therapies:*** Nanotechnology allows for the integration of multiple therapeutic modalities into single nanoscale platforms, enabling synergistic effects and overcoming treatment resistance. Combination therapies incorporating drugs, genes, imaging agents, and immunotherapeutic agents hold great promise for addressing complex diseases more effectively.
- ***Remote monitoring and therapy:*** Nanosensors and nanodevices can be deployed within the body for real-time monitoring of physiological parameters and disease progression. Additionally, remote-controlled nanorobots could perform targeted therapeutic interventions inside the body, offering new avenues for minimally invasive surgery and disease management.

While the potential of nanomedicine is vast, significant challenges remain in terms of safety, regulatory approval, scalability, and accessibility. Addressing these challenges will be crucial for realizing the full benefits of nanotechnology in healthcare.

References

Arya S. K., Prusty A. K., Singh S. P., Solanki P. R., Pandey M. K., Datta M., & Malhotra B. D. (2007). Biosensor for total cholesterol estimation using N-(2-aminoethyl)-3-aminopropyltrimethoxysilane self-assembled monolayer. Analytical and Bioanalytical Chemistry, 363(2), 210- 218.

Bleeker, E. A., de Jong, W. H., Geertsma, R. S., Groenewold, M., Heugens, E. H., & Koers-jacquemijns, M. (2013). Considerations on the EU definition of a nanomaterial: science to support policy making. Regulatory Toxicology and Pharmacology, 65, 119-125.

Boverhof, D. R., Bramante, C. M., Butala, J. H., Clancy, S. F., Lafranconi, M., & West, J. (2015). Comparative assessment of nanomaterial definitions and safety evaluation considerations. Regulatory Toxicology and Pharmacology, 73, 137-150.

Dessale M., Mengistu G., & Mengist H. M. (2022). Nanotechnology: A Promising Approach for Cancer Diagnosis, Therapeutics and Theragnosis. International Journal of Nanomedicine, 17, 3735–3749.

Fakruddin, M., Hossain, Z., & Afroz, H. (2012). Prospects and applications of nanobiotechnology: A medical perspective. Journal of Nanobiotechnology, 10, 31.

Fox K. E., Tran N. L., Nguyen T. A., Nguyen T. T., & Tran P. A. (2019). Biomaterials in Translational Medicine. Academic Press; Cambridge, MA, USA: 2019. Surface modification of medical devices at nanoscale—Recent development and translational perspectives; 163–189.

Loevestam G., Rauscher H., Roebben G., Sokull-Kluettgen B., Gibson P., Putaud J., & Stamm H. Considerations on a Definition of Nanomaterial for Regulatory Purposes. EUR 24403 EN. Luxembourg Publications Office of the European Union.

Luther, W. (2004) Industrial application of Nanomaterial-Chances and risks. Future Technologies Division of VDI Technologiezentrum GmbH. GrafRecke-Str., 84 40239 Düsseldorf Germany, 5-9.

Moshed, A. M. A., Sarkar, M. K. I., & Khaleque Md. A. (2017) The Application of Nanotechnology in Medical Sciences: New Horizon of Treatment. American Journal of Biomedical Sciences, 9, 1-14.

Oberdorster, G. (2010) Safety assessment for nanotechnology and nanomedicine: concepts of Nanotoxicology. Journal of Internal Medicine, 267, 89-105.

Pramanik, P. K. D., Solanki, A., Debnath, A., Nayyar, A., El-Sappagh S., & Kwak, K. S. (2020) Advancing Modern Healthcare With Nanotechnology, Nanobiosensors, and Internet of Nano Things: Taxonomies, Applications, Architecture, and Challenges. IEEE Access, 8, 65230–65266.

Saha, M. (2009) Nanomedicine: Promising Tiny Machine for the Healthcare in Future-A Review. Oman Medical Journal, 24(4), 242–247.

Tinkle, S., McNeil, S. E., Muhlebach, S., Bawa, R., Borchard, G., Barenholz, Y. C. (2014) Nanomedicines: Addressing the scientific and regulatory gap. Annals of the New York Academy of Sciences, 1313, 35-56.

10

Nanotechnology in Food Production Processing and Preservation

Navpreet Kaur

Department of Food Processing and Preservation, Guru Nanak College, Sri Muktsar Sahib, Punjab, India

Abstract

Nanotechnology has been successfully applied in various fields such as agriculture, biochemistry and medicine. However, its application in food systems has been new and is yet to be scrutinised. It offers a range of physical and chemical benefits that can be employed in production, development, fabrication, packaging, storage and food distribution systems. These novel nanomaterials such as nanocapsules, nanocomposites, nanoemulsions, nanoencapsulation, nano packaging, nanotubes and nanosensors have been used to improve food quality with better shelf life. However, there is concern about the safety of these nanomaterials, especially because of partial knowledge about the effects of these materials on human health. This chapter includes the different types of nanomaterials used in food preservation, their toxicity to humans and risk assessments related to these. The most fundamental questions regarding safety and risk posed by these high-performance materials still need to be addressed. The guidelines and regulations governing the use of nanomaterials should be restructured. However, the only answer to this is the synthesis of bio-inspired nanomaterials.

Keywords: *Food nanomaterials; Food safety; Bio-inspired nanomaterials; Bio-synthesized nanomaterials.*

1. Introduction to Nanotechnology

Nanotechnology has found its wide application in our everyday life and has been changing entire society for the last two decades (Nasrollahzadeh *et al.*, 2019). This all started with the United States Department of Agriculture publishing its first reports on its use in the agriculture and food industry in 2003 (Agriculture Organization, 2004). The word 'nano' derives its roots from the Greek word 'nanos' which means dwarf. Nanotechnology is a branch of applied science and technology that includes the use of nanomaterials and the design and implementation of devices at the nanoscale (Bhushan, 2017).

Nanomaterials (NPs) are materials of nano dimensions i.e. 1-100nm scale (Sudha *et al.*, 2018). This miniaturised version of materials leads to significant changes in surface-to-volume ratio thus impacting their overall surface chemistry (Sudha *et al.*, 2018; Pawar, 2020). The surface properties largely govern the availability of active sites and their interactions. These can also be tailored for specific functions such as size, shape, surface properties, aggregation state, solubility, structure and chemical composition imparting antibacterial and antioxidant properties to nanomaterials (Sudha *et al.*, 2018). The food industry has also welcomed the nano era to the growing need for healthier products with rich nutrition and low expenses (Ingale *et al.*, 2018). Nanomaterials are universally accepted to be in nano dimensions. However, their dimensional size might vary for food-based applications (Couto and Almeid, 2022). Foods are complex assemblies of nanosized elements with varying physical and chemical characteristics. The foods are colloidal mixtures of carbohydrates, lipids, proteins, vitamins and minerals. These complex structures and interactions of interfacial colloidal mixtures determine the overall stability of food items. Thus, fabricating nanomaterials, their assemblies and their interactions with each other with the help of nanotechnological tools provide a means to design nano foods from scratch.

The nanoparticles can be used for developing novel nanomaterial-based food products, improving food efficiency and food characteristics, as functional ingredients in food additives, food packaging materials, nutrient delivery and food safety (Pushparaj *et al.*, 2022). Amazingly, there is a range of products that have been already marketed and used in the food business. The majority of these products are designed as food-contacting materials but are not consumed by people except titanium dioxide and iron oxide, that has been used directly in food as pigment and colourant respectively (Sohal *et al.*, 2020).

However, despite the initial excitement the use of nanofoods has been slowed down. The fundamental reason is risk and toxicity, which is yet to be established (McClements and Xiao, 2017). The most of studies establishing the toxicity of these nanomaterials are in vitro and very little data for in vivo studies is available. The chronic effects are these nanomaterials especially metal nanoparticles have been well documented (Maharramov *et al.*, 2019.). There are also several gaps in establishing the toxicity of nanomaterials to human cells, tissues and organs, its chronic effects on the human body, migration of these nanomaterials to food, bioaccumulation of nanomaterials in organic systems, degradation or environmental fate of nanomaterials and their impact on ecosystems. The acceptance of these nano foods by the public is also an important aspect that is often ignored by researchers, manufacturers and authorities. The less scrutiny of the safety of these nanomaterials in food is also the root cause for legislation delays regarding nano foods.

2. Type of Nanomaterials

Nanomaterials can be categorized into organic and inorganic nanomaterials based on their composition as depicted in Figure 1. Among organic, it can further be classified into carbon-based and other organic-based nanomaterials. The inorganic nanomaterials are metal-based and metal oxide-based nanomaterials (Hasan, 2015).

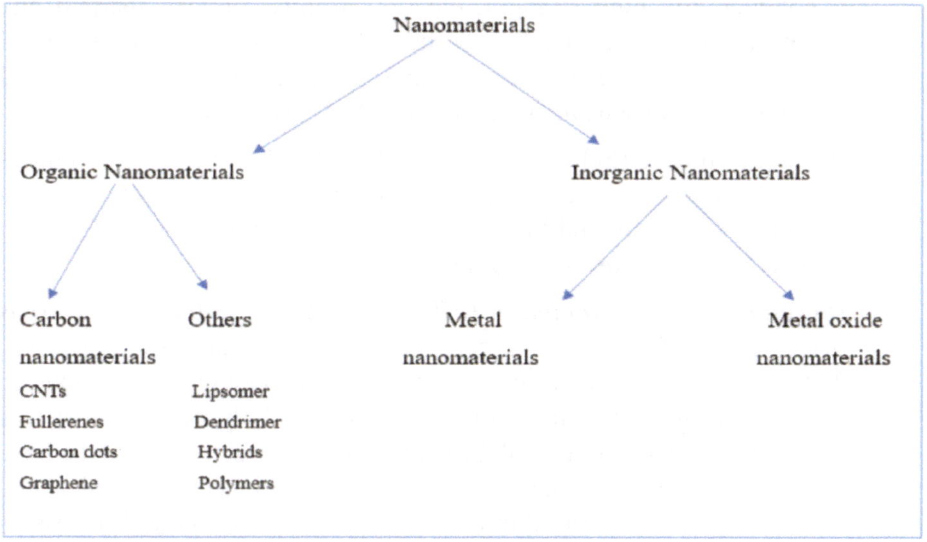

Fig. 1. Classification of nanomaterials based on composition.

A. ***Carbon-based Nanomaterials:*** This class comprises nanomaterials solely made from carbon. The carbon atom due to catenation property forms covalent bonds with other carbon atoms with different hybridization states (Sp, Sp^2 and Sp^3) (Rao and Dresselhaus, 2001; Vejpravová, 2021). The different hybridization states form a variety of morphological structures such as sheets, ellipsoids, hollow tubes or spheres (Coville, 2011; Mayeen, 2018). Various examples of carbon-based nanomaterials are graphene sheets (Gr), Carbon Nanotubes (CNTs), Fullerenes (C_{60}), carbon nanofibers, carbon onions and carbon black nanoparticles (Coville, 2011).

   - Fullerenes are formed by sixty carbon atoms forming a symmetrical closed cage structure like a soccer ball (Dresselhaus *et al.*, 1993). Fullerenes formed from C_{70} or C_{540} also have been described.

- Carbon black nanoparticles are grape-like aggregates formed by fused spherical particles (Lawrence *et al.*, 2014).
- Carbon quantum dots are quasi-spherical and discrete carbon structures whose size is below 10nm (Jelinek, 2017).
- Carbon-based nanoparticles (NPs) have three external nanoscale dimensions (Dizaj *et al.*, 2015). The sp^2-hybridised carbon bonds have unusual physicochemical properties at the nanoscale. These carbon-based nanoparticles have unique properties like electrical conductivity, high strength, electron affinity, optical, thermal and sorption properties. Carbon-based NPs are used in a wide range of fields such as drug delivery, energy storage, photovoltaic devices and environmental sensing applications (Dizaj *et al.*, 2015).
- Nanodiamonds (Schrand *et al.*, 2009) and carbon-based nano onions (Mykhailiv *et al.*, 2017) are carbon-based complex nanoparticles with low toxicity and biocompatibility. Thus, it can be used in drug delivery and tissue engineering applications.

B. ***Inorganic-based Nanomaterials:*** This class represents the nanomaterials made of metal, metal oxides, ceramics or semiconductors (Kang *et al.*, 2021). The metal nanoparticles are made from metal precursors and can be monometallic, bimetallic or polymetallic (Mody *et al.*, 2010; Dos Santo *et al.*, 2020; Couto and Almeida, 2022). These metallic NPs have localised surface plasmon resonance and thus possess unique optical and electrical properties. In addition, some metallic nanoparticles possess thermal, magnetic and biological properties making them increasingly important materials for the development of nanodevices that find their utility in numerous fields such as physical, chemical, biological and pharmaceutical fields (Mody *et al.*, 2010; Dos Santo *et al.*, 2020; Couto and Almeida, 2022). These nanomaterials can be synthesized from metals like Ag, Au, and Fe or metal oxides like ZnO, TiO_2 and Fe_3O_4, CeO_2.

NPs made from semiconductor materials possess the properties of both metals and non-metals. These NPs can be used in photocatalytic, optic and electronic devices (Terna *et al.*, 2021). Others are ceramic NPs made from inorganic solids such as carbonates, carbides, and phosphates and are usually found in amorphous, polycrystalline, dense, porous or hollow forms. The ceramic NPs have high stability and load capacity and thus can be used in biomedical applications (Khan *et al.*, 2019).

C. ***Organic-based Nanomaterials:*** This class of nanoparticles is made of carbohydrates, proteins, lipids, polymers or other organic compounds that form covalent bonds. The basis of these nanomaterials is covalent interactions that help the organic nanomaterials to form coveted structures such as micelles, dendrimers, micelles, protein complexes like ferritin, compact polymers and liposomes (Szakal *et al.*, 2014). These nontoxic and biodegradable nanomaterials are environment-friendly and in some cases even have hollow cores such as liposomes. Organic nanomaterials are sensitive to electromagnetic and thermal radiations such as light and heat. The non-covalent interactions also stabilize the 3D structures of these nanoparticles, although this adds to their labile nature (Szakal *et al.*, 2014; Babu and Tingirikari, 2023). Hence, organic nanomaterials can be easily cleared from the body. These are mostly used in the biomedical field for target drug delivery, cancer therapy and the food sector (Niknam *et al.*, 2024; Babu and Tingirikari, 2023).

3. Use of nanomaterials in production, development, fabrication, packaging, storage and food distribution systems

Food Nanotechnology has permeated into many aspects of production, development, fabrication, packaging, storage and food distribution systems. This novel technology has modernised various existing technologies in food systems and has attracted the attention of manufacturers, industrialists, scientists and consumers worldwide. Various elements added as additives to food or used in food packaging were found to exist on the nanometre scale e.g. TiO_2 NPs. A few novel applications of nanotechnology/nanomaterials in food have been listed in Figure 2.

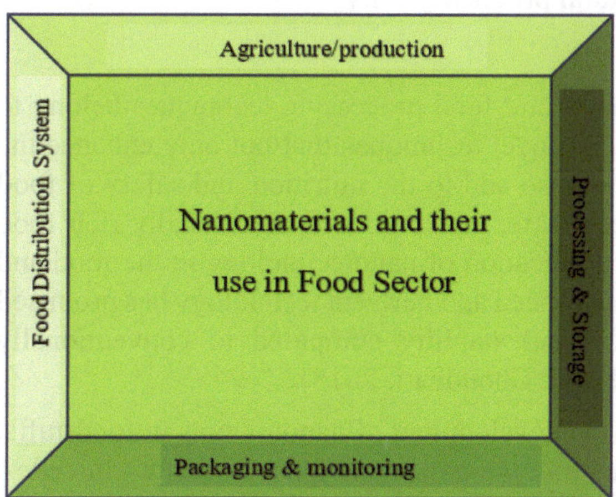

Fig. 2. Applications of nanomaterials in various fields of the food sector.

3.1. Food Production/Agriculture

Nanotechnology holds great potential for various challenges faced by agriculture. The key issues include the overuse of pesticides and fertilizers to increase yield. Pesticides and fertilizers are not biodegradable and tend to accumulate in living flora and fauna. They also have a long shelf-life and if biodegraded by some microbes, these generally are metabolised into harmful free radicals that are cancer-causing species. To regulate these burning issues in today's agriculture sector, nanotechnological strategies are currently being examined.

Utilization of nanotechnology in the agriculture sector might lower the cost of production inputs in chemical fertilizers and pesticides and also improve precision farming which allows precise control at the nanometer scale (Ingale and Choudhari, 2013). Precision farming is the process of enhancing crop yield by reducing the usage of fertilizers, pesticides and herbicides with the aid of advanced remote-sensing devices, thus identifying problems related to crops and the environment. The implementation of nanosensors enhances the yield by providing accurate information therefore impacting the decisions related to crop growth and productivity. This technique lowers the massive agricultural waste (bioaccumulation of toxins) and hence keeps a check on environmental pollution. Nanosensors or nano biosensors can also be used as smart sensors to detect the presence of pathogenic microorganisms and pests in the field or nutrient deficiency. Nanoencapsulation is yet another technique where pesticides are encapsulated in nanoparticles and are only released in the target insects' stomach, which minimises the possibility of environmental contamination. Application of nano barcodes and nano processing techniques can be used to monitor agricultural produce quality.

3.2. Food Processing

The next-generation food industry and food processing techniques belong to innovative products processed by novel techniques that not only enhance the shelf life and food attributes but also add to the nutrition and safety of food products. This demand for fresh, safe, authentic and nutritionally rich food by consumers has brought the application of nanotechnology in the food and food processing industry. This advanced and new-era technology has promised improved textures, consistency and stability compared to conventionally processed food products (Ingale and Choudhari, 2018).

One such application involves the development of nanotextures in foodstuffs. The most common mechanisms involve nanoemulsions, surfactant micelles, and emulsion bilayers (Ingale and Choudhari, 2018). A range of nanostructured foodstuffs have been developed such as creams, ice creams, yoghurts, mayonnaise and spreads.

A further application of nanomaterials includes the use of nano-encapsulated or nanosized colour or flavour additives, preservatives or supplements in products. These nanomaterials can be engineered for unique properties that offer great advantages for food ingredients. Efforts are being made to enhance water dispersion qualities without excessive use of surfactants or fats. The few nanomaterials listed below have been permitted and thus can be used as anti-caking agents, food flavour carriers and additives.

- TiO_2 can be used as a colour additive, provided it should be less than 1% by weight in food products (Ropers *et al.*, 2017; Younes *et al.*, 2021).
- Synthetic iron oxide can also be used as a colour additive but its concentration should be less than 0.1% in human-finished food (EFSA Panel, 2015; Aquilina *et al.*, 2016).
- The use of zinc oxide, iron oxide, aluminium oxide, cobalt oxide, silicon dioxide, manganese oxide, titanium nitride or carbon black as an additive or as a polymer production aid has been authorised EC 10/2011.
- Silver-silica can be used as an antimicrobial agent where silver blended into polymer should be less than 4ppm by weight (FCN No. 1235) (He *et al.*, 2019).
- Silicon dioxide (E551) can be used as a flavour carrier which should be less than 10,000 mg per kg of food. However, it is excluded from infant's food and the food of young children (authorised by EC 1334/2008).
- Silicon dioxide (E551), Copper oxide and iron oxide can be used as an anticaking agent and nutritional dietary supplement (He *et al.*, 2019).

3.3. Food Preservation/Storage and Monitoring

Once processed, the finished food products are preserved with different techniques to enhance the shelf life of food products. The two major food factors that deteriorate the food are self-oxidative reactions by enzymes in food and the metabolic action of various microorganisms in the food. The deterioration accelerates the process of discolouration, texture changes, off-odours, flavour trouble and rancidity. Nanoparticles can also immobilise the enzymes that aid in the dispersion of these particles throughout the food matrices and enhance the enzymatic activity. Nanomaterials can act as oxygen scavengers, thus lowering the oxidation-reduction reaction in a range of products like processed meat, beer, beverages, cooked pasta etc. Packaging films in collaboration with silicate nanoparticles reduces the oxygen entrance and prevents food spoilage (Babu, 2022).

The nanoparticles possess unique bactericidal properties (Hajipour *et al.*, 2012) and thus are used in food preservation (Hama *et al.*, 2018). Different nanoparticles like copper, silver, chitosan and metal oxide have been used to extend the shelf life of food products by monitoring the bacterial growth in food items (Kumar *et al.*, 2021). These nanoparticles also control the enzymatic and metabolic activity thus enhancing the shelf life of food products (Khan *et al.*, 2022).

In another application, nanosensors can be used as an advanced detection system for speedy detection of product spoilage. Thus, nanotechnology can prove to be cutting-edge technology for quality control at the source, and during the production chain as well. Nanobiosensor technology for quality assurance can prove to be quick, real-time, inexpensive and an alternative to conventional methodologies. The nanomaterials can also be used as biosensors to detect the presence of pesticides, pathogens and toxins in the food. In one such study, ZnO quantum dots (QD) were developed to detect a range of pesticides including aldrin, tetradifon, glyphosate and atrazine (Sahoo *et al.*, 2018). The pesticides with the Cl group interact with QD with high binding affinity. The ZnO QD could also be photocatalyzed pesticides forming the new trend of active packaging or smart packaging.

3.4. Food Packaging

The food packaging maintains the quality and safety of the product during its storage and transportation. Smart Packing lowers the loss of food nutrients and extends its shelf life by avoiding unfavourable factors such as aeration, spills, growth of microorganisms, light and other external factors. Materials used for food packaging are in direct contact with food products during manufacturing, transportation and storage (Kim *et al.*, 2014.). Active and Smart packaging systems with nanomaterials for food packaging have many advantages compared to conventional packaging materials. Among many nanomaterials, the most widely used is nano clay preferred for mechanical, thermal and barrier properties and low price (Tiwari *et al.*, 2021). These bionanomaterials are considered more eco-friendly as they reduce the use of nonbiodegradable plastics.

Nano clays are developed into a range of subclasses including bentonite, hectorite, halloysite, kaolinite and montmorillonite (Hong & Rhim, 2008; Saba *et al.*, 2016). Bentonite (Siddiqui and Alrumman, 2021) and montmorillonite (Bangar *et al.*, 2023) are listed as GRAS and in Effective Food Contact Substances (FCS) notifications by the U.S. FDA. A few examples of nano clay loaded with different substances are listed below:

- Nanocomposite membrane made of 1wt % bentonite clay/poly (vinyl alcohol) has been reported with enhanced permeance to water (Buruga *et al.*, 2019).
- 3 % nano clay loaded with woven carbon fibre with improved interlaminar fracture toughness has been developed (Siddiqui *et al.*, 2007).
- Nanocomposites made of epoxy/clay have been scrutinised for corrosion resistance (Zaghloul *et al.*, 2023).

In another application, antimicrobial packaging systems have been developed. Antimicrobial nanoparticle covering is included in the matrix of the packaging material. It reduces bacterial growth on and near the product. Antimicrobial nanoparticle sachets are also included in the package that disperses bioactive antimicrobial agents into the package (Khaneghah *et al.*, 2018). Various bakery and meat products that are prone to surface spoiling can be safely packed with antimicrobial packaging imbued nanoparticle layer (Birwal *et al.*, 2019). Edible food films also have been used to kill various microbes present on the meat surface (Walayat *et al.*, 2024).

Antimicrobials that are extracted from natural sources such as organic acids, plant extracts, essential oils, chitosan or nisin are generally recognised as safe compounds (Liao *et al.*, 2021). Various organic acids like citric acid, lactic acid and malic acid reduce the pH, disrupt the substrate transport and reduce the proton motive force (Papadochristopoulos *et al.*, 2021). In another research, chitosan has been used as a preservative and it inhibits the growth of many spoilage-causing yeast, moulds and pathogenic bacteria. The efficacy of chitosan depends on the type of application (Gautam *et al.*, 2022).

The recent interest of researchers in green packaging made of natural biopolymers, and bio-nanocomposites has generated great potential for enhancing the safety, stability and quality of food products (Ahari *et al.*, 2022). The biodegradable and fully compostable plastic packaging made up of corn starch was developed with the help of nanotechnology (Onyeaka *et al.*, 2022). These biodegradable polymers were also reported to enhance the organoleptic properties thus lowering the chances of toxicity generated by nano foods (Biswas *et al.*, 2022).

Edible coating of nanomaterials has also shown promising results for food preservation and storage. Fresh food products such as vegetables and fruits coated with nanomaterials remain active during storage and transportation. The coating of nanomaterials lowers the active respiration, hence lowering the post-harvest losses by extending the storage time and nutritional quality. The coating of food products not only lowers food respiration but also microbial activity in the products. A thin layer of nano-edible film of less than 100 μm enhances the mechanical properties and sensory perceptions, acts as a gas and moisture barrier and prevents microbial spoilage thus improving the storage life of fresh foods (Flores-Lopez *et al.*, 2016). These nanomaterials from natural extracts listed below are applied as nanoencapsulation and nanoemulsions on various fresh fruits and vegetables:

- Poly-ε-caprolactone was tested on fresh-cut "Red Delicious" apples (Zambrano-Zaragoza *et al.*, 2014).
- Nanoemulsion of Quinoa Protein and Chitosan was tested on fresh strawberries (Chaudhary *et al.*, 2020).
- Nanoemulsion with lemongrass essential oil was tested on fresh-cut Fuji apples (Hasan *et al.*, 2020).
- Bio-nano-hybrid of pectin and LDH salicylate was tested on fresh apricots (Saputra *et al.*, 2023).

3.5. Nanomaterials in food chain distribution systems

Nanotechnology can be employed from the beginning to the end of the food chain i.e. from farm to fork. The application of nanotechnology from such a vast environment can improve yield in production methods and the shelf life of food along with improvement in food quality in processing and storage with the promise of promoting the wellness of the ecosystem along with its components.

Various devices that are similar to human-based senses such as e-tongues, e-eyes and e-noses have been developed that can analyse different compounds in food matrices. These nanosensors have high efficiency and efficacy and thus detect a wide variety of analytes such as heavy metals, toxins, pathogens, nucleic acids and proteins (Poeta *et al.*, 2023). Hence, the quality checks in distribution systems can be quick, easy and real-time reading elevating the profits and lowering the risk of losses.

4. Safety and Risk Assessment Associated with Nanomaterials

The key issue associated with the use of nanomaterials is the safety of such materials. Researchers are concerned with the migration of these materials from packaging to food items and from food items to human systems. The current safety laws, safety testing protocols and workplace health procedures are inadequate to measure and assess the risk posed by nanofood. The various risks posed by these materials are listed

4.1. Toxicological fundamentals and risk assessments of exposure roots and interactions

Nanomaterials are either unintentionally introduced via migration in food materials or are intentionally added as additives or preservatives (Siddiqui and Alrumman, 2021). This increasing application of nanomaterials in the food sector has attracted consumer attention, especially regarding the safety of their use. Owing to the unique physiochemical properties of nanomaterials, the spread of nanomaterials in the environment and the human body is under question (Zhu *et al.*, 2013; El-Kady *et al.*, 2023.). The fate and behaviour of nanomaterials within the human systems and environment are difficult to predict due to the complexity of these systems. It is tedious to trace and monitor the distribution of nanomaterials. Although, various studies have suggested a holistic approach to understanding these nano-bio-eco interactions yet case by case studies are needed for the assessment.

4.2. Nanotoxicological Mechanisms

Another major apprehension is the effect of these nanomaterials on human health that are added as additives or other functional ingredients in food items. The direct contact of these nanomaterials can cause serious human health concerns. The nanomaterials generate reactive oxidative species (ROS) which is a mechanism major toxicological mechanism causing cellular damage and cell death (Samrot and Noel Richard Prakash, 2023). Overproduction of ROS can damage DNA and neurons and might lead to autophagy in the cells. The ROS are mutagenic species and can cause cancer within a biological system (Jomova *et al.*, 2023). Another adverse outcome is allergic reactions caused by exposure to food with nanomaterials (Chen *et al.*, 2023).

5. Rules and Regulations Associated with Nanomaterials

The nanomaterials behave differently from bulk materials of the conventional world. The protocols for nanoworld are rather distinct from than already set protocols. To standardize these protocols, both understanding and control of nanoscale interactions/properties/functions/toxicities are required. This further determines the beneficial and adverse effects, the transition of nanomaterials and finally the environmental fate of nanomaterials that has an impact on the ecosystem can be scrutinised. Hence, it is extremely difficult to set up a database for nanomaterials and thus formulate rules and regulations. Worldwide, the current legislation is in the initial stage of development and covers an extremely broad and general aspect regarding nanotechnology and nanomaterials. It needs evaluation and analysis when it comes to specific nanomaterials.

Every year several nano-based food products are introduced in the market globally. Food manufacturers have invested a lot in developing novel nanotechnology-based products in the food sector. However, most of these are in the field of food-contacting materials i.e. food packages, sensors and detectors while those consumed directly are limited (titanium dioxide and iron oxide as food additive or pigment). The food products that are directly consumed require the highest safety levels than the contacting materials or sensors. Hence, these are under strict control or evaluation and evaluated for their toxicity to human systems and their environmental impact. Otherwise, it is highly risky and these nanomaterials could bring new disasters to humans and the environment. This makes the regulation and legislation far more urgent. Also, the success and implication of nanotechnology and the marketing of nanofood products depend upon regulations and legislation laws. These laws are windows for public awareness. The legislation involves:

- Suggestions by academicians, researchers, scientists or government agencies.
- Reasonable suggestions are taken by government agencies and are proposed as a directive, known as legislation.
- The legislation is then passed by the parliament or legislative arm of the country and becomes law.
- The laws covering general or broad aspects related to a given matter are then enforced by regulators.

In India, nanotechnology-based food products are covered by various agencies like the Department of Biotechnology (DBT), The Ministry of Environment, Forest and Climate Change (MoEFCC), The Food Safety and Standards Authority of India (FSSAI) and The Indian Council of Agricultural Research (ICAR). The DBT issued the guidelines in 2020 that have been complied with to evaluate nano-based foods with significant advantages over existing food ingredients in terms of altered but beneficial properties or dimensions. The present guidelines cover nano-agri-input products (NAIPs), nano-agri-products (NAPs), nanocomposites and nanosensors made from nanomaterials. FSSAI is an autonomous body of the Ministry of Health & Family Welfare, Government of India. The NAP covers the nanofood along with FDA guidelines (FDA, 2014), (FDA, 2015) and the Food Safety and Standards Act 2006 with additional criteria that may be adopted by FSSAI.4.2.2.

6. Conclusion

Numerous techniques and technologies play an important role in agriculture, food processing and preservation and one such is nanotechnology in this new era. The benefits and uses of nanotechnology are innumerable. The use of nanomaterials in production enhances productivity by maximising profits and minimising investments in terms of fertilizers and pesticides. Thus, it directs the production towards precision farming.

The use of nanotechnology in food production is another benefit as certain preservatives or components that cannot be dissolved into aqueous systems can be done at nanoscales e.g. one such being the dissolution of turmeric nanoparticles in milk (Ipar *et al.*, 2022). Also, the use of biopolymers with nanomaterials has cut down the high cost and increased the overall performance when compared to conventional synthetic polymers.

The use of active and smart packaging with the use of nanomaterials has revolutionised packaging as well as the retail industry. The antimicrobial properties, enzyme immobilisation and oxygen-scavenging properties are key factors that led to their increased use in the food sector. Ecofriendly nanomembranes developed from natural sources are also gaining more focus in the food industry due to less toxicity and biodegradability.

The nano biosensors and their use in the food industry are another unmatched application as these diagnostic devices have high sensitivity and selectivity and assure food quality measurements along the real-time production lines. The use of edible films and coating on food items is also being utilized for improving the shelf life of food products.

Thus, for the food industry nanotechnology exhibits promising potential as demonstrated by laboratory trials. However, the practical application of nanotechnology remains uncertain. At the nanoscale, it is difficult to control the properties of nanomaterials, thus limiting their application at commercial scales. The absorption and delivery of these nanoparticles in human systems are still under study. Moreover, the behaviour of nanomaterials in both in vitro and in vivo studies is remarkably different thus even complicating the safety evaluations. There are also loopholes in regulation and legislation concerning these nanomaterials and their use in the food industry. This gap needs to be addressed so that safe, non-toxic and next-generation food items with the best quality and nutrition are brought to market.

References

Agriculture Organization, 2004. The State of Food and Agriculture 2003-04 (No. 35). Food & Agriculture Org.

Ahari, H., Golestan, L., Anvar, S.A.A., Cacciotti, I., Garavand, F., Rezaei, A., Sani, M.A. and Jafari, S.M. (2022). Bio-nanocomposites as food packaging materials; the main production techniques and analytical parameters. Advances in Colloid and Interface Science, 310, p.102806.

Aquilina, G., Azimonti, G., Bampidis, V., de Lourdes Bastos, M., Bories, G., Chesson, A., Cocconcelli, P.S., Flachowsky, G., Gropp, J., Kolar, B. and Kouba, M. (2016). Safety and efficacy of iron oxide black, red and yellow for all animal species. Efsa Journal, 14(6), 1-16.

Babu, P.J. and Tingirikari, J.M.R., (2023). A review on polymeric nanomaterials intervention in food industry. Polymer Bulletin, 80(1), 137-164.

Babu, P.J., 2022. Nanotechnology mediated intelligent and improved food packaging. International Nano Letters, 12(1), 1-14.

Bangar, S.P., Chaudhary, V., Akhila, P.P. and Sunooj, K.V., 2023. Recent functionality developments in montmorillonite as a nanofiller in food packaging. Trends in Food Science & Technology, p.104148.

Bhushan, B. ed., 2017. Springer handbook of nanotechnology. Springer.

Birwal, P., Rangi, P. and Ravindra, M.R. (2019). Nanotechnology applications in packaging of dairy and meat products. In Nanotechnology Applications in Dairy Science, pp. 3-26. Apple Academic Press.

Biswas, R., Alam, M., Sarkar, A., Haque, M.I., Hasan, M.M. and Hoque, M., 2022. Application of nanotechnology in food: processing, preservation, packaging and safety assessment. Heliyon, 8(11).

Buruga, K., Song, H., Shang, J., Bolan, N., Jagannathan, T.K. and Kim, K.H. (2019). A review on functional polymer-clay based nanocomposite membranes for treatment of water. Journal of Hazardous Materials, 379, p.120584.

Chaudhary, S., Kumar, S., Kumar, V. and Sharma, R. (2020). Chitosan nanoemulsions as advanced edible coatings for fruits and vegetables: Composition, fabrication and developments in last decade. International Journal of Biological Macromolecules, 152, 154-170.

Chen, J., Guo, Y., Zhang, X., Liu, J., Gong, P., Su, Z., Fan, L. and Li, G. (2023). Emerging nanoparticles in food: sources, application, and safety. Journal of Agricultural and Food Chemistry, 71(8), 3564-3582.

Couto, C. and Almeida, A., 2022. Metallic nanoparticles in the food sector: A mini-review. Foods, 11(3), p.402.

Coville, N.J., Mhlanga, S.D., Nxumalo, E.N. and Shaikjee, A. (2011). A review of shaped carbon nanomaterials. South African Journal of Science, 107(3), 1-15.

Dizaj, S.M., Mennati, A., Jafari, S., Khezri, K. and Adibkia, K. (2015). Antimicrobial activity of carbon-based nanoparticles. Advanced pharmaceutical bulletin, 5(1), p.19.

Dos Santos, C.A., Ingle, A.P. and Rai, M., 2020. The emerging role of metallic nanoparticles in food. Applied microbiology and biotechnology, 104, pp.2373-2383.

Dresselhaus, M.S., Dresselhaus, G. and Eklund, P.C., (1993). Fullerenes. Journal of materials research, 8(8), 2054-2097.

EC, N. and SE, S. (2011). Crop Conditions as of April 10, 2011.

EFSA Panel on Food Additives and Nutrient Sources added to Food (ANS) (2015). Scientific Opinion on the re□evaluation of iron oxides and hydroxides (E 172) as food additives. EFSA Journal, 13(12), p.4317.

El-Kady, M.M., Ansari, I., Arora, C., Rai, N., Soni, S., Verma, D.K., Singh, P. and Mahmoud, A.E.D. (2023). Nanomaterials: A comprehensive review of applications, toxicity, impact, and fate to environment. Journal of Molecular Liquids, 370, p.121046.

Flores-López, M.L., Cerqueira, M.A., de Rodríguez, D.J. and Vicente, A.A. (2016). Perspectives on utilization of edible coatings and nano-laminate coatings for extension of postharvest storage of fruits and vegetables. Food Engineering Reviews, 8, 292-305.

Gautam, G., Rani, R., Badwaik, L.S. and Mahanta, C.L. (2022). Chitosan□Based Films and Coatings. Biopolymer□Based Food Packaging: Innovations and Technology Applications, 110-146.

Hajipour, M.J., Fromm, K.M., Ashkarran, A.A., de Aberasturi, D.J., de Larramendi, I.R., Rojo, T., Serpooshan, V., Parak, W.J. and Mahmoudi, M. (2012). Antibacterial properties of nanoparticles. Trends in biotechnology, 30(10), 499-511.

Hamad, A.F., Han, J.H., Kim, B.C. and Rather, I.A., (2018). The intertwine of nanotechnology with the food industry. Saudi journal of biological sciences, 25(1), pp.27-30.

Hasan, S., (2015). A review on nanoparticles: their synthesis and types. Res. J. Recent Sci, 2277, p.2502.

Hasan, S.K., Ferrentino, G. and Scampicchio, M., (2020). Nanoemulsion as advanced edible coatings to preserve the quality of fresh□cut fruits and vegetables: a review. International journal of food science & technology, 55(1), pp.1-10.

He, X., Deng, H. and Hwang, H.M., 2019. The current application of nanotechnology in food and agriculture. Journal of food and drug analysis, 27(1), 1-21.

Hong, S.I. and Rhim, J.W. (2008). Antimicrobial activity of organically modified nano-clays. Journal of Nanoscience and Nanotechnology, 8(11), 5818-5824.

https://dbtindia.gov.in/sites/default/files/uploadfiles/Guidlines%20for%20Evaluation%20of%20Nano%20Based%20Agri%20Input%20and%20Food%20Products%20in%20India.pdf

https://eurlex.europa.eu/LexUriServ/LexUriServ.do?uri=OJ:L:2008:354:0034:0050:en:PDF

Ingale, A.G. and Chaudhari, A.N., (2013). Biogenic synthesis of nanoparticles and potential applications: an eco-friendly approach. Journal of Nanomed Nanotechonology, 4(165), pp.1-7.

Ingale, A.G. and Chaudhari, A.N. (2018). Nanotechnology in the food industry. Nanotechnology, food security and water treatment, pp.87-128.

Ipar, V.S., Singhal, R.S. and Devarajan, P.V. (2022). An innovative approach using microencapsulated turmeric oleoresin to develop ready-to-use turmeric milk powder with enhanced oral bioavailability. Food Chemistry, 373, p.131400.

Jelinek, R.2017). Carbon quantum dots. Carbon quantum dots. Springer International Publishing, Cham, pp.29-46.

Jomova, K., Raptova, R., Alomar, S.Y., Alwasel, S.H., Nepovimova, E., Kuca, K. and Valko, M., (2023). Reactive oxygen species, toxicity, oxidative stress, and antioxidants: Chronic diseases and aging. Archives of toxicology, 97(10), 2499-2574.

Kang, Y., Liu, J., Jiang, Y., Yin, S., Huang, Z., Zhang, Y., Wu, J., Chen, L. and Shao, L., (2021). Understanding the interactions between inorganic-based nanomaterials and biological membranes. Advanced Drug Delivery Reviews, 175, p.113820.

Khan, I., Saeed, K. and Khan, I., (2019). Nanoparticles: Properties, applications and toxicities. Arabian journal of chemistry, 12(7), 908-931.

Khan, J., Khurshid, S., Sarwar, A., Aziz, T., Naveed, M., Ali, U., Makhdoom, S.I., Nadeem, A.A., Khan, A.A., Sameeh, M.Y. and Alharbi, A.A. (2022). Enhancing bread quality and shelf life via glucose oxidase immobilized on zinc oxide nanoparticles—a sustainable approach towards food safety. Sustainability, 14(21), p.14255.

Khaneghah, A.M., Hashemi, S.M.B. and Limbo, S. (2018). Antimicrobial agents and packaging systems in antimicrobial active food packaging: An overview of approaches and interactions. Food and Bioproducts Processing, 111, 1-19.

Kim, Y.T., Min, B. and Kim, K.W., (2014). General characteristics of packaging materials for food system. In Innovations in food packaging, 13-35.

Kumar, A., Choudhary, A., Kaur, H., Mehta, S. and Husen, A., (2021). Metal-based nanoparticles, sensors, and their multifaceted application in food packaging. Journal of Nanobiotechnology, 19(1), p.256.

Lawrence, K., Baker, C.L., James, T.D., Bull, S.D., Lawrence, R., Mitchels, J.M., Opallo, M., Arotiba, O.A., Ozoemena, K.I. and Marken, F. (2014). Functionalized carbon nanoparticles, blacks and soots as electron□transfer building blocks and conduits. Chemistry–An Asian Journal, 9(5), 1226-1241.

Liao, W., Badri, W., Dumas, E., Ghnimi, S., Elaissari, A., Saurel, R. and Gharsallaoui, A., (2021). Nanoencapsulation of essential oils as natural food antimicrobial agents: An overview. Applied Sciences, 11(13), p.5778.

Maharramov, A.M., Hasanova, U.A., Suleymanova, I.A., Osmanova, G.E. and Hajiyeva, N.E., (2019). The engineered nanoparticles in food chain: Potential toxicity and effects. SN Applied Sciences, 1, 1-25.

Mayeen, A., Shaji, L.K., Nair, A.K. and Kalarikkal, N. (2018). Morphological characterization of nanomaterials. In Characterization of Nanomaterials, 335-364 Woodhead Publishing.

McClements, D.J. and Xiao, H. (2017). Is nano safe in foods? Establishing the factors impacting the gastrointestinal fate and toxicity of organic and inorganic food-grade nanoparticles. npj Science of Food, 1(1), p.6.

Mody, V.V., Siwale, R., Singh, A. and Mody, H.R. (2010). Introduction to metallic nanoparticles. Journal of Pharmacy and bioallied sciences, 2(4), 282-289.

Mykhailiv, O., Zubyk, H. and Plonska-Brzezinska, M.E. (2017). Carbon nano-onions: Unique carbon nanostructures with fascinating properties and their potential applications. Inorganica Chimica Acta, 468, pp.49-66.

Nasrollahzadeh, M., Sajadi, S.M., Sajjadi, M. and Issaabadi, Z. (2019). An introduction to nanotechnology. In Interface science and technology, 28, 1-27.

Niknam, Z., Zadeh, F.H., Toosi, S., Shoreh, N.K.H., Rasmi, Y. & Saleem, I., (2024). Organic-based nanomaterials for regenerative medicine. In Handbook of Nanomaterials, 2,359-400.

Onyeaka, H., Obileke, K., Makaka, G. and Nwokolo, N., (2022). Current research and applications of starch-based biodegradable films for food packaging. Polymers, 14(6), 1126.

Papadochristopoulos, A., Kerry, J.P., Fegan, N., Burgess, C.M. and Duffy, G., (2021). Natural anti-microbials for enhanced microbial safety and shelf-life of processed packaged meat. Foods, 10(7), 1598.

Pawar, S., (2020). Science of nanomaterials. Progress and Prospects in Nanoscience Today, p.1.

Poeta, E., Liboà, A., Mistrali, S., Núñez-Carmona, E. and Sberveglieri, V., (2023). Nanotechnology and E-Sensing for Food Chain Quality and Safety. Sensors, 23(20), p.8429.

Pushparaj, K., Liu, W.C., Meyyazhagan, A., Orlacchio, A., Pappusamy, M., Vadivalagan, C., Robert, A.A., Arumugam, V.A., Kamyab, H., Klemeš, J.J. and Khademi, T., (2022). Nano-from nature to nurture: A comprehensive review on facets, trends, perspectives and sustainability of nanotechnology in the food sector. Energy, 240, p.122732.

Rao, A.M. and Dresselhaus, M.S., (2001). Nanostructured forms of carbon: an overview. Nanostructured Carbon for Advanced Applications: Proceedings of the NATO Advanced Study Institute on Nanostructured Carbon for Advanced Applications Erice, Sicily, Italy July 19–31, 2000, 3-24.

Ropers, M.H., Terrisse, H., Mercier-Bonin, M. and Humbert, B., (2017). Titanium dioxide as food additive. Application of Titanium Dioxide, 10.

Saba, N., Jawaid, M. and Asim, M., (2016). Recent advances in nanoclay/natural fibers hybrid composites. Nanoclay reinforced polymer composites: natural fibre/nanoclay hybrid composites, 1-28.

Sahoo, D., Mandal, A., Mitra, T., Chakraborty, K., Bardhan, M. and Dasgupta, A.K., (2018). Nanosensing of pesticides by zinc oxide quantum dot: an optical and electrochemical approach for the detection of pesticides in water. Journal of agricultural and food chemistry, 66(2), 414-423.

Samrot, A.V. and Noel Richard Prakash, L.X., (2023). Nanoparticles induced oxidative damage in reproductive system and role of antioxidants on the induced toxicity. Life, 13(3), 767.

Saputra, A., Herpandi, H., Supriadi, A. and Saputra, D., (2023). The effects of nano-edible coating on shelf life, physicochemical, microbial and sensory properties in food preservation and horticulture: A mini review. In E3S Web of Conferences , 373, p. 04006). EDP Sciences.

Schrand, A.M., Hens, S.A.C. and Shenderova, O.A., (2009). Nanodiamond particles: properties and perspectives for bioapplications. Critical reviews in solid state and materials sciences, 34(1-2), 18-74.

Siddiqui, N.A., Woo, R.S., Kim, J.K., Leung, C.C. and Munir, A., (2007). Mode I interlaminar fracture behavior and mechanical properties of CFRPs with nanoclay-filled epoxy matrix. Composites Part A: Applied science and manufacturing, 38(2), 449-460.

Siddiqui, S. and Alrumman, S.A., (2021). Influence of nanoparticles on food: An analytical assessment. Journal of King Saud University-Science, 33(6), p.101530.

Sohal, I.S., DeLoid, G.M., O'Fallon, K.S., Gaines, P., Demokritou, P. and Bello, D., (2020). Effects of ingested food-grade titanium dioxide, silicon dioxide, iron (III) oxide and zinc oxide nanoparticles on an in vitro model of intestinal epithelium: Comparison between monoculture vs. a mucus-secreting coculture model. NanoImpact, 17, p.100209.

Sudha, P.N., Sangeetha, K., Vijayalakshmi, K. and Barhoum, A., (2018). Nanomaterials history, classification, unique properties, production and market. In Emerging applications of nanoparticles and architecture nanostructures (pp. 341-384). Elsevier.

Szakal, C., Roberts, S.M., Westerhoff, P., Bartholomaeus, A., Buck, N., Illuminato, I., Canady, R. and Rogers, M., (2014). Measurement of nanomaterials in foods: integrative consideration of challenges and future prospects. ACS nano, 8(4), 3128-3135.

Terna, A.D., Elemike, E.E., Mbonu, J.I., Osafile, O.E. and Ezeani, R.O., (2021). The future of semiconductors nanoparticles: Synthesis, properties and applications. Materials Science and Engineering: B, 272, 115363.

Tiwari, K., Singh, R., Negi, P., Dani, R. and Rawat, A., (2021). Application of nanomaterials in food packaging industry: A review. Materials Today: Proceedings, 46, 10652-10655.

Vejpravová, J., (2021). Mixed sp2–sp3 nanocarbon materials: A status quo review. Nanomaterials, 11(10), 2469.

Walayat, N., Campagnol, P.C., Zhang, W., Su, Z. and Lorenzo, J.M., (2024). Films with antimicrobial activity for meat products. Current Opinion in Food Science, 56, p.101127.

Younes, M., Aquilina, G., Castle, L., Engel, K.H., Fowler, P., Frutos Fernandez, M.J., Fürst, P., Gundert□Remy, U., Gürtler, R. and Husøy, T., (2021). Safety assessment of titanium dioxide (E171) as a food additive. Efsa Journal, 19(5), p.e06585.

Zaghloul, M.M.Y., Zaghloul, M.M.Y. and Fuseini, M., (2023). Recent progress in Epoxy Nanocomposites: Corrosion, structural, flame retardancy and applications—A comprehensive review. Polymers for Advanced Technologies, 34(11), 3438-3472.

Zambrano-Zaragoza, M.D.L.L., Mercado-Silva, E., Gutiérrez-Cortez, E., Cornejo-Villegas, M.A. and Quintanar-Guerrero, D., (2014). The effect of nano-coatings with α-tocopherol and xanthan gum on shelf-life and browning index of fresh-cut "Red Delicious" apples. Innovative Food Science & Emerging Technologies, 22, 188-196.

Zhu, M., Nie, G., Meng, H., Xia, T., Nel, A. and Zhao, Y., (2013). Physicochemical properties determine nanomaterial cellular uptake, transport, and fate. Accounts of chemical research, 46(3), 622-631.

11

Unveiling the Potential of Nanotechnology in Food Additives

***Manvik Joshi*[1] *and Kamalesh Kumar Meena*[2*]**

[1]*Department of Dairy and Food Chemistry, Maharana Pratap University of Agriculture and Technology, Udaipur-313001, Rajasthan, India*

[2*]*Department of Dairy and Food Microbiology, Maharana Pratap University of Agriculture & Technology, Udaipur-313001, Rajasthan, India*

Abstract

Nanotechnology has emerged as a revolutionary force in the food industry, providing innovative solutions to improve food quality, safety, and usefulness. Researchers have devised novel ways to numerous issues in food production and preservation by harnessing nanoparticles' unique traits, such as their high surface area-to-volume ratio and customizable physicochemical characteristics. Nanotechnology-enabled food additives provide a wide range of benefits, including improved texture, taste, color, and nutritional value, as well as increased shelf life and microbiological safety. Furthermore, nanoparticles are effective transporters for bioactive chemicals, allowing for targeted and regulated release within the food matrix. Despite these prospective benefits, worries about the safety and regulatory implications of nanotechnology in food additives remain. This chapter critically explores the current state of knowledge, emphasizing important accomplishments, problems, and future directions in the area. By encouraging interdisciplinary collaboration among scientists, engineers, regulators, and stakeholders, the full potential of nanotechnology in food additives can be realized, ensuring the production of safe, sustainable, and creative food products to satisfy the changing needs of global consumers.

Keywords: *Nanotechnology, Food additives, Safety, Innovation*

1. Introduction

The significance of food additives in enhancing food quality underscores the importance of developing sensitive and selective methods for their detection as improper use of these additives can pose potential threats to human

* *Correspondence: author e-mail: Kamleshrj14@gmail.com*

health. Nanomaterials, with their distinctive chemical and electrochemical properties, are increasingly utilized in the design of sensing systems for food additives, offering promising avenues for precise detection and monitoring. Nanotechnology represents a groundbreaking interdisciplinary approach revolutionizing the landscape of the food industry. It involves the manipulation and modification of organic materials at the molecular level, ushering in transformative advancements. This innovative methodology has profoundly influenced the way food is produced, enhancing its quality, safety, and sustainability. Through the utilization of nanomaterials, it plays a crucial role in elevating food safety by detecting pathogens and toxins and offering insights into nutritional content (Nile *et al.*, 2020). Government departments overseeing food administration must implement regulatory measures to ensure the accurate and safe commercialization of nanofoods. Additional research is necessary to discover safer methods of integrating nanotechnology into the food industry, to address concerns, and to ensure responsible usage (Joshi & Meena, 2023).

2. Nanotechnology

The progress of knowledge has given rise to several new technologies beneficial for food processing. Over the past few decades, the food industry has undergone a revolution due to the emergence of numerous scientific fields and technologies. Among these, nanotechnology stands out as a unique multidisciplinary approach involving the development of nano materials and their manipulation at the molecular level. This process entails the examination and control of matter at the molecular scale, generating nanomaterials with modified chemical properties (Francisco & García-estepa, 2018). When particles size is reduced to nano size, particles exhibit unique physicochemical and mechanical properties that differ vastly from their macroscale size. In today's context, nanotechnology is pivotal in bolstering food safety by aiding in the identification of pathogens and toxins, while also offering valuable insights into nutritional content. Its integration into food technology holds immense potential to transform the packaging of dairy and food items. Consequently the promising advancements of nanotechnology in food industry stand to elevate food safety, quality, and prolong the shelf life of products (Rajamalar *et al.*, 2011).

2.1 Nanomaterials

Nanomaterials are the materials which size is reduced to nanometer. Their typical dimension varied from 1-100 nm. Whether engineered or naturally occurring, have multiple applications across diverse fields owing to their specific properties. These materials often demonstrate enhanced strength conductivity, reactivity, or other advantageous features compared to their larger-scale counterparts. In the realm of nanotechnology, nanomaterials serve as essential building blocks for the creation of novel structures and devices. They play a pivotal role in a variety of applications, including nanomedicine, electronics, energy storage, catalysis, sensors, and more. The distinct characteristics of nanomaterials contribute significantly to innovations and progress in various scientific and industrial domains. Nevertheless, it is crucial to consider the potential environmental and health impacts associated with the utilization and disposal of nanomaterials, as ongoing research in this area continues to advance (Sahoo *et al.*, 2007). Processes such as homogenization and milling produce nanoparticles.

Nano components have unique ability to self-assemble into nano structures like micelles and nanofibers. This event brings about advancements in structural, physical, and chemical properties. Furthermore, these nanoparticles help to extend the shelf life of food products. As a result, nanotechnology has the potential to alter the food processing sectors while also positively impacting the subject of food science (Biswas *et al.*, 2022). In the application in food industry, noteworthy nanomaterials which can be classified in following forms (Table 1 & 2).

Table 1: Classification of nanomaterials

S. No.	Type of Nanomaterial	Size (nm)	Physical Description
1.	Nanoparticles	20–200	Composed of biodegradable polymers for antioxidant release
2.	Liposomes	100–400	Small sized, spherical synthetic vesicles mainly composed of lipid bilayers
3.	Micelles	10–100	Self-assembling particles amphiphilic particles
4.	Nanocapsules	10–1000	Encapsulate nucleic acids, like DNA, microRNA, and siRNA
5.	Nanoconjugates	10–1000	Polymers covalently bonded with medicinal molecules
6.	Dendrimers	3–20	Monodisperse macromolecules that can encapsulate medication

Table 2: Various types of nanomaterials with their applications in food industry

S. No.	Nanomaterials	Applications in food industry	References
1.	Metal based like oxides of zinc, silver, titanium, gold and palladium	Antioxidant and antimicrobial properties. Strengthen mechanical and barrier properties aid in food preservation. Increase heat stability. Metal based nanocomposites used in food products like milk and cheese. Observing and transforming fluctuations in light, humidity, heat, gases, and chemicals into electrical signals; identifying deviations in food coloration; detecting toxins like aflatoxin B1 in milk; and recognizing gases generated by spoilage.	(Meetoo, 2011; Mizielińska *et al.*, 2018; Ramachandraiah *et al.*, 2017; Roilo *et al.*, 2018; Thirumurugan *et al.*, 2013)
2.	Nano polymer & nano clay nanoparticles	Enhanced gas barrier properties are aimed at preserving food freshness and controlling both pathogenic and spoilage microorganisms. This is achieved through heightened antimicrobial activity, increased levels of total volatiles, bolstered antioxidant capabilities, and the presence of organic acids in food packaging materials, all contributing to extending the shelf life of products.	(López-Rubio *et al.*, 2019; Yotova *et al.*, 2013)
3.	Natural polymers like chitosan and cellulose based polysaccharides	Nano capsules containing epicatechin gallate are integrated into the packaging, aiming to prolong shelf life and inhibit pathogen growth. The deposited substances in packaging have shown effectiveness. This advanced composition for food packaging boasts distinctive characteristics, including oxygen scavenging capabilities.	(Liang *et al.*, 2017; Pradhan *et al.*, 2015; Thirumurugan *et al.*, 2013)
4.	Plasmon-coupled emission biosensors on the surface (with Au)	Intelligent food packaging (extending shelf life and control as well as identify pathogens)	(Nile *et al.*, 2020)
5.	Nanomaterials as sensors like Polyaniline incorporated with carbon black and iStrip	Intelligent food packaging (extending shelf life and control as well as identify pathogens).	(Li *et al.*, 2005)
6.	Nanocomposites and nanolaminates	Application of coatings on bakery goods, cheese, vegetables, meats, and fruits with inherent antimicrobial properties. Active packaging to eradicate typical insect infestations in packaged food items at retail stores and shops.	(Miranda *et al.*, 2022)

Table 3. Nano-techniques with their mechanisms and types of nanomaterials used

S. No.	Nano techniques	Mechanism	Types of nano-materials	Applied areas of utilization	References
1.	Nano-emulsion	Enhancing nutrient composition of food products, stabilizing properties, aid in improving the texture and uniformity of ice creams and reduces creaming.	Vegetable oil, sucrose, carbohydrates and proteins like alginate, pectin, carrageenan and ester gum	Homogenization, emulsification, food additives, fortification,	(Biswas *et al.*, 2022; Mohammed *et al.*, 2020)
2.	Nano-encapsulation	Enhanced stability, and preservation of volatile flavor and odor; control growth of microbes.	Nano capsules, nanoliposomes, nano-cuticles,	Encapsulation, Food additives Food preservation	(Boonlao *et al.*, 2022; Mohammad *et al.*, 2022; J *et al.*, 2022; Thakur *et al.*, 2021)

3. Food Safety Issues with Nano Food Additives

Even with the progress in food preservation technology, regulatory frameworks, global apprehensions about food safety endure. The advent of nanotechnology in food additives presents fresh hurdles in guaranteeing the safety of food items. While nanotechnology offers promising solutions for microbial and toxin detection, extending shelf life, and enhancing food packaging, there are notable food safety issues associated with its use. One significant concern is the potential for nanoparticles to migrate from food packaging materials into the food products themselves. This migration could lead to unintended health consequences if ingested by consumers. Additionally, the antimicrobial properties of nanoparticles, while beneficial for food preservation, raise questions about their long-term effects on human health and the environment. There is a need for thorough risk assessments to evaluate the safety of nanotechnology-based food additives and their potential impact on consumers and the environment (Dudefoi *et al.*, 2018).

Furthermore, the rapid and accurate detection methods enabled by nanotechnology may pose challenges in terms of ensuring the reliability and consistency of results. Issues such as false positives or negatives could undermine confidence in the safety of food products and compromise public health. Despite these challenges, the integration of nanotechnology into food safety practices offers significant potential benefits. By providing faster, more accurate, and cost-effective detection methods, nanotechnology has the potential to enhance the safety and quality of food products. However, it is essential to address the associated food safety issues through rigorous testing, regulation, and transparency in order to ensure the safe and responsible use of nanotechnology in food additives (Biswas *et al.*, 2022; Shabnam *et al.*, 2020).

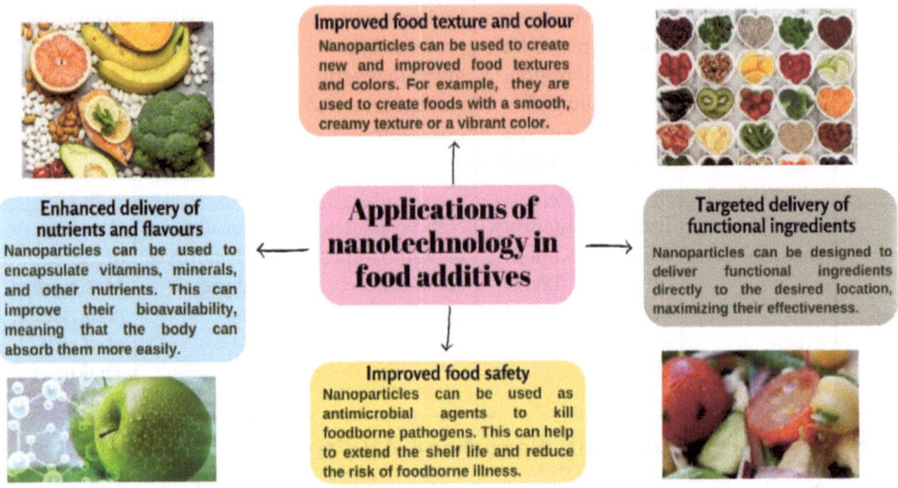

Fig. 1. Application of nanotechnology in food additives

4. Nanomaterials in Food Additives

Nanotechnology has emerged as a groundbreaking field with vast applications across diverse industries, notably in the realm of food processing. Throughout history, humans have tirelessly pursued methods to enhance food preservation and processing, and with the advent of nanotechnology, this quest has achieved unprecedented progress. By leveraging nanotechnology, the food industry has witnessed a paradigm shift, offering a myriad of benefits such as improved food quality, extended shelf life, heightened nutritional value, and efficient delivery of nutrients.

In the context of food processing, nanotechnology has revolutionized practices by significantly prolonging the shelf life of various food items and mitigating wastage attributed to microbial contamination. Moreover, it facilitates the development of "smart" additives and packaging solutions that can react to specific conditions within food, ensuring optimal freshness and safety for consumers. This technological advancement also addresses environmental concerns associated with traditional additives, paving the way for eco-friendly and sustainable alternatives. These innovations play a pivotal role in advancing food safety, quality, and sustainability amid global challenges like population growth, food security, and evolving consumer preferences. Nanotechnology in food additives offers numerous benefits such as improved food quality, increased shelf life, enhanced nutritional value, and efficient delivery of nutrients (Mishra *et al.*, 2018)

Furthermore, nano-delivery systems have emerged as effective carriers for essential food supplements, contributing to the creation of healthier food options. Additionally, nanotechnology offers promise in optimizing agricultural methods and pest control through the utilization of nanoagrochemicals and nanoscale materials. However, the widespread adoption of nanotechnology in food processing necessitates careful consideration of potential challenges and risks. Misuse of nanopesticides, for instance, could lead to pollution and harm to non-target organisms. To address such concerns, stringent regulations and precautionary measures must be implemented to ensure the safe and responsible application of nanotechnology in food processing.

Nanoadditives are revolutionizing the food industry, offering innovative solutions for enhancing the dispersibility of water-insoluble additives like colors, flavors, preservatives, and supplements without the need for surfactants. These nanoadditives boast a large surface area, facilitating improved flavor, nutritional value, and absorption of components into the body. Presently, a range of nanoadditives are available, including vitamins, antioxidants, colors, flavors, and preservatives. Moreover, nanosized materials such as selenium, iron, silver, magnesium, calcium, and silica are gaining traction as nanoadditives in food production (Thangavel & Thiruvengadam, 2014).

One notable example is nanosalt, which not only enhances flavor but also requires lesser quantities compared to normal salt. Nanoadditives like nanopolylysine are employed in oil to safeguard it from oxidation by infiltrating the gaps between phytoglycogen octenyl succinate nanoparticles. In response to increasing demand in the health and sports food markets, some countries are incorporating nanoingredients into food products (Srilekha *et al.*, 2022). Nanominerals are added to food for enhanced and quicker absorption in the stomach compared to larger minerals (Neethirajan & Jayas, 2011) Nutraceuticals, such as bioactive proteins, are also being integrated into food to elevate its nutritional value. Nano-based bioactive molecules are instrumental in boosting absorption in the gastrointestinal tract. Consequently, food and dairy products infused with nanoparticles, encompassing carotenoids, minerals, omega-3 fatty acids, probiotic bacteria species, lycopene, and vitamins, are gaining popularity among consumers. For instance, butter enriched with nanosized functional herbs like polysaccharides, pectins, inulin, carrots, and cryo powder of red beets has emerged as an innovative product in the market (Afroz *et al.*, 2012; Dhinesh kumar *et al.*, 2015)

The utilization of nano additives in food production heralds a new era of possibilities, offering solutions to long standing challenges in the industry. By leveraging nanotechnology, food manufacturers can improve the quality, safety, and nutritional content of their products while meeting evolving consumer demands. Nano additives represent a paradigm shift in food processing, enabling precise control over ingredient dispersibility and enhancing the overall consumer experience. However, as with any emerging technology, the integration of nano additives into food products necessitates careful consideration of safety, regulatory compliance, and ethical implications. One of the key advantages of nano additives is their ability to enhance the dispersibility of water-insoluble additives in food products. Traditionally, the dispersibility of such additives has been a challenge, often requiring the use of surfactants to achieve satisfactory results. Nano additives circumvent this need by virtue of their extremely small size and large surface area, which allows them to interact more effectively with other ingredients in the food matrix. In addition to improving the quality and nutritional content of food products, nanoadditives also offer potential benefits in food preservation and safety. For example, nanoscale silver particles have been shown to exhibit antimicrobial properties, which could help extend the shelf life of perishable foods and reduce the risk of foodborne illness (Ivanov & Rashevskaya, 2011).

The potential benefits of incorporating nano additives into food products are undeniable, as they offer improvements in flavor, nutritional value, and shelf life. However, along with these advantages come concerns about safety, regulatory oversight, and consumer acceptance. The use of nanoparticles in food raises important questions about their potential health risks upon ingestion and their environmental impact. Regulatory agencies are actively working to establish guidelines for the safe use of nanoadditives, but more research is needed to fully understand their long-term effects (Shabnam *et al.*, 2020).

Despite these challenges, nano additives represent a promising new frontier in food technology, offering innovative solutions to long standing issues in the food industry. As research and development in nanotechnology continue to progress, it is essential to prioritize safety and regulatory considerations. Education about the potential risks and benefits of nano additives is also crucial in garnering consumer acceptance. Ultimately, the successful integration of nanotechnology into food processing has the potential to revolutionize the industry, providing consumers with fresh, authentic, convenient, and flavorful food products while ensuring their safety and well-being (Samal, 2017).

5. Commercially Available Nanotechnology-based Products

Undoubtedly, nanotechnology has emerged as a solution to key challenges within the dairy and food sectors. Its integration holds the promise of a complete transformation in food production by elevating both safety standards and nutritional content. Nanotechnology's impact on prolonging shelf life and modifying food properties has opened up new horizons for processing industries (Mustafa & Andreescu, 2020). Leveraging nanotechnology in food processing offers numerous benefits, including enhanced biosensing capabilities, improved detection of foodborne pathogens and toxins, revolutionary advancements in food packaging, efficient delivery systems for bioactive compounds, and fortified functional ingredients. The primary focus of nanotechnology in the food domain revolves around crafting nanostructured food components and delivery mechanisms for nutrients and supplements (Chausali *et al.*, 2022). Challenges arise when integrating certain bioactive compounds into commercial products like functional foods and beverages, necessitating their encapsulation into nanoparticles to enhance bioavailability. Moreover, the utilization of micelles and liposomes, alongside nano-emulsions, emerges as a preferred choice due to their stability during processing. Within the dairy and food sectors, nanotechnology serves as a cornerstone for developing innovative food packaging materials that regulate food oxidation and prevent the emergence of undesirable flavors and textures (Mei & Wang, 2020).

Table 4: Some of the nanotechnology based commercial applications

S. No.	Institution	Nanotechnology based products	Applications
1.	Clemson University, USA	Nanoparticles	With help of nanoparticles contamination by *Campylobacter jejuni* was removed.
2.	USDA, USA	Precision farming	Real-time monitoring of crop growth and soil fertility.
3.	BASF, Germany	Nanocides	Nanoencapsulation of pesticides.
4.	Cornell University, USA	Nanofibers	Cotton waste was utilized for improving fabric strength.
5.	Nestle, USA	Nano-biosensors	Quality and safety monitoring of packaged food.
6.	NanoVic, Dingley, Australia	Nano-vaccines	Veterinary medicines produced by nanotechnology.

6. Technical Challenges and Future Perspectives

Nanotechnology offers vast potential for innovation within the food industry yet it faces numerous challenges that must be addressed before widespread implementation. One significant obstacle is the creation of edible delivery systems using economical processing techniques and formulations that are deemed safe for human consumption. The paramount concern revolves around maintaining the integrity of food products, particularly regarding the potential migration and release of nanoparticles from packaging materials into consumables. The unique behavior of materials at the nanoscale presents distinctive challenges, and our current technical proficiency in their analysis remains limited (Dupas *et al.*, 2007; Hannon *et al.*, 2015; Ishan Das Rastogi, 2012).

To advance practical applications and establish robust safety regulations, it is imperative to comprehensively comprehend the nanoscale functionalities and toxicities of nanomaterials. Chemical methods employed in nanoparticle synthesis not only pose potential adverse effects but also yield environmentally unfriendly by-products contributing to severe pollution concerns (Bajpai *et al.*, 2018; Dupas *et al.*, 2007). Hence, alongside considerations of popularity, technical advantages, and public demand, meticulous risk assessment programs, stringent regulatory frameworks, biosafety evaluations, and public health considerations must be integrated into the processing of nanotechnology-based food and dairy products.

To advance the practical applications of nanotechnology in food processing and establish robust safety regulations, it is imperative to comprehensively understand the functionalities and toxicities of nanomaterials at the nanoscale. Adverse effects of chemical methods used in nanoparticle synthesis, which can lead to environmentally unfriendly by-products and contribute to pollution. Therefore, Bajpai *et al.*, (2018) emphasize the need for comprehensive risk assessment programs, strict regulatory policies, and considerations of biosafety and public health during the processing of nanotechnology-based food and dairy products. However, further extensive research across multiple domains is needed to address safety concerns and challenges comprehensively. The challenges and considerations surrounding nanotechnology in the food industry underscore the need for a balanced approach that prioritizes both innovation and safety. As nanotechnology continues to evolve, it is essential to integrate rigorous risk assessment processes, regulatory frameworks, and biosafety evaluations into food processing practices. Only through a comprehensive understanding of nanomaterials and their impacts can the full potential of nanotechnology be realized while ensuring the safety and wholesomeness of food products (Ishan Das Rastogi, 2012; Nile *et al.*, 2020).

Addressing the migration and absorption of nanoparticles from food packaging materials into food products is critical for ensuring food safety in the industry. Nanomaterials exhibit complex behaviors at the nanoscale, making it challenging to fully understand their characteristics and toxicity. Key factors such as nanoparticle size, shape, surface chemistry, and interactions with food components influence migration, complicating the prediction of release mechanisms. Furthermore, gaps in knowledge regarding release mechanisms and the complexities of gastrointestinal absorption add layers of complexity to risk assessment. To overcome these challenges, further research is needed to understand nanoparticle behavior in food packaging materials and their interactions with food components. Advanced analytical techniques such as electron microscopy and spectroscopy can provide valuable insights, while systematic studies on nanoparticle fate in the gastrointestinal tract are crucial for assessing health risks. Regulatory agencies must establish robust frameworks for evaluating nanoparticle safety and conducting toxicity studies. Industry stakeholders also have a role in prioritizing consumer safety through risk assessments, quality control measures, and transparent communication. Addressing nanoparticle migration in food packaging requires a multidisciplinary approach involving scientific research, regulatory oversight, and industry collaboration. By gaining a deeper understanding of nanoparticle behavior and implementing appropriate safety measures, we can ensure the safe and responsible use of nanotechnology in food packaging, ultimately safeguarding consumer health (Jokar *et al.*, 2017; Sahoo *et al.*, 2007). The escalating presence of nanoparticles in food has sparked increasing apprehension, primarily due to their ability to infiltrate cells and organs, thereby triggering oxidative stress, protein denaturation, and DNA damage. Presently, a lack of standardized safety assessment methods for nanoparticles underscores the urgent need to devise regulations that align with the toxicity testing objectives of the modern era (McClements & Xiao, 2017).

7. Conclusion

In conclusion, nanotechnology has great promise for transforming the food industry by providing transformative solutions that improve food quality, safety, and utility. The study of nanotechnology in food additives demonstrates its significant impact on food development and consumer pleasure. Researchers have used nanoparticles' unique properties to develop new ways for solving a wide range of food production and preservation concerns. Nanotechnology-enabled food additives offer a variety of benefits, including improved texture and flavor, extended shelf life, and microbiological integrity. Furthermore, nanoparticles perform as excellent transporters for bioactive substances, allowing for precise and regulated release into the food matrix.

However, alongside these promising advancements, concerns persist regarding the safety and regulatory implications of nanotechnology in food additives. This abstract has critically examined the current landscape, shedding light on significant achievements, dilemmas, and future pathways in this domain. Leveraging collaborative synergy among stakeholders is crucial in realizing the full potential of nanotechnology to enhance food processing practices.

Over the past two decades, driven by advancements in nanoscience and nanotechnology, a multitude of sensitive detection techniques for food additives have emerged, capitalizing on nanomaterials' unique properties. These techniques underscore the versatility and utility of nanotechnology in addressing challenges across various domains, including food safety and quality assurance.

When integrating nanotechnology into the food industry, it is critical to address issues such as nanoparticle migration, environmental implications, and rigorous safety studies. By adopting joint efforts, rigorous research, and tight regulatory frameworks, the benefits of nanotechnology can be maximized while protecting both consumers and the environment. Finally, through responsible and sustainable adoption, nanotechnology has the ability to usher in a new era of safe, innovative, and ecologically sensitive food items that satisfy the changing requirements of global consumers.

References

Afroz, M., Karthikeyan, P., Ahmed, P., & Kumar, U. (2012). Application of nanotechnology in food and dairy processing: An overview. Pakistan Journal of Food Sciences, 22(1s).

Bajpai, V. K., Kamle, M., Shukla, S., Mahato, D. K., Chandra, P., Hwang, S. K., Kumar, P., Huh, Y. S., & Han, Y. K. (2018). Prospects of using nanotechnology for food preservation, safety, and security. Journal of Food and Drug Analysis, 26(4), 101–108. https://doi.org/10.1016/j.jfda.2018.06.011

Biswas, R., Alam, M., Sarkar, A., Haque, M. I., Hasan, M. M., & Hoque, M. (2022). Application of nanotechnology in food: processing, preservation, packaging and safety assessment. Heliyon, 8(11), 413–424. https://doi.org/10.1016/j.heliyon.2022.e11795

Boonlao, N., Ruktanonchai, U. R., & Anal, A. K. (2022). Enhancing bioaccessibility and bioavailability of carotenoids using emulsion-based delivery systems. Colloids and Surfaces B: Biointerfaces, 209, 112211. https://doi.org/10.1016/J.COLSURFB.2021.112211

Chausali, N., Saxena, J., & Prasad, R. (2022). Recent trends in nanotechnology applications of bio-based packaging. Journal of Agriculture and Food Research, 7. https://doi.org/10.1016/j.jafr.2021.100257

Dhineshkumar, V., Ramasamy, D., & Sudha, K. (2015). Nanotechnology application in food and dairy processing. International Journal of Farm Sciences, 5(3).

Dudefoi, W., Villares, A., Peyron, S., Moreau, C., Ropers, M. H., Gontard, N., & Cathala, B. (2018). Nanoscience and nanotechnologies for biobased materials, packaging and food applications: New opportunities and concerns. In Innovative Food Science and Emerging Technologies (Vol. 46, pp. 107–121). https://doi.org/10.1016/j.ifset.2017.09.007

Dupas, C., Houdy, P., & Lahmani, M. (2007). Nanoscience: Nanotechnologies and nanophysics. In Nanoscience: Nanotechnologies and Nanophysics. https://doi.org/10.1007/978-3-540-28617-2

Francisco, E. V. De, & García-estepa, R. M. (2018). Nanotechnology in the agrofood industry. Journal of Food Engineering, 238(November 2017), 1–11. https://doi.org/10.1016/j.jfoodeng.2018.05.024

Hannon, J. C., Kerry, J. P., Cruz-Romero, M., Azlin-Hasim, S., Morris, M., & Cummins, E. (2015). Assessment of the migration potential of nanosilver from nanoparticle-coated low-density polyethylene food packaging into food simulants. Food Additives and Contaminants - Part A Chemistry, Analysis, Control, Exposure and Risk Assessment, 33(1), 167–178. https://doi.org/10.1080/19440049.2015.1114184

Ishan Das Rastogi. (2012). Nanotechnology: Safety paradigms. Journal of Toxicology and Environmental Health Sciences, 4(1), 1–12. https://doi.org/10.5897/jtehs11.071

Ivanov, S. V., & Rashevskaya, T. A. (2011). The nanostructure's management is the basis for a functional fatty foods' production. Procedia Food Science, 1. https://doi.org/10.1016/j.profoo.2011.09.005

Jokar, M., Pedersen, G. A., & Loeschner, K. (2017). Six open questions about the migration of engineered nano-objects from polymer-based food-contact materials: a review. Food Additives and Contaminants - Part A Chemistry, Analysis, Control, Exposure and Risk Assessment, 34(3), 434–450. https://doi.org/10.1080/19440049.2016.1271462

Joshi, M., & Meena, K. K. (2023). Online ISSN : 2583-0376 Comprehending fundamental concepts and applications of nanotechnology in food processing : Current overview and future perspectives. 3(4), 15–23.

Li, Y., Cu, Y. T. H., & Luo, D. (2005). Multiplexed detection of pathogen DNA with DNA-based fluorescence nanobarcodes. Nature Biotechnology, 23(7), 885–889. https://doi.org/10.1038/nbt1106

Liang, J., Yan, H., Zhang, J., Dai, W., Gao, X., Zhou, Y., Wan, X., & Puligundla, P. (2017). Preparation and characterization of antioxidant edible chitosan films incorporated with epigallocatechin gallate nanocapsules. Carbohydrate Polymers, 171(1), 300–306. https://doi.org/10.1016/j.carbpol.2017.04.081

López-Rubio, A., Fabra, M. J., & Martínez-Sanz, M. (2019). Food packaging based on nanomaterials. Nanomaterials, 9(9), 3–5. https://doi.org/10.3390/nano9091224

McClements, D. J., & Xiao, H. (2017). Is nano safe in foods? Establishing the factors impacting the gastrointestinal fate and toxicity of organic and inorganic food-grade nanoparticles. Npj Science of Food, 1(1), 1–13. https://doi.org/10.1038/S41538-017-0005-1

Meetoo, D. D. (2011). Nanotechnology and the food sector: From the farm to the table. In Emirates Journal of Food and Agriculture (Vol. 23, Issue 5, pp. 1–13).

Mei, L., & Wang, Q. (2020). Advances in Using Nanotechnology Structuring Approaches for Improving Food Packaging. In Annual Review of Food Science and Technology (Vol. 11). https://doi.org/10.1146/annurev-food-032519-051804

Miranda, M., Sun, X., Marín, A., dos Santos, L. C., Plotto, A., Bai, J., Benedito Garrido Assis, O., David Ferreira, M., & Baldwin, E. (2022). Nano- and micro-sized carnauba wax emulsions-based coatings incorporated with ginger essential oil and hydroxypropyl methylcellulose on papaya: Preservation of quality and delay of post-harvest fruit decay. Food Chemistry: X, 13(1), 1–13. https://doi.org/10.1016/j.fochx.2022.100249

Mishra, P., Singh, A., Mishra, R. R., Ong, S. Y., & Prasad, S. M. (2018). Application of Nanotechnology to Enhance the Nutrient Quality of Food Crops and Agricultural Production. Nanomaterials in Plants, Algae, and Microorganisms, 1, 453–472. https://doi.org/10.1016/B978-0-12-811487-2.00020-7

Mizielińska, M., Kowalska, U., Jarosz, M., & Sumińska, P. (2018). A comparison of the effects of packaging containing nano ZnO or polylysine on the microbial purity and texture of cod (Gadus morhua) fillets. Nanomaterials, 8(3), 1–13. https://doi.org/10.3390/nano8030158

Mohammad, Z. H., Ahmad, F., Ibrahim, S. A., & Zaidi, S. (2022). Application of nanotechnology in different aspects of the food industry. Discover Food, 2(1), 1–21. https://doi.org/10.1007/s44187-022-00013-9

Mohammed, N. K., Muhialdin, B. J., & Meor Hussin, A. S. (2020). Characterization of nanoemulsion of Nigella sativa oil and its application in ice cream. Food Science and Nutrition, 8(6), 2608–2618. https://doi.org/10.1002/fsn3.1500

Mustafa, F., & Andreescu, S. (2020). Nanotechnology-based approaches for food sensing and packaging applications. RSC Advances, 10(33), 19309–19336. https://doi.org/10.1039/d0ra01084g

Neethirajan, S., & Jayas, D. S. (2011). Nanotechnology for the Food and Bioprocessing Industries. In Food and Bioprocess Technology (Vol. 4, Issue 1). https://doi.org/10.1007/s11947-010-0328-2

Nile, S. H., Baskar, V., Selvaraj, D., Nile, A., Xiao, J., & Kai, G. (2020). Nanotechnologies in Food Science: Applications, Recent Trends, and Future Perspectives. In Nano-Micro Letters (Vol. 12, Issue 1). Springer Singapore. https://doi.org/10.1007/s40820-020-0383-9

Pradhan, N., Singh, S., Ojha, N., Shrivastava, A., Barla, A., Rai, V., & Bose, S. (2015). Facets of nanotechnology as seen in food processing, packaging, and preservation industry. BioMed Research International, 2015(2), 1–17. https://doi.org/10.1155/2015/365672

Rajamalar, C. G., Chandrika, M., & Chellaram, D. C. (2011). Chemical Synthesis and Structural Elucidation of Novel Compounds - Schiff Bases. Biometrics and Bioinformatics, 3(10), 468–472.

Ramachandraiah, K., Gnoc, N. T. B., & Chin, K. B. (2017). Biosynthesis of Silver Nanoparticles from Persimmon Byproducts and Incorporation in Biodegradable Sodium Alginate Thin Film. Journal of Food Science, 82(10), 2329–2336. https://doi.org/10.1111/1750-3841.13865

Roilo, D., Maestri, C. A., Scarpa, M., Bettotti, P., & Checchetto, R. (2018). Gas barrier and optical properties of cellulose nanofiber coatings with dispersed TiO2 nanoparticles. Surface and Coatings Technology, 343(1), 21844–21853. https://doi.org/10.1016/j.surfcoat.2017.10.015

Sahoo, M., Vishwakarma, S., Panigrahi, C., & Kumar, J. (2007). Nanotechnology : Current applications and future scope in food. In Nanoscience: Nanotechnologies and Nanophysics (pp. 3–22). https://doi.org/10.1002/fft2.58

Samal, D. (2017). Use of Nanotechnology in Food Industry: A review. International Journal of Environment, Agriculture and Biotechnology, 2(4), 2270–2278. https://doi.org/10.22161/ijeab/2.4.90

Shabnam, Perwez, M., Mazumder, J. A., & Sardar, M. (2020). Phytonanotechnology: A new horizon for the food industry. In Phytonanotechnology: Challenges and Prospects. Elsevier Inc. https://doi.org/10.1016/B978-0-12-822348-2.00012-7

Srilekha, G. K. P., Tiwari, H., Mketo, N., & Lakkakula, J. (2022). Application of nanomaterials in the dairy industry. In Advances in Dairy Microbial Products. https://doi.org/10.1016/B978-0-323-85793-2.00018-7

Thangavel, G., & Thiruvengadam, S. (2014). Nanotechnology in food industry – a review. International Journal of ChemTech Research, 6(9), 4096–4101.

Thirumurugan, A., Ramachandran, S., & Shiamala Gowri, A. (2013). Combined effect of bacteriocin with gold nanoparticles against food spoiling bacteria - an approach for food packaging material preparation. International Food Research Journal, 20(4), 50–54.

Yotova, L., Yaneva, S., & Marinkova, D. (2013). Biomimetic nanosensors for determination of toxic compounds in food and agricultural products (review). Journal of the University of Chemical Technology and Metallurgy, 48(3), 215–227.

12

Microbial Mediated Nanoparticles to Clean Environmental Contaminants: A Green Nano-bioremediation Approach

Jyoti Gulia[1], Amit Lath[2], Nater Pal Singh[2*] and Anita Rani Santal[1*]

[1]*Department of Microbiology, Maharshi Dayanand University, Rohtak-124001, Haryana*
[2]*Centre for Biotechnology, Maharshi Dayanand University, Rohtak-124001, Haryana*

Abstract

The increase in population and global industrialization has led to environmental pollution, which is a significant threat to the health of the ecological system. Different measures are recommended to remediate polluted areas effectively. The applications of nanotechnology in medicine, cosmetics, sports, textiles, and pharmaceuticals have increased over the years. Another approach to remediate the adverse effects of pollutants on the environment is by utilizing nanotechnology for contaminant remediation. Biogenic nanoparticles are tiny particles produced by biological entities such as fungi, bacteria, yeast, and algae. These biogenic nanoparticles are efficient in removing pollutants like heavy metals, pharmaceutical waste, dyes, and hydrocarbon waste. The interactions between biofunctionalized nanoparticles and contaminants can aid in understanding the structural relationships and significantly contribute to environmental cleanup. This chapter aims to provide a thorough analysis of recent developments, reported efforts, and approaches to use nanoparticles for effective nano-bioremediation.

Keywords: *Nano-bioremediation, contaminants, nanoparticles, micro-organisms, environment.*

1. Introduction

The modern environment is highly contaminated with a wide range of dangerous pollutants, including heavy metals, dyes, and waste products from various industries that contain synthetic chemicals resulting from human activities (Santal *et al.*, 2016 and 2022). Increased industrial development and

* *Corresponding authors e-mail: anita.micro@mdurohtak.ac.in; npsinghcbt@mdurohtak.ac.in*

agricultural growth have led to the release of many chemicals into the land, water, and air. The primary reasons for environmental contamination are the reckless use of pesticides, veterinary drugs, and hazardous substances in various consumer goods, as well as the discharge of industrial waste into water bodies. Unidentified systemic pollutants can't be broken down naturally and can cause harm to living organisms when they enter the food chain. These pollutants can accumulate in higher species via biomagnification and pose a significant risk to life (Huang *et al.*, 2021). These hazardous substances can endure for extended periods in environmental matrices due to their elemental composition. Effluents containing polyethylenes, petroleum, and trace metals are detrimental to the environment. The ecology is negatively affected by such effluents. Pollutants known as heavy metals are found in the Earth's crust. They are found in rocks as ores and are extracted as minerals. Heavy metals can be released into the environment by high exposure levels (Ali *et al.*, 2019). Numerous of these contaminants are carcinogenic to people and the environment in which they are found. Heavy metal absorption causes accumulation in the kidney, liver, and brain. Animals may also experience development retardation, cancer, neurological system damage, or even death (Priydarshanee *et al.*, 2022). Because heavy metals impede plant development, nutrient absorption, and physiological metabolic processes, they lower the quality and quantity of food that is produced. Chemical, biological, and physical methods are being used to clean up soil contaminated with heavy metals. Microbial enzymes must be used in a bioremediation process to convert hydrocarbons into less hazardous substances (Bala *et al.*, 2022). Ongoing studies are exploring the potential of genetically modified microorganisms to eliminate xenobiotic compounds like naphthalene, petroleum, toluene, and benzene (Hussain *et al.*, 2018). Several factors impact the efficiency of bioremediation, such as the temperature of the environment, whether or not there is oxygen present or anaerobic conditions exist, and the accessibility of nutrients. Human activities are the main cause of the introduction of environmental pollutants that can be harmful to humans, animals, and plants. Bioremediation is one of the most cost-effective and environmentally beneficial biotechnological advancements (Alori *et al.*, 2022). Bioremediation is a recovery approach that can be adapted to many types of contaminants. However, it may not be effective in cases where the number of pollutants, xenobiotics, or refractory chemicals is significant. This can result in treatment inefficiencies and longer recovery periods that are not sustainable (Mapelli *et al.*, 2017). Because of this, the issue has drawn attention from people all around the world, requiring in-depth research to create practical mitigation plans for contaminated areas. Within this framework, the advancement of nanotechnology represents a novel approach to advance bioremediation beyond its current constraints. This integrated method has a wider range of potential uses due to the reduced expenses and minimal harm to the environment. It can be applied to treat pollutants in wastewater and groundwater, sediments contaminated with heavy metals and hydrocarbons, and both organic and inorganic compounds in soil (Briffa *et al.*, 2020). This approach combines the advantages of nanotechnology, which provides high surface area and reactivity, with the biological processes mediated by microorganisms to degrade, detoxify, or sequester pollutants.

2. Sources of Contaminants

The main sources of contaminants are the pharmaceutical industry, pesticides, artificial sweeteners, hormones, fertilizer industry, solvents for cleaning, antibacterial, ointments, clothing adhesives, foam for cushions, cosmetics, beauty care, sun protection, personal care products, drinks, and packaged food (Richardson and Kimura, 2017; Guo *et al.*, 2023). The existence of emerging micropollutants such as medications, cosmetics, endocrine disruptors, fluorinated chemical products, and nanoparticles in our surroundings is a significant threat to the health and welfare of both humans and marine creatures (Bala *et al.*, 2022).

2.1 Pharmaceutical sector

Despite their growing concern as contaminants, pharmaceuticals have been detected in trace concentrations in water sources worldwide (Mishra *et al.*, 2023). In nutrition, diagnosis, treatment, and prevention, pharmaceuticals are a class of substances that are commonly used. The pharmaceutical industry is one of the major developing contaminants. Antibiotics, hormones, illegal medications, and compounds that resemble hormones are among the other serious developing pollutants. These pollutants are significant because they contaminate freshwater supplies, are widely found in aquatic habitats, and may have negative consequences on the environment and public health (Hejna *et al.*, 2022; Khan and Barros 2023). The three main ways that pharmaceuticals end up in the atmosphere are through agriculture, human waste, and the disposal of unwanted medications. A vast array of drugs linked to wastewater discharge was discovered in samples of surface and groundwater. Through the use of manure and purines as organic fertilizers in intensive animal husbandry, medications can also indirectly affect the environment. This is because they can enter living things through the food supply chain and be absorbed by them (Fabbri *et al.*, 2023; Gaw *et al.*, 2014). These microorganisms have a poor elimination rate of approximately 10 percent and a complicated chemical composition, which makes it difficult for them to eliminate some types of organic substances. Various types of drugs are most frequently found in drinking water and wastewater (Wydro *et al.*, 2023). First-class drugs such as acetaminophen, enalapril, and ibuprofen were mostly removed during the course of treatment. Although second-class reagents like lipid regulators haven't been eliminated, their concentrations are far lower than those of significant reactants after therapy. Sewage water effluents, as opposed to influential effluents, contained a group of compounds that were not found in influential effluents, such as antibiotics, macrolide, and anxiolytics. In wastewater effluent, last-class medications such as sulfamethazine, paroxetine, and simvastatin were never found in detectable concentrations.

2.2 Personal care products (PCPs)

Products classified as personal care are a grouping of typical chemical items used for cleaning and cosmetics. Cosmetics, skincare, hair care, cleaning supplies, and fragrances are among the personal care items. Because skincare and personal care products are used daily, large amounts of them are constantly emitted into the environment (Hawash *et al.*, 2023). These products are used extensively around the world. The amount of emerging contaminants released into the environment is rising due to individual actions; this is a reflection of the exponential growth in the use of a wide range of commodities, including personal care and prettiness (Mishra *et al.*, 2023; Xiao *et al.*, 2023). Nonylphenol and octylphenol ethoxylates are two types of alkylphenol polyethoxylates (APEOs), which are frequently utilized as surfactants in PCPs, whereas alkylphenol itself is the breakdown product of APEOs. Detergents, disinfectants, surface cleaners, and many PCPs include APEOs (Patrolecco *et al.*, 2023; Ginebreda *et al.*, 2016). A portion of PCPs also include antimicrobials because of their aseptic qualities. The four PCP antibacterial compounds that are most frequently employed are 1,4-dichlorobenzene, triclocarban, orthophenylphenol, and triclosan. Toothpaste, soaps, deodorants, and detergents frequently include these kinds of antimicrobials (Khalid and Abdollahi, 2021). Bisphenols are utilized in the manufacturing of plastics, such as polycarbonates and epoxy resins. These are only present because they have leaked out of plastic containers or are degrading; they are not the intended constituents in PCPs (Abraham and Chakraborty, 2020). The most prevalent Bisphenol A among PCPs was found in various cosmetic products (Khalid and Abdollahi, 2021). Some personal care products contain several new pollutants such as microplastics, polydimethylsiloxane, and UV filters. Other pollutants found in personal care products include butylated hydroxyanisole, insect repellents, fragrance contaminants, disinfectant contaminants (such as triclosan and hydroxytoluene), preservatives (diethyl phthalate), ZnO nanoparticles, octinoxate benzophenone, methoxycinnamate, parabens, and butylparaben. Inexpensive perfumes like polycyclic musks are used in items such as shampoos, washing powder, hair products, laundry detergents, and cosmetics procedures (Abedi *et al.*, 2018). Applying it mostly to the skin's outer layer enhances the likelihood that it will be found in higher concentrations throughout the environment without changing metabolism (Rathi *et al.*, 2021). Fragrances are a blend of volatile organic compounds (VOCs) that react and combine with dust and other pollutants to create hazardous byproducts or air pollutants that may cause irritation or allergies worse than those of the original material. For instance, PCP terpenes may react with indoor ozone to produce formaldehyde and other secondary pollutants (Steinemann, 2016). Fragrances can cause atopic asthma and worsen asthmatic sufferers' symptoms. Highly absorbed aromas cause more dermatitis, hypersensitivity, and skin sensitization. In children, allergic contact dermatitis is also frequently observed. A common scent and possible allergy in PCPs is geraniol (Basketter and Kimber, 2015).

2.3. Pesticides

Pesticides are a broad class of substances with physiochemical properties that are used to control or stop the growth of dangerous weeds, insects, and microorganisms. Herbicides, fungicides, insecticides, and bactericides are the four categories into which generalized pesticides fall. Runoff from the application site carries the pollutants into maritime areas, where they may bioaccumulate in plants and other living things, contingent on the chemical's solubility (Gani and Kazmi, 2017). When newly discovered contaminants are found in the agricultural ecology, they will encounter comparable outcomes and transit mechanisms seen by numerous current pollutant categories. Soil debris can be contaminated by emerging contaminants, which can also be absorbed by roots, dissolved in groundwater, or traveled to surface water. They can also be contaminated chemically, physically, or biologically. Among the most widely used insecticides were endosulfan, hexachlorocyclohexane, and dichlorodiphenyltrichloroethane (DDT). The following items are also available: bentazone, dimethoate, phorate, methyl parathione, diazinon, alachlor, terbuthylazine, simazine, atrazine, phosphamidone, and cyanazine (Kock *et al.*, 2021).

2.4. Dyes

Environmental laws and regulatory organizations are growing more concerned about the destructive nature of toxic substances, such as synthetic dyes made for industrial purposes and hazardous wastewater effluents that contain colors. The best way to prevent pollution is to abide by strict environmental laws and policies (Ardila-Leal *et al.*, 2021). Chemical substitution is an additional workable solution to this problem. To attain comparable functionality, it is characterized as "the substitution or reduction of hazardous substances in products and processes by technological or organizational methods" (Bilal *et al.*, 2018). The manufacturing sector uses a large number of structurally varied synthetic dyes, including metal-complex, acidic, diazo, basic, azo, dispersion, reactive, and anthraquinone. The release of these colors and/or the often-toxic breakdown products into large bodies of water presents a serious risk to both human health and the environment (Pavithra *et al.*, 2019). Various methods have been created and utilized to remove dye-based contaminants from industrial wastewater, such as chemical treatment, ion-pair extraction, biosorption, coagulation, and flocculation. Al-Tohamy *et al.* (2022) have researched and implemented these techniques. However, it has been demonstrated that these treatments are costly and time-consuming, which means that the entire treatment process can raise further concerns about contamination. Because of this, the fight against these environmental pollutants greatly worries the public and depends on the development of ever-more-efficient technology (Manisalidis *et al.*, 2020; Bilal *et al.*, 2018).

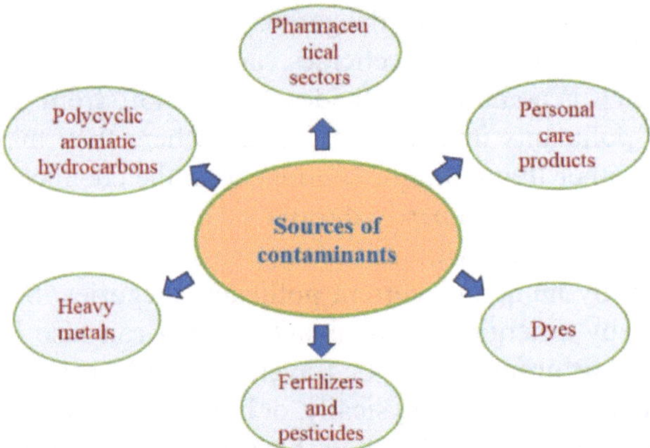

Fig. 1: Sources of contaminants

2.5 Heavy metals

The primary businesses that release heavy metals into the air, groundwater, and soil are mining and manufacturing. Sewage irrigation, industrial waste, sludge farm application, and the release of various exhausted gases due to the acceleration of urbanization and industrialization are the primary causes of heavy metal contamination (Sharma *et al.*, 2023). The health of the soil deteriorates when industrialization grows because it contaminates rivers and releases waste into the sewer system or the areas around cities. As per Dhaliwal *et al.* (2020), heavy metals present in the atmosphere are a result of fuel burning, waste materials generated by industries like dye and textile, waste from electroplating, metal coating, cycles and spare parts, mining and smelting processes, sewage sludge, and the use of chemical fertilizers. According to Bolan *et al.* (2014), the parent rock material, aerosol particles from burning fossil fuels, landfilling, applying organic material, and pollutants in fertilizers and other sources all influence the composition of heavy metals in agricultural soil. Each type of contaminant presents unique challenges and requires specific strategies for detection, monitoring, and remediation to mitigate their environmental and health impacts effectively.

3. Nano-bioremediation

The amalgamation of nanotechnology and bioremediation significantly contributes to the degradation of environmental pollutants and accelerates the pace of progress (Pete *et al.*, 2021). Nanoparticles produced from biological materials such as algae, fungi, bacteria, and plants are utilized to eliminate impurities such as metalloids, poisonous metal ions, and biotic and abiotic contaminants. Two major applications of utilizing eco-friendly nanomaterials as a sensor for monitoring environmental issues and as a means to eliminate impurities from wastewater (Roy *et al.*, 2021). The main reason nanotechnology is being used in bioremediation is because of its ability to produce particles that are small (less than 100 nm), have a significant surface area, and have stable chemical characteristics (Vázquez-Núñez *et al.*, 2020). A nanoparticle's physio-chemical characteristics can be altered by its size, which varies from 1 to 100 nm. There are two types of nanoparticles: inorganic and organic. A nanocrystal is a single crystalline nanoparticle that is also proven to be amorphous. Different structures, such as platelets, cubes, spheres, and tetrads, are displayed by nanoparticles (Gautam *et al.*, 2021). The biological technique is recommended for the synthesis process because of its low cost, toxicity, environmental safety, and quick synthesis. Biological systems can synthesize molecules that possess different physical and chemical characteristics. These characteristics comprise surface chemistry, zeta potential, chemical composition, and photocatalytic properties (Khan *et al.*, 2019). High surface and catalytic activity are displayed by bio-mediated nanoparticles (Castro *et al.*, 2014).

3.1. Synthesis of nanoparticles

The synthesis of nanoparticles was accomplished with the help of physical, chemical, and biological techniques. Without any control, larger particles in a physical process break down into smaller particles. In contrast, chemical techniques including laser pyrolysis, plasma spraying, and sol-gel are used to combine the smaller particles into nanoparticles. In the process of reducing metal ions, biogenic substances release biomolecule proteins, which lead to the formation of nanoparticles and capping through a biological mechanism (Kumari *et al.*, 2023). Living organisms produce bioactive substances such as peptides, polyphenols, and amino acids, which facilitate the processes of bio-precipitation and bio-reduction. These processes are the fundamental mechanisms responsible for the production of bio-mediated nanoparticles (Park *et al.*, 2011). Microorganisms like bacteria, yeast, fungus, and actinomycetes are known to release a potent chemical that can oxidize or reduce metal ions and generate nanoparticles. Nanoparticles were produced via several biochemical processes, including complexolysis, redoxolysis, alkylation, and acidolysis (Zhang *et al.*, 2011; Sathiyanarayanan *et al.*, 2017). The drawback of synthesizing microbe-mediated nanoparticles is their high expense and requirement for an ideal microbial growth environment. One way to synthesize nanoparticles intracellularly is to damage the microbial cell (Singh *et al.*, 2015).

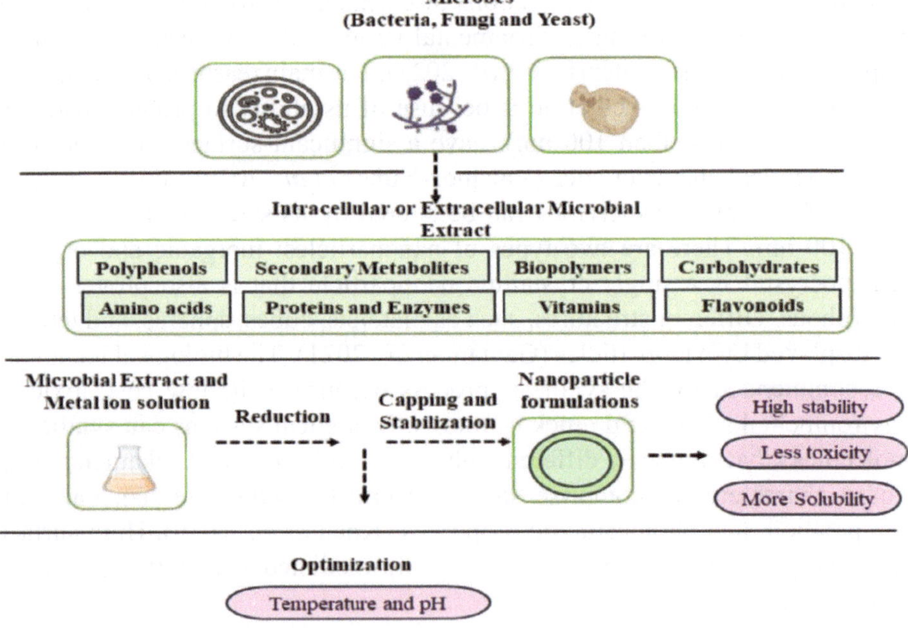

Fig. 2: Synthesis of nanoparticles

3.1.1. Synthesis of microbe-mediated nanoparticles

Microorganisms and nanotechnology work together to develop a great wastewater treatment bioremediation project that removes heavy metal ions. Bacterial and fungal enzymes can act as reductive agents for metal complex salt, resulting in the synthesis of metallic nanoparticles. Enzymes produced by microorganisms, both extracellular and intracellular, aid in the synthesis of metallic nanoparticles. Microbial cell walls contain amide groups and metal cations that work together to actively reduce ions through the synthesis of intracellular enzymes. This leads to the development of metallic nanoparticles (Mahanty *et al.*, 2020; Kapoor *et al.*, 2021). Extracellular enzymes act as reducing agents to facilitate the production of nanoparticles. Various types of enzymes, including glucosidase, acetyl xylan esterase, NADH, and cellobiohydrolase D, are produced by fungal species and aid in the production of metallic nanoparticles (Subbaiya *et al.*, 2017; Ovais *et al.*, 2018).

3.1.1.1. Bacteria-mediated nanoparticles in bioremediation of contaminants

In bioleaching, biomineralization, and bioaccumulation, bacteria play a crucial role. To solubilize the metal ions, the oxidation or reduction process modifies their oxidation state. Bacteria can convert harmful soluble inorganic ions into insoluble metal nanoparticles that are not harmful (Pande *et al.*, 2022). The process of synthesis occurs either intracellularly or extracellularly, depending on certain physiochemical conditions, including the pH, temperature, concentration of metal salts, and bacterial load of the solution. The biomolecules in the bacterial cell wall aid in the reduction of the metal ions during the extracellular process. In contrast, during the intracellular process, the functional groups in the bacterial cell wall draw in metal ions and combine with protein through electrostatic contact (Fang *et al.*, 2019). The bacterial cell wall is capable of capturing metal ions due to the presence of negatively charged groups within its structure. This characteristic enables the bacterial cell wall to entrap the metal ions effectively. The process consists of complexation, oxidation/reduction, precipitation, biosorption to cell walls, and trapping in extracellular capsules. Bacteria such as *Bacillus* sp., *Pseudomonas cepacian*, *Micrococcus luteus*, *Streptomyces coelicolor*, *Bacillus subtilis*, *Burkholderia cepacia* MCMB-821, and *Enterobacter* sp. are efficient in eliminating copper ions and other pollutants from wastewater (Wani *et al.*, 2007; Soare *et al.*, 2019). *Klebsiella oxytoca* GS-4-08 was utilized in fermentative conditions to create palladium nanoparticles (bio-Pd). These nanoparticles were subsequently employed to decrease the presence of an azo dye. The results of this experiment may be beneficial in developing and operating bioreactors to remove azo dye from wastewater (Wang *et al.*, 2017). Methylene blue is a harmful organic compound that can cause significant harm to aquatic life. Its toxicity makes it difficult to remove through biological means. Two kinds of bio-nanoparticles, namely BioMnOx and biopalladium (Bio-Pd), are effective in removing various micro-pollutants from sewage treatment plant effluent. These nanoparticles were derived from bacteria called *Pseudomonas putida* MnB6 and *Shewanella oneidensis*, respectively. In laboratory tests, the nanoparticles were used in membrane bioreactors (MBR) to eliminate pollutants through oxidation or reduction techniques. The BioMnOx-MBR method was able to significantly reduce the levels of various drugs up to the level of 50%. The use of Bio-Pd as a nano-catalyst is effective in eliminating highly persistent iodinated contrast media (ICM) and dehalogenated compounds through innovative reductive techniques, as demonstrated by Forrez *et al.* (2011). Wang *et al.* (2017) have reported that biogenic selenium nanoparticles synthesized from *Citrobacter freundii* Y9 can be used to bioremediate soil contaminated with Hg. The bacterium was isolated from an anaerobic sulfate-reducing bioreactor.

Various bacteria involved in the synthesis of nanoparticles have been mentioned in Table 1. The formation of silver and gold nanoparticles was also actively facilitated by bacterial species such as *P. aeruginosa*, *B. megaterium*, *E. coli*, *B. subtilis*, *Staphylococcus aureus*, *Lactobacillus* sp., and *Enterobacter* sp. (Sinha and Khare, 2012; Ameen *et al.*, 2020). Nanoparticles synthesized from iron and copper oxide were effective in bioremediation of arsenic from the environment. Copper-resistant *Escherichia* sp. SINT7 was used to synthesize copper nanoparticles. It was demonstrated that the biogenic nanoparticles broke down both textile effluent dyes at a lower concentration of 25 mg/L. At a concentration of 100 mg/L, these reductions were reduced to 76.84, 83.90, 62.32, and 31.08%, respectively (Norman *et al.*, 2020). Iron-sulfur nanoparticles were synthesized by Cheng *et al.* (2019) without the use of extra sulfur. These nanoparticles demonstrated the ability to degrade Naphthol Green B dye via the transfer of electrons extracellularly. The utilization of *Pseudoalteromonas* sp. CF10-13 in the synthesis process offers an environmentally safe method for biodegradation. The in-situ production of nanoparticles led to a reduction in the generation of hazardous gases and metal complexes. Utilizing biogenic particles presents a more advanced technology to employ in the purification of industrial effluents. In addition to generating nanoparticles directly from microorganisms, there are numerous other methods in which microorganisms can contribute to enhancing nanotechnology. For example, microorganisms can supply catalytic enzymes that, combined with nanoparticles, assist in the purification of effluents.

3.1.1.2. Fungi-mediated nanoparticles in bioremediation of contaminants

Fungi are abundant in nature and have the highest degree of metal tolerance as well as the ability to concentrate metal ions within them. Because of its high surface area, which aids in the effective synthesis of the nanoparticles, the fungal mycelium secretes proteins. Different fungal species have unique growth and nanoparticle manufacturing mechanisms (Fouda *et al.,* 2022). By secreting different enzymes, polypeptides, proteins, and other metabolic interactions, synthesis, stabilization, and immobilization were accomplished. This also maintained and preserved their propensities as nanocatalysts and nano-adsorbents (Yadav *et al.*, 2015). Like other microbes, fungi can generate nanoparticles both inside and outside of their cells. Numerous little nanoparticles that are created elsewhere in the environment are formed by the fungi (Table 1). Particle nucleation particular to a species is linked to the size constraint. Additionally, intracellular synthesis has many important benefits. Removing diminished metals from the ecosystem, such as copper and platinum, is necessary to address the effects of ecological degradation. Because they can eliminate the contaminants from the sample, fungi that generate intracellularly would be ideal to use. Because they are made outside of cells, nanoparticles often have a wide range of useful applications (Boroumand Moghaddam *et al.*, 2015). Fungi are commonly considered extracellular species because of their capacity to trap nanoparticles through a range of released compounds. There is increased extracellular synthesis. Extracellular gold nanoparticles were created by employing *Fusarium oxysporum.* It preserves the produced gold nanoparticles' extended fastness because of its protein binding capability property (Adebayo *et al.*, 2021). Heavy metal ions are also eliminated by *Cladosporium resinae*, *Aureobasidium pullulans*, *Trametes versicolor*, *Aspergillus niger*, *Rhizopus arrhizus*, *Funaliatrogii*, *Ganoderma lucidum*, and *Penicillium* species. Iron oxide nanoparticles mediated by *Aspergillus tubigensis* (STSP. 25) can remove 90% of hazardous metals from wastewater, including As, Cu, Ni, Pb, and Zn (Mahanty *et al.*, 2020). *Cladosporium cladosporioides* can synthesize silver nanoparticles that range in width from 30 to 60 nm. AgNPs showed a spherical morphology and constant size in FE-SEM images (Hulikere and Joshi, 2019). It is well known that certain fungi, such as *Agaricus bisporus*, *Cladosporium cladosporioides*, and *Gliomastixmurorum*, aggressively eliminate copper ions from their surroundings. The nickel ions found in wastewater are actively removed by the production of *Hypocrealixii* nickel oxide nanoparticles (Agarwal and Singh, 2017; Torimiro *et al.*, 2021).

Table 1 Microbial synthesis of nanoparticles for bioremediation

Sr. No.	Source	Nanoparticle synthesized	References
		Bacteria	
1	*Bacillus methylotrophicus*	Silver	Wang *et al.* (2016)
2	*Bacillus amyloliquefaciens*	Titanium dioxide	Khan and Fulekar (2016)
3	*Pseudoalteromonas* sp CF10-13	Iron-sulfur	Cheng *et al.* (2019)
4	*Escherichia* sp. SINT7	Copper	Norman *et al.* (2020)
5	*Desulfovibrio vulgaris*	Palladium	Martins et al. (2017)
6	*Citrobacter freundii* Y9	Selenium	Wang *et al.* (2017)
7	*Pseudomonas putida* MnB6	BioMnOx	Forrez *et al.* (2011)
8	*Shewanella oneidensis*	Palladium	Forrez *et al.* (2011)
		Fungi	
1	*Penicillium oxalicum* GRS-1	Silver	Rose *et al.* (2019)
2	*Aspergillus fumigatus* BTCB10	Silver	Shahzad *et al.* (2019)
3	*Trichoderma longibrachaitum*	Silver	Elamawi *et al.* (2018)
4	*Rhizopus stolonifer*	Silver	Abdel-Rahim *et al.* (2017)
5	*Fusarium keratoplasticum*	Silver	Mohmed *et al.* (2017)
6	*Fusarium oxysporum*	Silver	Hamedi *et al.* (2017)
7	*Duddingtonia flagans*	Silver	Costa Silva *et al.* (2017)
8	*Cladosporium cladosporioides*	Silver	Hulikere and Joshi (2019)

3.2. Characteristics of nanoparticles

Nanoparticles have unique properties and attributes that make them useful for a wide range of applications. To ensure that they are suitable for the intended use, it is necessary to perform a thorough characterization and evaluation of these particles using a variety of tools and techniques that can provide the required data. The most commonly used techniques to investigate nanoparticles include electron microscopy (SEM/TEM), dynamic light scattering (DLS), Fourier transform infrared spectroscopy (FT-IR), energy-dispersive X-ray spectroscopy (EDS), atomic force microscopy (AFM), Raman spectroscopy (RS), and scanning/transmission electron microscopy (SEM/TEM). These techniques are utilized to analyze various properties of nanoparticles, such as their size, shape, structure, surface characteristics, and interactions with other materials (Hasan *et al.*, 2018).

3.2.1. Electron Microscopy

For the investigation of nanoparticles' size, shape, and surface, transmission electron microscopy (TEM), scanning electron microscopy (SEM), and scanning tunneling microscopy (STM) are commonly used. An electron cannon is used in SEM to fire an electron beam that is directed vertically through the microscope and strikes the samples under the lens system. After the beam strikes the sample, X-rays, and electrons are expelled from it. Detectors collect the dispersed electrons and X-rays to create a 3D image of the sample, providing information about the NPs' size, shape, aggregation, and dispersion. SEM offers details on the NPs' two-dimensional size, shape, localization, dispersal, and aggregation. An electromagnetic lens in TEM concentrates an extremely thin electron beam onto a thin portion of the material. As electrons from the beam pass through the specimen, they scatter or land on a fluorescent screen at the microscope's bottom to offer information on the NPs' size, shape, localization, dispersal, and aggregation (Shedbalkar *et al.*, 2014). The contrast to produce an image of the specimen is based on the difference in electron densities. By utilizing selected area electron diffraction (SAED), the electron beam can be focused on a specific portion of the sample, and the scattered electrons can be used to generate a diffraction pattern. This technique enables the characterization of NP crystal structure using TEM. An atomic resolution image of the surface is reconstructed by applying voltage, which causes electrons on the sample surface to be removed and create an electrical current. The primary purpose of STM is to describe the topography of NPs. These methods provide a good means of determining the morphological characteristics of inorganic NPs. There are significant limitations to the application of these techniques for organic NPs (or NPs coated with biological components) because they necessitate complex sample preparations. Sample dehydration from sample preparation for these procedures may result in sample shrinkage and aggregation, among other effects (Kumar and Dixit, 2017). It was discovered that the NPs aggregated into tiny aggregates, each containing 5-7 NPs. The NPs' crystal structure was ascertained concurrently using the selected area electron diffraction (SAED) method. It was discovered that the bulk of the NPs were single-crystalline cubic materials that were mostly orientated in their (111) orientation.

3.2.2. DLS

This method is frequently used to analyze the size and size distribution of NPs. This method relies on the Strokes-Einstein equation and uses the Brownian motion of NPs in suspension to evaluate light interference. It also correlates the diffusion coefficient of NP velocity with their size to make the assessment. The polydispersity index, which is the result of an autocorrelation function, displays the size distribution range of NPs (Kumar and Dixit, 2017). The values of the polydispersity index range from 0 to 1, where 0 denotes a homogeneous population and 1 denotes one that is extremely diverse. This technique allows for the examination of non-spherical nanoparticles using multistage DLS. Photon correlation spectroscopy (PCS) is a term used interchangeably with this method. DLS was utilized to measure various biogenic nanoparticles to determine their size and size distribution profile. The size distribution of Ag nanoparticles generated by *Trichoderma koningii* fungi was determined to be between 14 and 34 nm, with an average size of approximately 25 nm. The polydispersity index for these nanoparticles was 0.681, indicating polydispersion. DLS was also used to investigate the size expansion of biogenic MnO_2 nanoparticles over time. This demonstrated that three minutes after the reaction started, their size was 7.5 nm (Tripathi *et al.*, 2013).

3.2.3. FTIR

A technique that utilizes infrared light to a sample and observes the radiation that is either absorbed or transmitted is based on this method. The fingerprint of each sample is unique and provides information about its composition, including the bonds it contains, its polarity, and state of oxidation. This method is mainly used to assess organic materials, such as identifying the functionalization of nanoparticles or the chemical composition of their surface. When high purity is required, it is also utilized to detect pollutants (Manor *et al.*, 2012). For instance, biogenic NPs typically use FTIR to identify the probable functional groups on the NP surface, which are responsible for the NPs' stability and reduction. The confirmation of different proteins involved the reduction/stabilization process provided by the presence of protein residues on the surface of NPs produced by *Streptomyces* sp. NH28. Other nanoparticles produced by *Penicillium chrysogenum* PTCC 5031, and Se NPs produced by *Azospirillum thiophilum*, are examples of microbe-mediated NP synthesis results (Barabadi *et al.*, 2018).

3.2.4. Zeta Potential Analysis

The measurement of zeta potential is used to find the NP surface charge in colloidal solutions. The Stern layer is a thin layer that forms on the surface of NPs as a result of counter-ions being drawn to their surface charge. This layer moves through the solution with the NPs as they diffuse. The NP zeta potential is the electric potential at this layer's boundary. Zeta potential analyzers are the tools used to measure this potential. Higher absolute values of zeta potential suggest more stable NPs. Zeta potential levels are indicative of NP stability (Sikora *et al.*, 2015). The zeta potential serves as a useful gauge for the stability of nanoparticles (NPs); those having a zeta potential of more than +30 mV or less than - 30 mV are regarded as stable. The relative instability of AgNPs generated by *Aspergillus tubingensis* is indicated by their zeta potential of +8.48 (Ballottin *et al.*, 2016). Another crucial factor affecting zeta potential readings is the sample's pH; the greater the pH, the lower the zeta potential value. SeNPs can demonstrate varying potential values depending on the organism used for their synthesis. This indicates that having different zeta potential values for the same type of NPs based on the organism employed for their synthesis is not limited to silver nanoparticles (Menon *et al.*, 2019).

4. Mechanisms Involved in the Removal of Contaminants Using Nanoparticles

Abiotic and biotic processes are combined in the integrated notion of nano-bioremediation, which involves microorganisms and NPs. The first stage involves the introduction of nanoparticles (NPs) into a contaminated system and is followed by adsorption, dissolution, absorption, and photochemical reactions. Microbial introduction mechanisms like bioaccumulation, biostimulation, and biotransformation are included in the second phase. Hydrocarbons are first adsorbed onto the surfaces of the NPs, which raises the concentration of hydrocarbons around the microorganisms that are part of the bioremediation process. Hydrocarbons can also be adsorbed onto nanoparticles to improve their bioavailability to microorganisms and facilitate their breakdown. For example, when multiwalled carbon nanotubes (CNT) were combined with fungus, pyrene was first adsorbed on the surface of the CNTs, and then the cell broke down intracellularly. The overexpression of genes associated with ribosomes and oxidative phosphorylation, which are in charge of energy production and metabolic rate, respectively, was caused by the adsorption of pyrene by these multi-walled carbon nanotubes (Zhou *et al.*, 2021). Two beneficial degradation processes were proposed by another study that used iron nanoparticles. The first, high surface area and surface charge iron nanoparticles facilitate PAH adsorption on the NP surface, draw bacterial cells, and encourage bacterial adherence, all of which enhance the interaction between bacteria and hydrocarbons. The second is the rise in bacterial growth that NPs enabled, which led to a notable increase in the contaminant's breakdown and the production of more surface-active chemicals (Parthipan *et al.*, 2022). Moreover, certain nanoparticles can react more readily with hydrocarbons due to their special chemical characteristics. For instance, oxidation reactions that convert hydrocarbons into less toxic chemicals can be facilitated by nanoparticles containing metal catalysts. In addition, nanoparticles can be engineered to release substances that promote the growth of microorganisms involved in bioremediation, such as nutrients. This may contribute to increasing the bioremediation process's efficacy and activity. One such instance is polyvinylpyrrolidone (PVP) coated iron-oxide NPs, which, in addition to being effective oil collectors, improved the nutrients' bioavailability for the bacteria that break down oil by adsorbing and accumulating nitrate and phosphorus (Alabresm *et al.*, 2018). The biotransformation of petroleum pollutants is ultimately the responsibility of the enzymatic system. According to a study that used magnetic nanoparticles and the bacteria, a complex multicomponent enzyme system made up of hydrolases, decarboxylases, dehydrogenases, oxygenases, and dioxygenases was responsible for mediating the degradation. Ultimately, the β-ketoadipate route facilitates the biotransformation of products into Kreb cycle intermediates (Chauhan *et al.*, 2023). These are all potential processes that contribute to the degradation of petroleum pollutants by the application of nano-bioremediation techniques.

5. Conclusion and Future Prospects

Metal nanoparticles can be produced in an environmentally friendly, economical, and non-toxic manner using microbial synthesis techniques as an alternative to more conventional methods like physical and chemical ones. Actinomycetes, fungi, bacteria, and other microorganisms have all been shown to be capable of producing different metals and metal oxides. Various characterization techniques have been employed to ascertain the dimensions, configuration, and morphology of nanoparticles that have been biosynthesized. Because of these particles' versatility, biocompatibility, and ease of manufacture, microbial nanoparticles have piqued the interest of researchers worldwide. For a variety of uses, scientists can alter existing microbes to create metabolic pathways for the manufacture of nanoparticles with tunable sizes and shapes. Despite the many advantages, the use of recombinant organisms to fabricate nanoparticles is currently uncommon because of concerns about environmental safety and economic viability.

References

Abdel Rahim, K., Mahmoud, S.Y., Mohamed Ali, A., Almaary, K.S., Mustafa, M.A., & Moussa, S (2017) Extracellular biosynthesis of silver nanoparticles using *Rhizopus stolonifer*. Saudi Journal of Biological Sciences, 24:208–216. https://doi.org/10.1016/j.sjbs.2016.02.025

Abedi, G., Talebpour, Z., & Jamechenarboo, F. (2018). The survey of analytical methods for sample preparation and analysis of fragrances in cosmetics and personal care products. TrAC Trends in Analytical Chemistry, 102, 41-59. https://doi.org/10.1016/j.trac.2018.01.006

Abraham, A., & Chakraborty, P. (2020). A review on sources and health impacts of bisphenol A. Reviews on Environmental Health, 35(2), 201-210. https://doi.org/10.1515/reveh-2019-0034

Adebayo, E. A., Azeez, M. A., Alao, M. B., Oke, A. M., & Aina, D. A. (2021). Fungi as veritable tool in current advances in nanobiotechnology. Heliyon, 7(11). https://doi.org/10.1016/j.heliyon.2021.e08480

Agarwal, M., & Singh, K. (2017) Heavy metal removal from wastewater using various adsorbents: a review. Journal of Water Reuse and Desalination, 7(4), 387–419. https://doi.org/10.2166/wrd.2016.104

Alabresm, A., Chen, Y. P., Decho, A. W., & Lead, J. (2018). A novel method for the synergistic remediation of oil-water mixtures using nanoparticles and oil-degrading bacteria. Science of the Total Environment, 630, 1292-1297. https://doi.org/10.1016/j.scitotenv.2018.02.277

Ali, H., Khan, E., & Ilahi, I. (2019). Environmental chemistry and ecotoxicology of hazardous heavy metals: environmental persistence, toxicity, and bioaccumulation. Journal of Chemistry, 2019. https://doi.org/10.1155/2019/6730305

Alori, E. T., Gabasawa, A. I., Elenwo, C. E., & Agbeyegbe, O. O. (2022). Bioremediation techniques as affected by limiting factors in soil environment. Frontiers in Soil Science, 2, 937186. https://doi.org/10.3389/fsoil.2022.937186

Al-Tohamy, R., Ali, S. S., Li, F., Okasha, K. M., Mahmoud, Y. A. G., Elsamahy, T., ... & Sun, J. (2022). A critical review on the treatment of dye-containing wastewater: Ecotoxicological and health concerns of textile dyes and possible remediation approaches for environmental safety. Ecotoxicology and Environmental Safety, 231, 113160. https://doi.org/10.1016/j.ecoenv.2021.113160

Ameen, F. A., Hamdan, A. M., & El-Naggar, M. Y. (2020). Assessment of the heavy metal bioremediation efficiency of the novel marine lactic acid bacterium, *Lactobacillus plantarum* MF042018. Scientific reports, 10(1), 314. https://doi.org/10.1038/s41598-019-57210-3

Ardila-Leal, L. D., Poutou-Piñales, R. A., Pedroza-Rodríguez, A. M., & Quevedo-Hidalgo, B. E. (2021). A brief history of colour, the environmental impact of synthetic dyes, and removal by using laccases. Molecules, 26(13), 3813. https://doi.org/10.3390/molecules26133813

Bala, S., Garg, D., Thirumalesh, B. V., Sharma, M., Sridhar, K., Inbaraj, B. S., & Tripathi, M. (2022). Recent strategies for bioremediation of emerging pollutants: a review for a green and sustainable environment. Toxics, 10(8), 484. https://doi.org/10.3390/toxics10080484

Ballottin, D., Fulaz, S., Souza, M. L., Corio, P., Rodrigues, A. G., Souza, A. O., ... & Tasic, L. (2016). Elucidating protein involvement in the stabilization of the biogenic silver nanoparticles. Nanoscale Research Letters, 11, 1-9. https://doi.org/10.1186/s11671-016-1538-y

Barabadi, H., Kobarfard, F., & Vahidi, H. (2018). Biosynthesis and characterization of biogenic tellurium nanoparticles by using Penicillium chrysogenum PTCC 5031: A novel approach in gold biotechnology. Iranian journal of pharmaceutical research: IJPR, 17(Suppl2), 87.

Basketter, D., & Kimber, I. (2015). Fragrance sensitisers: Is inhalation an allergy risk. Regulatory Toxicology and Pharmacology, 73(3), 897-902. https://doi.org/10.1016/j.yrtph.2015.09.031

Bilal, M., Rasheed, T., Iqbal, H. M., & Yan, Y. (2018). Peroxidases-assisted removal of environmentally-related hazardous pollutants with reference to the reaction mechanisms of industrial dyes. Science of the total environment, 644, 1-13. https://doi.org/10.1016/j.scitotenv.2018.06.274

Bolan, N., Kunhikrishnan, A., Thangarajan, R., Kumpiene, J., Park, J., Makino, T., ... & Scheckel, K. (2014). Remediation of heavy metal (loid) s contaminated soils–to mobilize or to immobilize. Journal of hazardous materials, 266, 141-166. https://doi.org/10.1016/j.jhazmat.2013.12.018

Boroumand Moghaddam, A., Namvar, F., Moniri, M., Md. Tahir, P., Azizi, S., & Mohamad, R. (2015). Nanoparticles biosynthesized by fungi and yeast: a review of their preparation, properties, and medical applications. Molecules, 20(9), 16540-16565. https://doi.org/10.3390/molecules200916540

Briffa, J., Sinagra, E., & Blundell, R. (2020). Heavy metal pollution in the environment and their toxicological effects on humans. Heliyon, 6(9). https://doi.org/10.1016/j.heliyon.2020.e04691

Castro, L., Blázquez, M. L., González, F. G., & Ballester, A. (2014). Mechanism and applications of metal nanoparticles prepared by bio-mediated process. Reviews in Advanced Sciences and Engineering, 3(3), 199-216. https://doi.org/10.1166/rase.2014.1064

Chauhan, P., Imam, A., Kanaujia, P. K., & Suman, S. K. (2023). Nano-bioremediation: an eco-friendly and effective step towards petroleum hydrocarbon removal from environment. Environmental Research, 231, 116224. https://doi.org/10.1016/j.envres.2023.116224

Cheng, S., Li, N., Jiang, L., Li, Y., Xu, B., & Zhou, W. (2019). Biodegradation of metal complex Naphthol Green B and formation of iron–sulfur nanoparticles by marine bacterium *Pseudoalteromonas* sp CF10-13. Bioresource Technology, 273, 49-55. https://doi.org/10.1016/j.biortech.2018.10.082

Costa Silva, L. P., Oliveira, J. P., Keijok, W. J., da Silva, A. R., Aguiar, A. R., Guimarães, M. C. C. & Braga, F. R. (2017). Extracellular biosynthesis of silver nanoparticles using the cell-free filtrate of nematophagous fungus *Duddingtonia flagrans*. International Journal of Nanomedicine, 6373-6381. https://doi.org/10.2147/IJN.S137703

Dhaliwal, S. S., Singh, J., Taneja, P. K., & Mandal, A. (2020). Remediation techniques for removal of heavy metals from the soil contaminated through different sources: a review. Environmental Science and Pollution Research, 27(2), 1319-1333. https://doi.org/10.1007/s11356-019-06967-1

Elamawi, R. M., Al-Harbi, R. E., and Hendi, A. A. (2018). Biosynthesis and characterization of silver nanoparticles using *Trichoderma longibrachiatum* and their effect on phytopathogenic fungi. Egyptian Journal of Biological Pest Control. 28:28. https://doi.org/10.1186/s41938-018-0028-1

Fabbri, E., Valbonesi, P., & Moon, T. W. (2023). Pharmaceuticals in the marine environment: occurrence, fate, & biological effects. In Contaminants of Emerging Concern in the Marine Environment (pp. 11-71). Elsevier. https://doi.org/10.1016/B978-0-323-90297-7.00008-1

Forrez I, Carballa M, Fink G, Wick A, Hennebel T, Vanhaecke L, Ternes T, Boon N, Verstraete W (2011) Biogenic metals for the oxidative and reductive removal of pharmaceuticals, biocides and iodinated contrast media in a polishing membrane bioreactor. Water Research, 45(4):1763–1773. https://doi.org/10.1016/j.watres.2010.11.031

Fouda, A., Hassan, S. E. D., Eid, A. M., Abdel□Rahman, M. A., & Hamza, M. F. (2022). Light enhanced the antimicrobial, anticancer, and catalytic activities of selenium nanoparticles fabricated by endophytic fungal strain, *Penicillium crustosum* EP-1. Scientific reports, 12(1), 11834. https://doi.org/10.1038/s41598-022-15903-2

Gani, K. M., & Kazmi, A. A. (2017). Contamination of emerging contaminants in Indian aquatic sources: first overview of the situation. Journal of Hazardous, Toxic, and Radioactive Waste, 21(3), 04016026. https://doi.org/10.1061/(ASCE)HZ.2153-5515.0000348

Gautam, M., Kim, J. O., & Yong, C. S. (2021). Fabrication of aerosol-based nanoparticles and their applications in biomedical fields. Journal of Pharmaceutical Investigation, 51(4), 361-375. 51:361–375 https://doi.org/10.1007/s40005-021-00523-1

Gaw, S., Thomas, K.V., & Hutchinson, T.H. (2014). Sources, impacts and trends of pharmaceuticals in the marine & coastal environment. Philosophical Transactions of the Royal Society B: Biological Sciences, 369(1656), 20130572. https://doi.org/10.1098/rstb.2013.0572

Ginebreda, A., Pérez, S., Rivas, D., Kuzmanovic, M., & Barceló, D. (2016). Pollutants of emerging concern in rivers of catalonia: occurrence, fate, and risk. Experiences from surface water quality monitoring: the EU water framework directive implementation in the Catalan river basin district (Part I), 283-320. https://doi.org/10.1007/698_2015_348

Guo, T., Pan, K., Chen, Y., Tian, Y., Deng, J., & Li, J. (2023). When aerobic granular sludge faces emerging contaminants: A review. Science of The Total Environment, 167792. https://doi.org/10.1016/j.scitotenv.2023.167792

Hamedi, S., Ghaseminezhad, M., Shokrollahzadeh, S., & Shojaosadati, S. A. (2017). Controlled biosynthesis of silver nanoparticles using nitrate reductase enzyme induction of filamentous fungus and their antibacterial evaluation. Artificial cells, Nanomedicine, and Biotechnology, 45(8), 1588-1596. https://doi.org/10.1080/21691401.2016.1267011

Hawash, H. B., Moneer, A. A., Galhoum, A. A., Elgarahy, A. M., Mohamed, W. A., Samy, M., ... & Attia, N. F. (2023). Occurrence and spatial distribution of pharmaceuticals and personal care products (PPCPs) in the aquatic environment, their characteristics, and adopted legislations. Journal of Water Process Engineering, 52, 103490. https://doi.org/10.1016/j.jwpe.2023.103490

Hejna, M., Kapuścińska, D., & Aksmann, A. (2022). Pharmaceuticals in the aquatic environment: A review on eco-toxicology and the remediation potential of algae. International Journal of Environmental Research and Public Health, 19(13), 7717. https://doi.org/10.3390/ijerph19137717

Huang, W., Song, B., Liang, J., Niu, Q., Zeng, G., Shen, M., ... & Zhang, Y. (2021). Microplastics and associated contaminants in the aquatic environment: A review on their ecotoxicological effects, trophic transfer, and potential impacts to human health. Journal of Hazardous Materials, 405, 124187. https://doi.org/10.1016/j.jhazmat.2020.124187

Hulikere, M., and Joshi, C. (2019) Characterization, antioxidant and antimicrobial activity of silver nanoparticles synthesized using marine endophytic fungus *Cladosporium cladosporioides*, Process Biochemistry 82, 199–204. https://doi.org/10.1016/j.procbio.2019.04.011

Hussain, I., Aleti, G., Naidu, R., Puschenreiter, M., Mahmood, Q., Rahman, M. M., ... & Reichenauer, T. G. (2018). Microbe and plant assisted-remediation of organic xenobiotics and its enhancement by genetically modified organisms and recombinant technology: a review. Science of the total environment, 628, 1582-1599. https://doi.org/10.1016/j.scitotenv.2018.02.037

Kapoor, R. T., Salvadori, M. R., Rafatullah, M., Siddiqui, M. R., Khan, M. A., & Alshareef, S. A. (2021). Exploration of microbial factories for synthesis of nanoparticles–a sustainable approach for bioremediation of environmental contaminants. Frontiers in Microbiology, 12, 658294. https://doi.org/10.3389/fmicb.2021.658294

Khalid, M., & Abdollahi, M. (2021). Environmental distribution of personal care products and their effects on human health. Iranian journal of pharmaceutical research: IJPR, 20(1), 216. https://doi.org/10.22037/ijpr.2021.114891.15088

Khan, A. H. A., & Barros, R. (2023). Pharmaceuticals in water: risks to aquatic life and remediation strategies. Hydrobiology, 2(2), 395-409. https://doi.org/10.3390/hydrobiology2020026

Khan, I., Saeed, K., & Khan, I. (2019). Nanoparticles: Properties, applications and toxicities. Arabian journal of chemistry, 12(7), 908-931. https://doi.org/10.1016/j.arabjc.2017.05.011

Khan, R., and Fulekar, M. (2016). Biosynthesis of titanium dioxide nanoparticles using Bacillus amyloliquefaciens culture and enhancement of its photocatalytic activity for the degradation of a sulfonated textile dye Reactive Red 31. Journal of Colloid Interface Science, 475, 184–191. https://doi.org/10.1016/j.jcis.2016.05.001

Köck-Schulmeyer, M., Ginebreda, A., Petrovic, M., Giulivo, M., Aznar-Alemany, Ò., Eljarrat, E., ... & Barceló, D. (2021). Priority and emerging organic microcontaminants in three Mediterranean river basins: Occurrence, spatial distribution, and identification of river basin specific pollutants. Science of the Total Environment, 754, 142344. https://doi.org/10.1016/j.scitotenv.2020.142344

Kumar, A., & Dixit, C. K. (2017). Methods for characterization of nanoparticles. In Advances in nanomedicine for the delivery of therapeutic nucleic acids (pp. 43-58). Woodhead Publishing. https://doi.org/10.1016/B978-0-08-100557-6.00003-1

Kumari, S., Raturi, S., Kulshrestha, S., Chauhan, K., Dhingra, S., András, K., ... & Singh, T. (2023). A comprehensive review on various techniques used for synthesizing nanoparticles. Journal of Materials Research and Technology. https://doi.org/10.1016/j.jmrt.2023.09.291

Mahanty, S., Chatterjee, S., Ghosh, S., Tudu, P., Gaine, T., Bakshi, M., ... & Chaudhuri, P. (2020). Synergistic approach towards the sustainable management of heavy metals in wastewater using mycosynthesized iron oxide nanoparticles: biofabrication, adsorptive dynamics and chemometric modeling study. Journal of Water Process Engineering, 37, 101426. https://doi.org/10.1016/j.jwpe.2020.101426

Manisalidis, I., Stavropoulou, E., Stavropoulos, A., & Bezirtzoglou, E. (2020). Environmental and health impacts of air pollution: a review. Frontiers in Public Health, 8, 505570. https://doi.org/10.3389/fpubh.2020.00014

Manor, J., Feldblum, E. S., Zanni, M. T., & Arkin, I. T. (2012). Environment polarity in proteins mapped noninvasively by FTIR spectroscopy. The Journal of Physical Chemistry Letters, 3(7), 939-944. https://doi.org/10.1021%2Fjz300150v

Mapelli, F.; Scoma, A.; Michoud, G.; Aulenta, F.; Boon, N.; Borin, S.; Kalogerakis, N.; Daffonchio, (2017) D. Biotechnologies for Marine Oil Spill Cleanup: Indissoluble Ties with Microorganisms. Trends in Biotechnology 35, 860–870. https://doi.org/10.1016/j.tibtech.2017.04.003

Martins, M., Mourato, C., Sanches, S., Noronha, J. P., Crespo, M. B., & Pereira, I. A. (2017). Biogenic platinum and palladium nanoparticles as new catalysts for the removal of pharmaceutical compounds. Water Research, 108, 160-168. https://doi.org/10.1016/j.watres.2016.10.071

Menon, S., Agarwal, H., Kumar, S. V., & Rajeshkumar, S. (2019). Biomemetic synthesis of selenium nanoparticles and its biomedical applications. In Green Synthesis, Characterization and Applications of Nanoparticles (pp. 165-197). Elsevier.

Mishra, R. K., Mentha, S. S., Misra, Y., & Dwivedi, N. (2023). Emerging pollutants of severe environmental concern in water and wastewater: A comprehensive review on current developments and future research. Water-Energy Nexus.

Mishra, R. K., Mentha, S. S., Misra, Y., & Dwivedi, N. (2023). Emerging pollutants of severe environmental concern in water and wastewater: A comprehensive review on current developments and future research. Water-Energy Nexus. https://doi.org/10.1016/j.wen.2023.08.002

Mohamed, A., Fouda, A., S Elgamal, M., EL-Din Hassan, S., I Shaheen, T., & S Salem, S. (2017). Enhancing of cotton fabric antibacterial properties by silver nanoparticles synthesized by new Egyptian strain Fusarium keratoplasticum A1-3. Egyptian journal of chemistry, 60(Conference Issue (The 8th International Conference of The Textile Research Division (ICTRD 2017), National Research Centre, Cairo 12622, Egypt., 63-71. https://doi.org/10.21608/ejchem.2017.1626.1137

Noman, M., Shahid, M., Ahmed, T., Niazi, M. B. K., Hussain, S., Song, F., & Manzoor, I. (2020). Use of biogenic copper nanoparticles synthesized from a native *Escherichia* sp. as photocatalysts for azo dye degradation and treatment of textile effluents. Environmental Pollution, 257, 113514. https://doi.org/10.1016/j.envpol.2019.113514

Ovais, M., Khalil, A. T., Islam, N. U., Ahmad, I., Ayaz, M., Saravanan, M., ... & Mukherjee, S. (2018). Role of plant phytochemicals and microbial enzymes in biosynthesis of metallic nanoparticles. Applied Microbiology and Biotechnology, 102, 6799-6814. 018). https://doi.org/10.1007/s00253-018-9146-7

Park, Y., Hong, Y. N., Weyers, A., Kim, Y. S., & Linhardt, R. J. (2011). Polysaccharides and phytochemicals: a natural reservoir for the green synthesis of gold and silver nanoparticles. IET Nanobiotechnology, 5(3), 69-78. https://doi.org/10.1049/iet-nbt.2010.0033

Parthipan, P., Cheng, L., Dhandapani, P., Elumalai, P., Huang, M., & Rajasekar, A. (2022). Impact of biosurfactant and iron nanoparticles on biodegradation of polyaromatic hydrocarbons (PAHs). Environmental Pollution, 306, 119384. https://doi.org/10.1016/j.envpol.2022.119384

Patrolecco, L., Rauseo, J., Ademollo, N., Polesello, S., Vardè, M., Pizzini, S., & Spataro, F. (2023). Some organic compounds in potable water: the PFASs, EDCs and PPCPs issue. In Current Trends and Future Developments on (Bio-) Membranes (pp. 183-228). Elsevier. https://doi.org/10.1016/B978-0-12-824103-5.00001-2

Pavithra, K. G., & Jaikumar, V. J. J. O. I. (2019). Removal of colorants from wastewater: A review on sources and treatment strategies. Journal of Industrial and Engineering Chemistry, 75, 1-19. https://doi.org/10.1016/j.jiec.2019.02.011

Pete, A. J., Bharti, B., & Benton, M. G. (2021). Nano-enhanced bioremediation for oil spills: a review. ACS ES&T Engineering, 1(6), 928-946. https://doi.org/10.1021/acsestengg.0c00217

Priyadarshanee, M., Mahto, U., & Das, S. (2022). Mechanism of toxicity and adverse health effects of environmental pollutants. In Microbial biodegradation and bioremediation (pp. 33-53). Elsevier. https://doi.org/10.1016/B978-0-323-85455-9.00024-2

Rathi, B. S., Kumar, P. S., & Show, P. L. (2021). A review on effective removal of emerging contaminants from aquatic systems: Current trends and scope for further research. Journal of hazardous materials, 409, 124413. https://doi.org/10.1016/j.jhazmat.2020.124413

Rose, G. K., Soni, R., Rishi, P., and Soni, S. K. (2019). Optimization of the biological synthesis of silver nanoparticles using *Penicillium oxalicum* GRS-1 and their antimicrobial effects against common food-borne pathogens. Green Process Synthesis 8, 144–156. https://doi.org/ 10.1515/gps-2018-0042

Roy, A., Sharma, A., Yadav, S., Jule, L. T., & Krishnaraj, R. (2021). Nanomaterials for remediation of environmental pollutants. Bioinorganic Chemistry and Applications, 2021. https://doi.org/10.1155/2021/1764647

Santal, A. R., Rani, R., Kumar, A., Sharma, J. K., & Singh, N. P. (2022). Biodegradation and detoxification of textile dyes using a novel bacterium *Bacillus* sp. AS2 for sustainable environmental cleanup. Biocatalysis and Biotransformation, 42(1), 41-55. https://doi.org/10.1080/10242422.2022.2113518

Santal, A. R., Singh, N. P., & Saharan, B. S. (2016). A novel application of *Paracoccus pantotrophus* for the decolorization of melanoidins from distillery effluent under static conditions. Journal of Environmental Management, 169, 78-83. https://doi.org/10.1016/j.jenvman.2015.12.016

Sathiyanarayanan, G., Dineshkumar, K., & Yang, Y. H. (2017). Microbial exopolysaccharide-mediated synthesis and stabilization of metal nanoparticles. Critical reviews in microbiology, 43(6), 731-752. https://doi.org/10.1080/1040841X.2017.1306689

Shahzad, A., Saeed, H., Iqtedar, M., Hussain, S. Z., Kaleem, A., and Abdullah, R. (2019). Size-controlled production of silver nanoparticles by *Aspergillus fumigatus* BTCB10: likely antibacterial and cytotoxic effects. Journal of Nanomaterials. 2019:5168698. https://doi.org/10.1155/2019/5168698

Sharma, J. K., Kumar, N., Singh, N. P., & Santal, A. R. (2023). Phytoremediation technologies and their mechanism for removal of heavy metal from contaminated soil: An approach for a sustainable environment. Frontiers in Plant Science, 14, 1076876. https://doi.org/10.3389/fpls.2023.1076876

Sikora, A., Bartczak, D., Geißler, D., Kestens, V., Roebben, G., Ramaye, Y., ... & Minelli, C. (2015). A systematic comparison of different techniques to determine the zeta potential of silica nanoparticles in biological medium. Analytical methods, 7(23), 9835-9843. https://doi.org/10.1039/C5AY02014J

Singh, R., Shedbalkar, U. U., Wadhwani, S. A., & Chopade, B. A. (2015). Bacteriagenic silver nanoparticles: synthesis, mechanism, and applications. Applied microbiology and biotechnology, 99, 4579-4593. https://doi.org/10.1007/s00253-015-6622-1

Sinha, A., & Khare, S. K. (2012). Mercury bioremediation by mercury accumulating Enterobacter sp. cells and its alginate immobilized application. Biodegradation, 23, 25-34. https://doi.org/10.1016/j.biortech.2010.12.040

Steinemann, A. (2016). Fragranced consumer products: exposures and effects from emissions. Air Quality, Atmosphere & Health, 9(8), 861-866. https://doi.org/10.1007/s11869-016-0442-z

Subbaiya, R., Saravanan, M., Priya, A. R., Shankar, K. R., Selvam, M., Ovais, M., ... & Barabadi, H. (2017). Biomimetic synthesis of silver nanoparticles from *Streptomyces atrovirens* and their potential anticancer activity against human breast cancer cells. IET nanobiotechnology, 11(8), 965-972. https://doi.org/10.1049/iet-nbt.2016.0222

Torimiro, N., Daramola, O. B., Oshibanjo, O. D., Otuyelu, F. O., Akinsanola, B. A., Yusuf, O. O., ... & Omole, R. K. (2021). Ecorestoration of heavy metals and toxic chemicals in polluted environment using microbe-mediated nanomaterials. International Journal of Environmental Bioremediation & Biodegradation, 9(1), 8-21. https://doi.org/10.12691/ijebb-9-1-2

Tripathi, R. M., Gupta, R. K., Shrivastav, A., Singh, M. P., Shrivastav, B. R., & Singh, P. (2013). *Trichoderma koningii* assisted biogenic synthesis of silver nanoparticles and evaluation of their antibacterial activity. Advances in Natural Sciences: Nanoscience and Nanotechnology, 4(3), 035005. https://doi.org/10.1088/2043-6262/4/3/035005

Vázquez-Núñez, E., Molina-Guerrero, C. E., Peña-Castro, J. M., Fernández-Luqueño, F., & de la Rosa-Álvarez, M. G. (2020). Use of nanotechnology for the bioremediation of contaminants: A review. Processes, 8(7), 826. https://doi.org/10.3390/pr8070826

Wang, X., Zhang, D., Pan, X., Lee, D.J., Al-Misned, F.A., Mortuza, M.G., & Gadd, G.M., (2017) Aerobic and anaerobic biosynthesis of nano-selenium for remediation of mercury contaminated soil. Chemosphere, 170:266–273. https://doi.org/10.1016/j.chemosphere.2016.12.020

Wang, C., Kim, Y. J., Singh, P., Mathiyalagan, R., Jin, Y., & Yang, D. C. (2016). Green synthesis of silver nanoparticles by *Bacillus methylotrophicus*, and their antimicrobial activity. Artificial Cells, Nanomedicine, and Biotechnology, 44, 1127–1132 https://doi.org/10.3109/21691401.2015.1011805

Wydro, U., Wołejko, E., Luarasi, L., Puto, K., Tarasevičienė, Ž., & Jabłońska-Trypuć, A. (2023). A Review on Pharmaceuticals and Personal Care Products Residues in the Aquatic Environment and Possibilities for Their Remediation. Sustainability, 16(1), 169. https://doi.org/10.3390/su16010169

Xiao, R., Ou, T., Ding, S., Fang, C., Xu, Z., & Chu, W. (2023). Disinfection by-products as environmental contaminants of emerging concern: a review on their occurrence, fate and removal in the urban water cycle. Critical Reviews in Environmental Science and Technology, 53(1), 19-46. https://doi.org/10.1080/10643389.2022.2043101

Yadav, A., Kon, K., Kratosova, G., Duran, N., Ingle, A. P., & Rai, M. (2015). Fungi as an efficient mycosystem for the synthesis of metal nanoparticles: progress and key aspects of research. Biotechnology Letters, 37, 2099-2120. https://doi.org/10.1007/s10529-015-1901-6

Zhang, X., Yan, S., Tyagi, R. D., & Surampalli, R. Y. (2011). Synthesis of nanoparticles by microorganisms and their application in enhancing microbiological reaction rates. Chemosphere, 82(4), 489-494. https://doi.org/10.1016/j.chemosphere.2010.10.023

Zhou, G., Jia, X., Zhang, X., & Li, L. (2022). Multi-walled carbon nanotube-modified hydrothermal carbon: A potent carbon material for efficient remediation of cadmium-contaminated soil in coal gangue piling site. Chemosphere, 307, 135605. https://doi.org/10.1016/j.chemosphere.2022.135605

13

Nanoremediation: An Approach of Nanotechnology for Remediation of Heavy Metals

Vandna Kumari and Narendra Kumar*

Department of Food Safety and Quality Assurance, College of Dairy and Food Technology (RAJUVAS), Bassi, Jaipur, Rajasthan

Abstract

The hazardous toxic metals like cadmium (Cd), arsenic (As), mercury (Hg), lead (Pb), and chromium (Cr), which have detrimental health effects, are naturally discharged into the atmosphere in various ways viz., industrial wastes, anthropogenic sources, etc. Heavy metal pollution of the atmosphere has become a frightening ecological concern that has created serious threats to the ecosystem. These metals have been identified as group I, human carcinogens due to their persistence in the surroundings and bio-magnification in biological systems. Thus, the removal of heavy metals from the atmosphere becomes a serious matter from the biological and environmental point of view. In the past following techniques adopted for toxic heavy metals remediation were classified are bioadsorption, ion exchange, membrane filtration, flotation, and solvent extraction techniques. Moreover, among the abovementioned techniques, precipitation of heavy metals such as insoluble hydroxides, sulfides, and carbonates by electroplating is widely used (in about 75% of incidences) to treat wastewater. More recently, studies have illustrated that various forms of nanomaterials prepared by several techniques, showed remarkable potential for the remediation of innumerable contaminants including toxic heavy metals in comparison to their bulk counterparts making them a topic of importance to researchers in various fields. Nanotechnology has impacted water purification from different approaches, such as (i) Nanofiltration membranes (ii) Nanocatalysts conditions (iii) Magnetic nanoparticles, and (iv) Nanosensing. The tiny nanometals and nanometal oxides showed enormous potential in almost every extent of industrial and biotechnological domains, including their potential multipurpose approach to deal with water impurities. This chapter

[*] *Corresponding authors e-mail: prof.narenfsqa@gmail.com*

provides innovative facts on the application of nanotechnology for the remediation of toxic metals. It highlights myriad nanotechnology approaches that have been employed for the remediation of heavy metals, existing details on their synthesis techniques, adsorption processes, their factors, and the environmental and health impact of nanomaterials. Finally, it provides innovative research challenges and future trends of nanomaterials as well as novel remediation strategies for heavy metal removal.

Keywords: *Nanotechnology, heavy metals, nanoremediation, nanosensing*

1. Introduction

Rapid industrialization and urbanization in recent times have led to the emission of around 10 million tons of untreated hazardous chemicals every year into the environment from various sources such as metallurgical, mining, nuclear energy, and chemical manufacturing units (Bhatt *et al.* 2022). These are mainly pesticides, antibiotics, chlorinated phenols, steroids, azo dyes, and heavy metals are among the pollutants that contaminate the environment. Among these xenobiotics, heavy metals cause serious concern due to their toxicity, lethal effect on long-term exposure, and persistency in the environment which passes across the food chain (Lall *et al.*, 2022). Heavy metals (HMs) are defined either due to their high atomic weight or because of their high density, among those that are more common in our everyday life and have a density greater than 5 g/cm3 are Titanium, Vanadium, Chromium, Manganese, Cadmium, Tin, Platinum, Gold, Mercury, Lead, Nickel, etc. (Briffa *et al.* 2020). Pollution due to these chemicals ranks second among the most hazardous pollutions and more lethal than chief hazardous pollutants notably sulfur dioxide, carbon dioxide, and pesticides (Ahmad *et al.*, 2021). These metals enter our body in four proposed ways i) ingestion of contaminated food ii) inhalation from the atmosphere iii) drinking contaminated water iv) direct skin contact from agriculture, pharmaceutical, manufacturing, and industrial areas which leads to bioaccumulation which causes biological and physiological complications as shown in Figure 1 (Kumar *et al.*, 2018; Masindi and Muedi., 2018; Walker *et al.*, 2012).

Remediation of heavy metals requires effective management practices and must be implemented because of existing climate change challenges, food and energy security, and sustainability for environmental protection. To conquer the mentioned challenges in a proficient manner recent technological advancements have been made, Bioremediation is one such strategy that has been acknowledged combining biological methods with other remediation techniques such as biophysical, biochemical, and physiochemical (Patel *et al.*, 2020; Kumar[2] *et al.*, 2018).

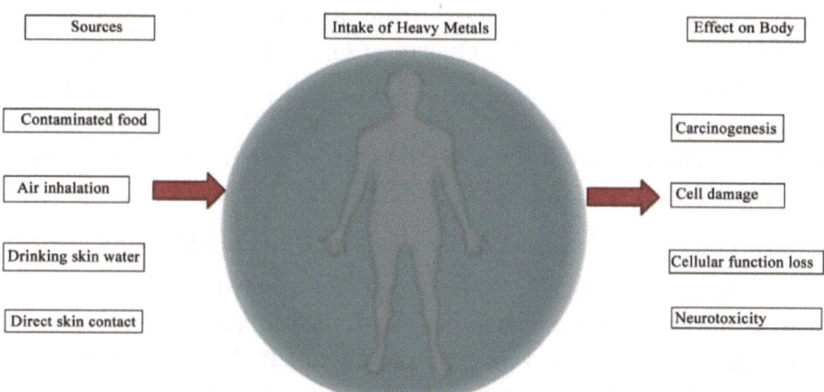

Fig. 1: Impact of heavy metals on the human body

It is an economical and efficient method using biological agents such as bacteria, fungi, protists, or their enzymes to degrade hazardous pollutants in less cyanogenetic forms and minimize chemical and biological sludge (Kumar *et al.*, 2020; Van Dillewijn *et al.*, 2007). Bioremediation comprises various other strategies such as bioaugmentation which is a natural attenuation process by using indigenous microorganisms (Kumar *et al.*, 2017), the addition of nutrients externally for a stimulated process known as biostimulation, the use of genetically modified organisms in the remediation process, involving the use of some plants and their enzymes (phytoremediation) and biomineralization process of biodegradation of organic substances into inorganic components (Joutey *et al.*, 2013).

Bioremediation in combination with nanotechnological advancements makes a more efficient technique with wider applications including heavy metal contamination of air, water, and soil by reducing the formation of intermediate compounds during the process, and mitigating their adverse environmental impacts (Bharagava *et al.*, 2020; Yogalakshmi *et al.*, 2020). Advancements in nanotechnologies and the application of novel nanomaterials with particles of two or three dimensions greater than 1 nm size integrated with biological processes to accelerate and promote the removal of toxic contaminants from the environment is termed nano bioremediation (Koul and Taak., 2018). The methods used are mainly surface plasmon resonance and surface-enhanced scattering, using different forms of nanoparticles such as silicon oxide, zinc oxide, and platinum used for the identification and removal of environmental contaminants (Mourdikoudis *et al.*, 2018). Moreover, for in-situ remediation of pesticides, heavy metals, antibiotics, and various other environmental pollutants gold, manganese oxide, and ferric oxide nanoparticles are being utilized along with green nanomaterials synthesized from microorganisms such as iron nanoparticles (Bolade *et al.*, 2020; Guerra *et al.*, 2018; Mourdikoudis *et al.*, 2018). Interactions among microorganisms and their enzymes, nanomaterials, and environment pollutants mainly heavy metals, to identify bioremediation processes has created a forum for a wide range of applications for ensuring safe ecosystems for current and future generations.

2. Nanomaterials and Nanoparticles Used in Bioremediation

For effective bioremediation of contaminated environments with pollutants, a proper interaction between nanoparticles (NPs) and living organisms is essential which depends on depend on a variety of parameters including size and shape, surface coating, chemical nature of the NMs and type of contaminant, organism, media, pH, and temperature during the process (Tan *et al.*, 2018). Various elements are used for the production of nanoparticles of different chemical composition, shape, and size which are essentially more mobile and reactive. Earlier physical and chemical methods were used to generate nanoparticles for HMs remediation but biosynthesis of nanoparticles using microorganisms has emerged as a sustainable approach for nanobioremediation as shown in Figure 2.

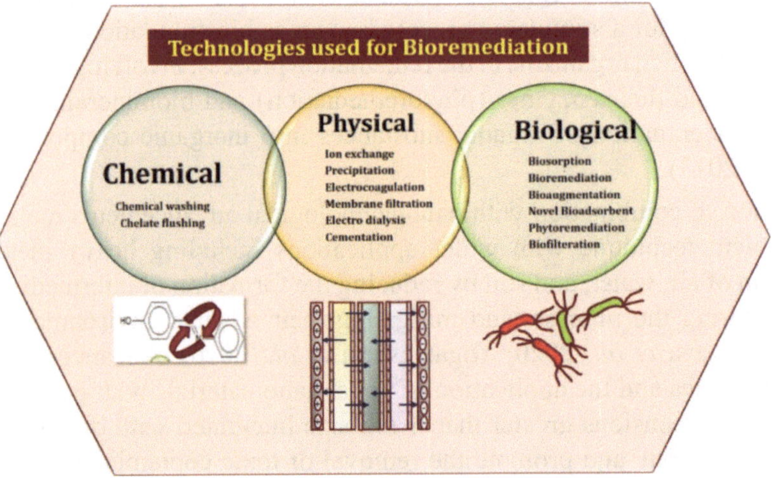

Fig. 2: Methods for remediation of toxic and HMs from the enviroment

In nanoparticles organic and inorganic groups are found, organic nanoparticles involve carbon nanoparticles (fullerenes), whereas magnetic nanoparticles, natural metal nanoparticles (such as gold and silver), and semiconductive nanoparticles (such as titanium and zinc oxide) are involved in inorganic nanoparticles.

The term "nanomaterial" is usually used to describe fabrics with at least one dimension specifically intended between 1 and 100nm. The selection criteria of the type of NM may be applied for more specificity such as (a) nano iron and its derivatives; (b) dendrimers; (c) carbon-based NMs; (d) single-enzyme NPs; (e) engineered polymeric; (f) biogenic uraninite; and (g) metals other than iron (Rizwan *et al.*, 2014). Nanomaterials play a crucial role in different fields including environmental remediation of heavy metals, agriculture, electronics, healthcare, cosmetics, nano-sensors, water filtration, etc. Nowadays novel nanomaterials are constantly applied with superior performance in environmental pollution cleanup. Metal-organic frameworks (MOFs) having various functional groups and flexible structures, have emerged as a new field of capturing various heavy metal ions (Li *et al.*, 2020). Nanoscale zero-valent irons (nZVI), are another having high mobility, reactivity, low toxicity, and abundant surface-active sites that are being used as effective adsorbents to eliminate heavy metal ions (Zou *et al.*, 2016). Similarly, MXenes with large specific surface areas and excellent mechanical properties are used for nanoparticle generation material.

2.1. Bacterial Nanoremediation

Bacterial cells are capable of synthesizing nanoparticles such as nanoparticles such as gold, silver, cadmium, platinum, titanium, etc using their enzymes and are capable of catalyzing specific reactions that lead to an inorganic nanoparticle synthesis ranging from 100–200nm in size which are a relatively pure state, free from other cell proteins (Patel *et al.*, 2020). Microbial enzymes such as Mn peroxidase, alcohol dehydrogenase, horseradish peroxidase, laccase, lipase, catalase, pectate lyase, and glucose oxidase facilitate bioremediation along with nanoparticles. They promote the production of biosurfactants in microorganisms on immobilized heavy metals and other pollutants by increasing the solubility of hydrocarbon and the degradation of pollutants in the environment (Kumari and Singh, 2016). Many researchers have proved that when microbial enzymes are immobilized on a NP matrix enhances their application on a broader range of temperature, pH, enhanced enzyme substrate contacts, and degradation capacity (Dutta *et al.*, 2013). Application of immobilized cells of *Micrococcus luteus* and *Pseudomonas sp.* on crude oil has been shown to enhance sorption and consumption capability by coating cells with magnetic nanoparticles (Shan *et al.*, 2005). Application of zero-valent iron nanoparticles (nZVI) along with compost from organic wastes causes in situ immobilization of As and Cr and also a reduction in the concentration of aliphatic hydrocarbons by up to 60% in soil (Galdames *et al.*, 2017). It has been recently found that the combined use of carbon nanotubes (CNT) and Arthrobacter sp. is emerging as a new method to degrade polychlorinated biphenyls (PCBs), however, high CNT concentrations have been found to reduce the biodegradation rate by inhibiting bacterial growth and microbial activity, while low CNT concentrations increased the biodegradation rate by stimulating bacterial growth and the overexpression of degradation genes (Zhang *et al.*, 2004). Hou *et al.* (2014) found that immobilized cells of *Rhodococcus rhodochrous* in magnetic NPs by using k-carrageenan and Fe3O4 NPs were able to degrade 2-chlorophenol, 4-chlorophenol, 2,3-dichlorophenol, and their mixture; with 30% higher removal efficiency compared to free cells. Similarly various studies have been conducted using different types of NPs to increase the microbial degradation of hazardous environmental pollutants as given in Table 1.

Table 1: Nanobioremediation technology involved in Heavy Metal and xenobiotics degradation

Metals	Biogenic source	Function	References
Ag_2O	*Paeonia emodi*	97.78% methylene blue (MB) was degraded	Shah *et al.* (2019)
Pd	*Chlorella vulgaris*	Removal of 99.8% of trivalent Arsenic from aqueous solutions	Arsiya *et al.* (2017)
Pd	*Spirulina platensis*	Lead removal from aqueous solution with 12% to 90% efficiency	Sayadi *et al.* (2018)
Iron oxide	*Aspergillus tubingensis*	Removal of Pb (II), Ni (II), Cu (II), and Zn (II) from wastewater	Mahanty *et al.* (2020)
Se	*Citrobacter freundii Y9*	Removal of Mercury from environmental sample with 45.8–57.1%	Wang *et al.* (2017)
TiO2/Ag	*Chlorella vulgaris*	Improved the photocatalytic reduction of Cr(VI) and organic pollutants treatment	Wang *et al.* (2017b)
Fe_2O_3/ FeS	*Vaccinium foribundum*	Removal of copper and zinc from water with high efficiencies (>99%)	Abril *et al.* (2018)
Si	*Actinomycetes*	Decolorization of industrial effluent by photocatalytic degradation upto 80%	Mohanraj *et al.* (2020)
NiO	*Microbacterium sp.*	Nickel removal efficiency of 95% from industrial effluents	Sathyavathi *et al.* (2014)
CdS	*Pseudomonas aeruginosa JP-11*	88.7% Cadmium removal efficiency	Raj *et al.* (2016)
Pd	*E. coli MC4100*	Reduction of Cr(VI) to Cr(III) with 85% efficiency	Deplanche *et al.* (2010)
Fe	*Eichhornia crassipes*	89.9% of Cr(VI) was removed by adsorption and reduction mechanism	Wei *et al.* (2017)

2.2. Algae in nanoparticle-based bioremediation

Bioremediation of heavy and toxic metals by phyto-nanotechnology has recently emerged as an effective technique as this is an energy-efficient and environmentally friendly method involving the synthesis of nanoparticles without using any toxic substances (Jeffryes *et al.*, 2015). Various groups of algae (Cyanophyceae, Rhodophyceae, Phaeophyceae, Chlorophyceae) are involved in the biosynthesis of different metallic nanoparticles such as gold, palladium, and silver. Fabrication of NPs of different metals using microalgae which are considered to be nano-bio factories is the fastest method among all the biosynthesizing approaches because of their large-scale bioproduction and can be achieved by using dead or living dried biomass (Zhou *et al.*, 2021). The algal species selected for the biosynthesis of NPs should have important attributes such as ease of harvesting, handling, and disruption of cells, faster growth rate, lower risk of toxicity, easy scalability, and cost-effectiveness (Sharma *et al.*, 2015).

The biosynthesis of silver nanoparticles (AgNPs) and gold nanoparticles (AuNPs)using various strains of brown algae such as (*Sargassum polycystum, Cystophora moniliformis, Gelidiella acerosa, Dictyota bartayresianna, Ecklonia cava, Sargassum wightii,* and *Cystoseira baccata*) as well as red algae species such as *Kappaphycus alvarezii, Kappaphycus* sp., *Galaxaura elongata, and K. alvarezii* was studied by various researchers (Azizi *et al.*, 2014; Castro-Longoria *et al.*, 2012). Silver and gold nanoparticles can also be obtained in different shapes and sizes using blue-green algae (BGA) species such as *Spirulina platensis, Oscillato riawillei,*

Microchaete diplosiphon, Phormidium valderianum, *S. platensis* and other species (Husain *et al.*, 2015). A study conducted by He and Chen, (2014) to find the efficiency of different algal biosorbents for the removal of metals such as Au(I) and Au(III) complexes reported the role of *Chlorella vulgaris*, a green microalga, in the bioaccumulation of heavy metals. Similarly, the biosorption activity of heavy metals cadmium, lead, nickel, copper, and zinc was observed using *Sargassum sp.,* the biomass of a marine brown macroalga, and biosynthesis of AuNPs after a few months (Sheng *et al.*, 2004). Furthermore, the bioremediation of toxic compounds using algal nanomaterials is also found to be an efficient technique. For example, aqueous extracts of *Sargassum tenerrimum* and *Turbinaria conoides* used for the biosynthesis of AuNPs showed higher catalytic activity for a reduction in aromatic nitro compounds organic dye molecules (sulforhodamine 101 and rhodamine B) and (p-nitroaniline and 4-nitrophenol) (Ramakrishna *et al.*, 2016). Degradation of hazardous chemicals such as methylene blue dyes, methyl orange, and other toxic compounds was enhanced by photocatalytic activity using AgNPs synthesized by freshwater microalgae *Chlorella pyrenoidosa,* and *Chlorella ellipsoidea* (Borah *et al.*, 2020; Aziz *et al.*, 2015). Algal-based NPs (Ag, Au, and Cu) have found a wide range of applications such as heavy metal and toxic compound reduction in different environments as shown in Table 2.

Table 2: Different algal and fungal nanoparticles application in Bioremediation.

Microorganisms	Nano-particles	Application	References
		Algal NPs	
Chlorella vulgaris	Ag	Wastewater treatment	Karthikeyan *et al.* (2015)
Chlorococcum sp. MM11	Fe	Remediation of toxic Cr (VI)	Subramaniyam *et al.* (2015)
Scenedesmus sp. 24	Cd-S	Sequestration of cadmium from cadmium polluted water	Jena *et al.*, 2015
Scenedesmus quadricauda	Au, Ag, Cu	Degrade polycyclic aromatic hydrocarbons	Agarwal *et al.* (2019)
Turbinaria conoides	Au	Higher catalytic activity for reduction of aromatic nitro compounds and organic dye molecules	Ramakrishna *et al.* (2016)
Microchaete sp. NCCU-342	Ag	Degrade azo dye and methyl red	Husain *et al.* (2019)
Ulva lactuca	Ag	Photocatalytic degradation of methyl orange dye	Kumar *et al.* (2013)
		Fungal NPs	
Trichoderma koningiopsis	Cu	Treatment of wastewater	Salvadori *et al.* (2014)
Rhizopus oryzae	Au	Strong adsorption capacity for organophosphorous pesticides	Das *et al.* (2009)
Pycnoporus sanguineus	Au	Degradation of 4- nitroaniline	Shi *et al.* (2015)
Trichoderma viride	Au	Degradation of para-nitro Phenol compounds	Mishra *et al.* (2014)
Fusarium solani	Ag	Textile industry	El-Rafie *et al.* (2010)
Fusarium solani	Ag	Silver ions resistance and biotransformation	El Sayed and El-Sayed, (2020)
Alternaria alternata	Ag	Enhancement in antifungal activity of fluconazole against *Phoma glomerata*	Gajbhiye *et al.* (2009)

Phytophthora infestans	Ag	Antibacterial activity	Ray *et al.* (2011)
Fusarium oxysporum	Cd-S	Diagnostic tool	Raheman *et al.* (2011)
Pleurotus djamor	Ag	Anticancer activity	Raman *et al.* (2015)
Aspergillus clavatus	Ag	Antibacterial activity	Verma *et al.* (2011)

* Table content adopted from Bhatt *et al.* (2022).

2.3. Fungi in nanoparticle-based bioremediation

Fungi are a diverse group of multicellular eukaryotic microorganisms heterotrophic in nature and include yeasts and molds, they are resistant to extreme environments and can survive in high metal concentrations which makes them favorable for bioremediation processes (Duhan *et al.*, 2017; Dixit *et al.*, 2015). The bioremediation process using fungal biosynthesized NPs for the removal of heavy metals and toxic compounds from different environmental sources is known as myco-nanotechnology or mycosynthesis of metal nanoparticles is emerging as a new green technology of remediation process (Honary *et al.*, 2012). This is because fungi produce more amount of protein than bacteria which directly results in the production of more nanoparticles with well-defined dimensions (Mohanpuria, Rana, & Yadav, 2008). The beneficial effects of fungi for the synthesis of metal nanoparticles include ease of cultivation, more metal tolerant, higher scalability and metal uptake, better control over morphology, more stability, rapid growth, a large surface area, and cost-effective technology which makes them ideal for industrial-scale production (Yadav *et al.*, 2015). Various fungal groups can biosynthesis metal NPs with significant applications in bioremediation as shown in table 2. Some potential groups include Fusarium, Aspergillus Verticillium, and Penicillium, (Mishra *et al.*, 2021b). Various studies have been conducted on mycosynthesis of metals NPs such as gold–silver alloys, cadmium telluride, titanium, selenium, tellurium, silver, platinum, zirconium, palladium, etc. (Gholami-Shabani *et al.*, 2016). Silver nanoparticles have been reported to be produced intracellularly by various groups of fungi with variable sizes for example, *F. oxysporum, Trichoderma reesei, Coriolus versicolor*, Phoma sp, *Trichoderma viride*, Neurospora crassa; *Phaenerochate chrysosporium* and Aspergillus flavus (Jain *et al.*, 2011; Bhainsa and D'souza, 2006; Vahabi *et al.*, 2011; Sanghi and Verma, 2009; Fayaz *et al.*, 2010). Similarly, the biosynthesis of gold nanoparticles was also reviewed through interactions with the *Cylindrocladium floridanum* fungus, *Verticillium sp.*, *Yarrowia lipolytica*, and *Aspergillus niger* (Soni and Prakash, 2012; Kitching *et al.*, 2015; Narayanan and Sakthivel, 2011). Furthermore, various studies have been conducted for the production of several other fungal metal NPs such as silica, platinum, magnetite, selenium, and cadmium nanoparticles, and their application in bioremediation. (Castro-Longoria *et al.*, 2012; Bharde *et al.*, 2006; Zare *et al.*, 2013). The main mechanism proposed for bioremediation by using metallic nanoparticles synthesized by fungi is the biosorption of heavy metals which convert the toxic form of compounds into less toxic form (Gaur *et al.*, 2014). For example, *Arbuscular mycorrhizal* fungi showed efficient adsorption, and immobilization ability of heavy-metal ions mitigating heavy-metal stress (Upadhyaya *et al.*, 2010).

Recently, fungi from extreme environments have been explored and studied for wider applications of bioremediation of heavy metals and hydrocarbons due to their ability to survive in extreme conditions (Damare *et al.*, 2012). For example, Marine fungi from mangrove areas were found to tolerate high lead and copper concentrations (Thatoi *et al.* 2013; Gazem and Nazareth, 2013). In a similar study, Bonugli-Santos *et al.*, (2012) observed enzyme-based bioremediation for Remazol Brilliant Blue-R dye using three basidiomycetes obtained from marine sponges and anthraquinone dye Reactive Blue 4 using Cerrena unicolor.

3. Nanotechnology and Water Pollution

Nanoparticles and nanomaterials synthesized by microorganisms have increasingly wider applications, especially the removal of various organic and inorganic compounds as well as heavy metals from polluted environments (Figure 3).

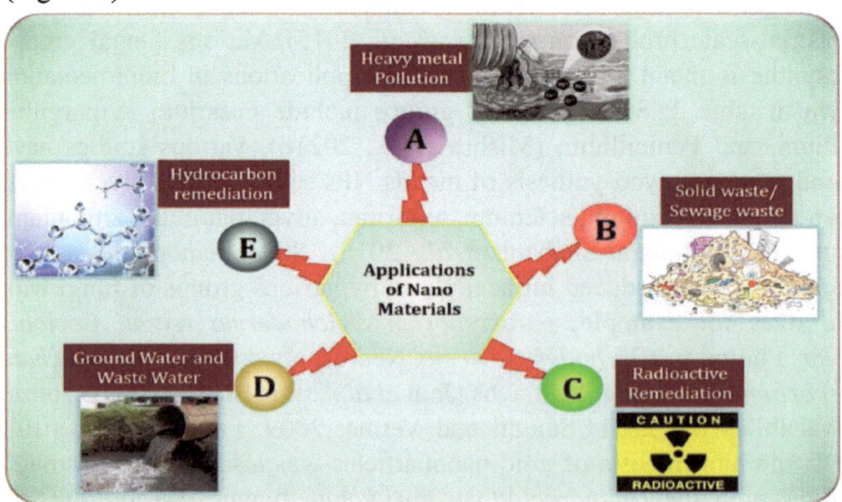

Fig. 3: Application of biosynthesized NPs/NMs in heavy metal polluted environment.

The conventional techniques of wastewater treatment from industrial effluents to reduce the effect of toxicity include electrochemical, advanced oxidation processes, and valorization techniques (Gupta and Shukla, 2020). But with the development and advancement of nanotechnology and nanoscience have made wider and effective methods of the remediation of water pollutants in comparison to the conventional methods as these are cost-effective, have superior chemical properties, and have high surface area-to-volume ratio (Baruah *et al.*, 2019). Moreover, biosynthesis of these NMs using microorganisms has opened new avenues for sustainable and environmentally friendly methods of toxic and heavy metal remediation from water sources as shown in Figure 4.

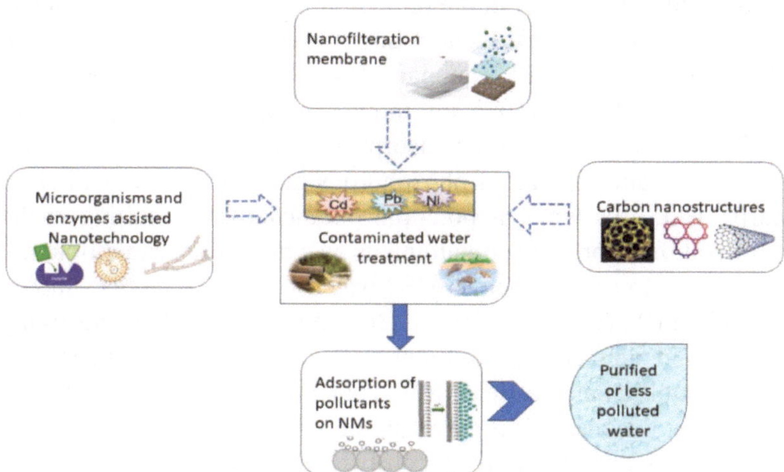

Fig. 4: Microbial-assisted nanotechnology for the treatment of HMs polluted water

Due to the smaller size of NMs their physical, chemical, and biological properties get enhanced which makes them suitable for a wider range of applications. These NPs mainly include carbon-based, metal, and metal oxide-based nanoparticles which are used as adsorbents to remove organic and inorganic pollutants from industrial wastewater (Kumari *et al.*, 2019). Carbon nanotubes (CNTs), activated carbon, graphene, and, fullerene are examples of carbon-based nanoparticles among them CNTs are efficient adsorbents for toxic chemicals from industrial wastewater (Kumari *et al.*, 2019). These are exceptional macromoles that can be used as single-walled nanotubes (SWNTs) and multi-walled nanotubes (MWNTs). For example, multi-walled carbon nanotubes (MWCNTs) prepared from Fe-Ni supported on activated carbon using the chemical vapor deposition (CVD) technique showed a good range (2–10) of adsorption characteristics for metronidazole and levofloxacin, respectively (Kariim *et al.*, 2020). A nanocomposite made with activated carbon-modified nanomagnets was also used for the removal of fluoride ions from wastewater (Takmil *et al.*, 2020).

Gold and copper NPs are effective in the nano remediation of wastewater systems and the degradation of harmful chemicals into less toxic forms. AgNPs have shown the removal of phosphorus, and sulfur elements from polluted water as well as anti-microbial effects against pathogenic bacteria. The main mechanism involves the initial, de-polymerization of the cell due to penetration of the NPs followed by the formation of ROS which interferes with DNA replication and energy production by the cell (Slavin *et al.*, 2017). Metal oxide NPs and nano-adsorbents of magnetic iron, aluminum, and titanium have been found efficient in the remediation of various heavy metals such as lead, arsenic, copper, nickel, cadmium, mercury and chromium from wastewater. For instance, iron oxide nanoparticles imbedded in hyperbranched polyglycerol were found to remove aluminum, nickel, and copper, from wastewater in 35s without affecting the adsorption efficiency of nanoparticles due to organic content and amount of phosphorous in water (Almomani *et al.* 2020). Similarly, TiO_2 NPs work as potential photocatalysts resulting in the complete solubilization of organic contaminants as well as they can efficiently remove a wide range of viruses, such as MS2 bacteriophages, Herpes simplex virus, and poliovirus (Panahi *et al.*, 2018). Biosynthesis of NPs using living cells from bacteria and plant cells act as reductive agents for the metal complex salt and generate metallic nanoparticles which can attain higher strength in an aqueous environment because of co-precipitation or by adding proteinaceous and bioactive elements onto the outer face of the nanoparticles. Mahanty *et al.* (2020) biofabricated iron oxide nanoparticles from *Aspergillus tubingensis* (STSP 25) which were able to remove more than 90% of heavy metals [Pb (II), Ni (II), Cu (II), and Zn (II)] from wastewater with a regeneration capability of up to five cycles. In another study, alginate-encapsulated bacteria were applied for the precipitation of the Cd^{2+} contaminated water (Xu *et al.*, 2021). Microbially synthesized NPs contain various functional groups that interact with pollutants and play a role in the bioremediation of toxic and heavy metals. The removal of organic/inorganic compounds from dairy wastewater using a *Leptolyngbya consortium* has been investigated (Tsolcha *et al.*, 2018). *Kluyveromyces marxianus* and *Lactococcus lactis* were involved in the bioremediation of organic matter and heavy metals, such as Ni and Cd, with 75% removal efficiency (Cherni *et al.*, 2020). Enzymes immobilized on NPs become more stable due to resistance in unfolding resulting in less vulnerability to diffusional constraints and can be used in multiple cycles, with enhanced kinetic characteristics (Ding *et al.*, 2015). For example, use of immobilized Laccase is widely used for the treatment of industrial effluents and biodegradation using magnetic nanoparticles such as Fe_3O_4 and chitosan composite (Zhang *et al.*, 2020).

4. Application of NPs in Soil Pollution

Contamination of soil with heavy metals or toxic compounds is alarming and becoming a threat to food security due to its long-term persistency and resistance to degradation. Accumulation of heavy metals results in changes in structure, function and microbiome of soil which causes loss in its fertility (Saleem *et al.*, 2020). To restore soil fertility and remove of contaminants various physico-chemical and biological methods have been employed such as thermal desorption, soil stabilization, soil excavation, chemical extraction, phytoremediation, etc (Raffa *et al.* 2021; Song *et al.* 2017). These techniques were not able to be applied in a large scale due to cost and time constrain, generation of secondary pollutants, and unsatisfactory results (Ullah *et al.* 2015). Nowadays remediation techniques have been shifted towards more sustainable, eco-friendly, and cost-effective methods along with precision in the result. Nanotechnology in combination with other conventional techniques is emerging as most effective and potent method in the future for mitigating heavy metal pollution (Yang et al. 2019). Metals such as silver, iron, gold, palladium, titanium, and copper, etc. are used as nano-materials as they are less expensive and have higher removal efficiency to remove heavy metals (Yu *et al.* 2021; Kaur and Roy, 2021; Yang *et al.* 2019). Due to the lesser size of nanoparticles, they can penetrate deeper into the soil and form a complex with inorganic and organic components of soil systems causing biological, chemical, and physical transformations. Various biotic and abiotic factors affect the interaction of NPs and heavy metals interaction which affect the uptake, translocation and accumulation of NPs into various plant organs. For example, Manzoor *et al.* (2021) found that the adsorption of cadmium from soil by FeO NPs is largely affected due to their properties such as reactivity, electrostatic attraction, large surface area, and capping molecules. Several novel NMs such as graphene oxide (GO) with ultra-large surface area found to form a conjugate with heavy metals like Cr (VI) due to their functional group increasing their adsorption (Wang *et al.*, 2017a, b). Use of nanobioremediation which is an integration of various NMs and NPs with the biological process including soil microbiome is emerging as a promising, cost effective and eco friendly method for the removal of organic and inorganic pollutants from soil (Bhatt *et al.* 2021; Singh *et al.* 2020). Various species of soil microbiome including bacteria, fungi, and algae or their cell extracts, biosynthesize NPs like zinc, silver, gold, copper, etc. through extracellular or intracellular routes (Jacob *et al.* 2021; Elegbede *et al.* 2018). These NPs decontaminate the toxic compounds from soil by transforming or inducing, immobilizing, synthesis of microbial enzymes having antioxidant or reducing properties which catalyse the process (Singh *et al.* 2020). However, various physical, chemical and biological factors affect the biosynthesis of metallic nanoparticles by microorganisms in attaining a controlled size, shape, dispersity, and reactivity (Anandaradje *et al.* 2020). Bacteria in soil ecology could proliferate in an environment contaminated with toxic metal as they have developed protective mechanisms such as biotransformation, biomineralization, and bioreduction to alleviate metal toxicity (Narayanan and Sakthivel 2010). Thus, these bacterial cells because of their exceptional metal-binding abilities of the bacterial cells and presence of S-layers can be used in a biofactory for the synthesis of different types of metallic NPs such as, platinum, gold, cadmium sulfide, silver, palladium, titanium, titanium dioxide, magnetite, etc. Palladium (Pd) nanoparticles biosynthesized by *Clostridium pasteurianum* have shown a significant reduction in carcinogenic Cr by transforming its hexavalent form to insoluble trivalent form (Alexakis 2016; Chidambaram *et al.* 2010). Polymeric mediated synthesis of CdS nanoparticles prepared from Gram-negative bacterium *Pseudomonas aeruginosa* efficiently removed Cd from a Cd-contaminated environment (Raj *et al.* 2016). In another study lead sulfide (PbS) nanoparticles synthesized from yeast *Rhodosporidium diobovatum*, were able to bio-transform the toxic Pb (II) ions into less toxic forms (Seshadri *et al.* 2011).

5. Nanotechnology and Air Pollution

Rapid industrialization and urbanization as well as various anthropogenic factors result in the release of various harmful and toxic gaseous compounds directly into the air in different forms such as oxides and dioxide form of carbon, nitrogen, sulfur, ozone, suspended airborne particles, and volatile organic compounds (VOCs) etc (Daly 2007). The emission of these gases causes alteration in the natural composition of the atmosphere, moreover adversely affects natural flora and fauna as well as human health by causing various types of diseases that can be fatal, such as respiratory, and cardiovascular diseases. The most important gaseous pollutants that contribute to global warming are greenhouse gases (GHG) which include carbon dioxide, methane, nitrous oxide, and fluorinated gases (Ibrahim *et al.* 2016). Removal and monitoring of GHG is a major challenge for the scientific community as these gases have long-term persistence and high environmental risk. Nanotechnology is emerging as a revolution to face the challenges of environmental remediation by synthesizing nanoparticle/ nanomaterials and applying them as adsorbents, catalysts, membranes, and sensors (Zhao 2009).

Carbon-based materials (CNMs) have been found very effective as CO_2 low-temperature adsorbents due to their large surface, high pore structure modification ability, and surface functionalization (Upendar *et al.* 2012). Application of carbon nanotube (CNTs) in CO_2 adsorption with enhanced performance in the presence of moisture and adsorption of different types of natural and inorganic compounds can be done in form of single-walled nanotubes (SWNTs) and as multi-walled nanotubes (MWNTs). Carbon nanotubes in combination with 3-aminopropyl-triethoxysilane (CNTs-APTS) create abundant amine groups on the surface which act as chemical sites for CO_2 adsorption even at low temperature range (20–100 °C) (Su *et al.* 2009). Adsorption of carbon dioxide at high temperature can be performed using Calcium (Ca)-based nano-adsorbents based on the reversible carbonation reaction of calcium oxides (CaO). Coating these nano calcium carbonates using titanium dioxide (TiO_2) can prevent problem of sintering during the carbonation/calcination cycles (Abanades and Alvarez 2003). Thermal decomposition of methane (CH4) to produce hydrogen has been reported using metallic nickel nanoparticles as catalysts and TiO_2 coated with stainless steel web net was also found efficient for photocatalytic degradation of CH4 (Wang and Lua 2012; Merajin *et al.* 2013). Similarly, titanate nanotubes (TNTs) and their derivatives have been widely reported for the photocatalytic oxidation of NOx. Such as increasing the surface area, and the amount of crystalline, sodium content of TNTs increased the removal efficiency of both NO and NO_2 (Nguyen and Bai, 2015). Elimination of SO_2 from the environment have been done using nanomaterials either by desulfurization of fossil fuel or by its removal directly from the emission source by various technologies, such as adsorption processes and catalytic oxidations (Rodriguez *et al.* 2010; Wu *et al.* 2011; Saleh *et al.* 2015). Moreover, nanoparticles can also be applied to control and eradicate the emission of VOCs such as formaldehyde (HCHO), by electrospun polyacrylonitrile (PAN)-based carbon nanofiber (CNF) membrane with higher microporosity and abundant nitrogen-containing functional groups (Lee *et al.*, 2010a, b). The application of various nanoparticles to treat different types of air pollutants is shown in Table 3.

Table 3: Application of various NPs for air pollutant remediation

Nanoparticles	Pollutant	Effect	Reference
Titanium dioxide nanoparticles (TNPs)	VOCs (ethylene)	Enhanced photocatalytic degradation of ethylene was done by Sol-gel method synthesized Novel TNPs with superior characteristics	Hussain *et al.* (2010)
Titanium dioxide nanoparticles (TNPs)	(SO_2)	Enhanced SO_2 adsorption in the absence as well as the presence of light irradiation	Baltrusaitis *et al.* (2011)
Single-walled carbon nanotubes (SWCNTs)	(N_2), (CH_4), (CO), (CO_2)	Maximum adsorption uptake was reported in the case of CO_2 followed by CO, CH_4, N_2 and H_2 by SWCNTs	Lithoxoos *et al.* (2010)
Silica nanoparticle (SiNPs)	Pb	Due to the large surface and the negative-charged groups in the SiNPs adsorption capacity of atmospheric lead was enhanced	Yang *et al.* (2013c)
Zinc oxide (ZnO) and zirconium hydroxide Nanoparticles ($Zr(OH)_4$)	(NO_2), (SO_2)	Chemisorption of SO_3 and NO_3 results due to exposure of ZnO and $Zr(OH)_4$ powders to SO_2 and NO_2	Singh *et al.* (2012)

* Table content is adopted from Ibrahim *et al.*, (2016).

6. Regulations of Nanotechnologies Applied in Bioremediation

Although nanotechnology is revolutionizing the current and future applications of conventional techniques, there is a need to assess the unintended risk associated with the production and application of nanomaterials for the remediation of human health and the environment. This concern related to biosafety is mainly related to a lack of knowledge and understanding of the fate and behavior of nanoparticles (Sun *et al.*, 2019). There is a need for validated protocols related to the effect of NMs on human health, their bioaccumulation, and their persistence in our food chain. Currently, there are many international organizations and institutions such as the European Observatory for Nanomaterials (EON), USEPA, ISO Technical Committee TC 229 "Nanotechnologies" and the OECD Working Party on Manufactured Nanomaterials (WPMN) and have established international cooperation to increase the application of available regulations (Rasmussen *et al.*, 2016; Kica, and Wessel, 2017). Risk assessment research of nanotechnology for establishing the potential impacts of nanoparticles on human health and the environment is crucial for balancing the technology's benefits and potential unintended consequences. Multi-criteria Decision Analysis (MDA) is one of the most reliable monitoring frameworks for nanoparticle risk assessment which connects efficiency details with decision criteria and permits quantification and visualization of the specific tradeoffs (Linkov *et al.*, 2007).

7. Conclusion and Future Prospects

Heavy metal environmental pollution is a global threat due to rapid urbanization and various anthropogenic activities. Remediation of these pollutants by nanotechnology is revolutionizing the conventional and existing technology due to nano size/shape, larger surface area, high reactivity, and surface charge which make them easy to apply and more efficient. The application of nanotechnology in combination with microorganisms is paving the way for a sustainable, environmentally friendly, and cost-effective manner. As discussed in this chapter many bacteria, fungi, and algae act as "biofactories" for the synthesis of various NPs and are applied for decontaminating and bioremediating HMs and toxic compound contamination from soil, water, and other environments with greater effectiveness. Moreover, enzyme-based nanotechnology derived from plants and microorganisms is emerging as a green and long-term approach for nano bioremediation which should be commercialized and pursued further to explore its maximum potential. Future research should focus on studying the detailed mechanisms involved in the biosynthesis of NPs using microorganisms, their wider application and the exploration of mixed microbial cultures for HM and toxic pollutant remediation at a large scale. Also, for the safe implementation and application of nanoparticles/nanomaterials extensive field trials as well as in vitro and in vivo trials of magnetic nanoparticles should be evaluated to establish their higher efficiency, reliability, and reproducibility. Furthermore, Life cycle assessment (LCA) of various metal NPs after application in the environment should be made to overcome the limitations of these particles such as their potential toxicity, recycling ability, production cost, and interaction with the production media. The use of novel nanoparticles and nanomaterials, and their safe and wider application by industrial collaboration can pave the way for the development of efficient and green HM nanoremediation approaches to achieve food safety/security goals at a global scale.

References

Abanades, J.C. & Alvarez, D. (2003). Conversion limits in the reaction of CO_2 with lime. Energy & Fuels, 17(2), 308-315.

Abril, M., Ruiz, H. & Cumbal, L.H., (2018). Biosynthesis of multicomponent nanoparticles with extract of mortiño (Vaccinium floribundum Kunth) berry: application on heavy metals removal from water and immobilization in soils. Journal of Nanotechnology/

Agarwal, P., Gupta, R. & Agarwal, N. (2019). Advances in synthesis and applications of microalgal nanoparticles for wastewater treatment. Journal of Nanotechnology.

Ahmad, B., Zaid, A., Zulfiqar, F., Bovand, F. & Dar, T.A. (2023). Nanotechnology: A novel and sustainable approach towards heavy metal stress alleviation in plants. Nanotechnology for Environmental Engineering, 8(1), 27-40.

Alexakis, D. (2016). Human health risk assessment associated with Co, Cr, Mn, Ni and V contents in agricultural soils from a Mediterranean site. Archives of Agronomy and Soil Science, 62(3), 359-373.

Almomani, F., Bhosale, R., Khraisheh, M. & Almomani, T., 2020. Heavy metal ions removal from industrial wastewater using magnetic nanoparticles (MNP). Applied Surface Science, 506, p.144924.

Anandaradje, A., Meyappan, V., Kumar, I. & Sakthivel, N. (2020). Microbial synthesis of silver nanoparticles and their biological potential. Nanoparticles in medicine, 99-133.

Aziz, N., Faraz, M., Pandey, R., Shakir, M., Fatma, T., Varma, A., Barman, I., & Prasad, R., 2015. Facile algae-derived route to biogenic silver nanoparticles: synthesis, antibacterial, and photocatalytic properties. Langmuir, 31(42), 11605–11612.

Azizi, S., Ahmad, M.B., Namvar, F., & Mohamad, R. (2014). Green biosynthesis and characterization of zinc oxide nanoparticles using brown marine macroalga Sargassum muticum aqueous extract. Materials Letter, 116, 275–277.

Baltrusaitis, J., Jayaweera, P.M. & Grassian, V.H. (2011). Sulfur dioxide adsorption on TiO2 nanoparticles: influence of particle size, coadsorbates, sample pretreatment, and light on surface speciation and surface coverage. The Journal of Physical Chemistry C, 115(2), 492-500.

Baruah, A., Chaudhary, V., Malik, R., & Tomer, V. K., 2019. Nanotechnology based solutions for wastewater treatment. Nanotechnology in Water Wastewater Treatment, 337–368.

Bhainsa, K.C. and D'souza, S.F., (2006). Extracellular biosynthesis of silver nanoparticles using the fungus Aspergillus fumigatus. Colloids and surfaces B: Biointerfaces, 47(2),160-164.

Bharagava, R.N., Saxena, G. & Mulla, S.I., 2020. Introduction to industrial wastes containing organic and inorganic pollutants and bioremediation approaches for environmental management. Bioremediation of Industrial Waste for Environmental Safety: Volume I: Industrial Waste and Its Management, 1-18.

Bharde, A., Rautaray, D., Bansal, V., Ahmad, A., Sarkar, I., Yusuf, S.M., Sanyal, M. and Sastry, M. (2006). Extracellular biosynthesis of magnetite using fungi. Small, 2(1), 135-141.

Bhatt, P., Pandey, S.C., Joshi, S., Chaudhary, P., Pathak, V.M., Huang, Y., Wu, X., Zhou, Z. & Chen, S. 2022. Nanobioremediation: A sustainable approach for the removal of toxic pollutants from the environment. Journal of Hazardous Materials, 427, 128033.

Bolade, O.P., Williams, A.B. & Benson, N.U. (2020). Green synthesis of iron-based nanomaterials for environmental remediation: A review. Environmental Nanotechnology, Monitoring & Management, 13, p.100279.

Bonugli-Santos, R.C., Durrant, L.R. and Sette, L.D. (2012). The production of ligninolytic enzymes by marine-derived basidiomycetes and their biotechnological potential in the biodegradation of recalcitrant pollutants and the treatment of textile effluents. Water, Air, & Soil Pollution, 223, 2333-2345.

Borah, D., Das, N., Das, N., Bhattacharjee, A., Sarmah, P., Ghosh, K., Chandel, M., Rout, J. *et al.* (2020). Alga□mediated facile green synthesis of silver nanoparticles: Photophysical, catalytic and antibacterial activity. Applied Organometallic Chemistry, 34(5), p.e5597.

Briffa, J., Sinagra, E. & Blundell, R. (2020). Heavy metal pollution in the environment and their toxicological effects on humans. Heliyon, 6(9).

Castro-Longoria, E., Moreno-Velasquez, S.D., Vilchis-Nestor, A.R., Arenas-Berumen, E. and Avalos-Borja, M. (2012). Production of platinum nanoparticles and nanoaggregates using Neurospora crassa. Journal of microbiology and biotechnology, 22(7), 1000-1004.

Cherni, Y., Botta, C., Kasmi, M., Franciosa, I., Cocolin, L., Chatti, A., Trabelsi, I. & Elleuch, L. (2020). Mixed culture of Lactococcus lactis and Kluyveromyces marxianus isolated from kefir grains for pollutants load removal from Jebel Chakir leachate. Water Environment Research, 92(12), 2041-2048.

Chidambaram, D., Hennebel, T., Taghavi, S., Mast, J., Boon, N., Verstraete, W., van der Lelie, D. & Fitts, J.P. (2010). Concomitant microbial generation of palladium nanoparticles and hydrogen to immobilize chromate. Environmental Sci. & Technology, 44(19),7635-7640.

Daly, A. & Zannetti, P. (2007). An introduction to air pollution–definitions, classifications, and history. Ambient Air Pollution P Zannetti, D Al-Ajmi and S Al-Rashied, The Arab School for Science and Technology and The EnviroComp Institute, 1-14.

Damare, S., Singh, P. & Raghukumar, S. (2011). Biotechnology of marine fungi. Biology of Marine fungi, 277-297.

Das, S.K., Das, A.R. & Guha, A.K. (2009). Gold nanoparticles: microbial synthesis and application in water hygiene management. Langmuir, 25(14), pp.8192-8199.

Deplanche, K., Caldelari, I., Mikheenko, I.P., Sargent, F. & Macaskie, L.E. (2010). Involvement of hydrogenases in the formation of highly catalytic Pd (0) nanoparticles by bioreduction of Pd (II) using Escherichia coli mutant strains. Microbiology, 156(9), pp.2630-2640.

Ding, S., Cargill, A.A., Medintz, I.L. & Claussen, J.C. (2015). Increasing the activity of immobilized enzymes with nanoparticle conjugation. Current opinion in biotechnology, 34, 242-250.

Dixit, R., Wasiullah, X., Malaviya, D., Pandiyan, K., Singh, U.B., Sahu, A., Shukla, R., Singh, B.P., Rai, J.P., Sharma, P.K. & Lade, H. (2015). Bioremediation of heavy metals from soil and aquatic environment: an overview of principles and criteria of fundamental processes. Sustainability, 7(2), 2189-2212.

Duhan, J.S., Kumar, R., Kumar, N., Kaur, P., Nehra, K. & Duhan, S. (2017). Nanotechnology: The new perspective in precision agriculture. Biotechnology reports, 15, 11-23.

Dutta, N., Mukhopadhyay, A., Dasgupta, A.K. & Chakrabarti, K. (2013). Nanotechnology enabled enhancement of enzyme activity and thermostability: study on impaired pectate lyase from attenuated Macrophomina phaseolina in presence of hydroxyapatite nanoparticle. PLoS One, 8(5), p.e63567.

El Sayed, M.T. & El-Sayed, A.S. (2020). Biocidal activity of metal nanoparticles synthesized by Fusarium solani against multidrug-resistant bacteria and mycotoxigenic fungi. Journal of Microbiology and Biotechnology, 30(2), p.226.

Elegbede, J.A., Lateef, A., Azeez, M.A., Asafa, T.B., Yekeen, T.A., Oladipo, I.C., Adebayo, E.A., Beukes, L.S. & Gueguim□Kana, E.B., 2018. Fungal xylanases□mediated synthesis of silver nanoparticles for catalytic and biomedical applications. IET nanobiotechnology, 12(6), 857-863.

El-Rafie, M.H., Mohamed, A.A., Shaheen, T.I. & Hebeish, A. (2010). Antimicrobial effect of silver nanoparticles produced by fungal process on cotton fabrics. Carbohydrate Polymers, 80(3), 779-782.

Fayaz, M., Tiwary, C.S., Kalaichelvan, P.T. and Venkatesan, R., 2010. Blue orange light emission from biogenic synthesized silver nanoparticles using Trichoderma viride. Colloids and Surfaces B: Biointerfaces, 75(1), pp.175-178.

Fernández, P.M., Viñarta, S.C., Bernal, A.R., Cruz, E.L. & Figueroa, L.I. (2018). Bioremediation strategies for chromium removal: current research, scale-up approach and future perspectives. Chemosphere, 208, 139-148.

Gajbhiye, M., Kesharwani, J., Ingle, A., Gade, A. & Rai, M. (2009). Fungus-mediated synthesis of silver nanoparticles and their activity against pathogenic fungi in combination with fluconazole. Nanomedicine: Nanotechnology, Biology and Medicine, 5(4), 382-386.

Galdames, A., Mendoza, A., Orueta, M., de Soto García, I.S., Sanchez, ′ M., Virto, I. & Vilas, J.L. (2017). Development of new remediation technologies for contaminated soils based on the application of zero-valent iron nanoparticles and bioremediation with compost. Resource-Efficient Technologies. 3, 166–176.

Gaur, N., Flora, G., Yadav, M. & Tiwari, A. (2014). A review with recent advancements on bioremediation-based abolition of heavy metals. Environmental Science: Processes & Impacts, 16(2), 180-193.

Gazem, M.A. & Nazareth, S. (2013). Sorption of lead and copper from an aqueous phase system by marine-derived Aspergillus species. Annals of microbiology, 63, 503-511.

Gholami-Shabani, M., Shams-Ghahfarokhi, M., Gholami-Shabani, Z. & Razzaghi-Abyaneh, M. (2016). Microbial enzymes: current features and potential applications in nanobiotechnology. Advances and Applications through Fungal Nanobiotechnology, 91-127.

Guerra, F.D., Attia, M.F., Whitehead, D.C., & Alexis, F. (2018). Nanotechnology for environmental remediation: materials and applications. Molecules, 23, 1760.

Gupta, G. K., & Shukla, P. (2020). Insights into the resources generation from pulp and paper industry wastes: challenges, perspectives and innovations. Bioresource Technology, 297:122496.

Gupta, G.K. & Shukla, P. (2020). Insights into the resources generation from pulp and paper industry wastes: challenges, perspectives and innovations. Bioresource technology, 297, p.122496.

He, J. & Chen, J.P. (2014). A comprehensive review on biosorption of heavy metals by algal biomass: materials, performances, chemistry, and modeling simulation tools. Bioresource technology, 160, 67-78.

He, J., & Chen, J.P. (2014). A comprehensive review on biosorption of heavy metals by algal biomass: materials, performances, chemistry, and modeling simulation tools. Bioresoure Technology, 160, 67–78.

Honary, S., Barabadi, H., Gharaei-Fathabad, E. & Naghibi, F. (2012). Green synthesis of copper oxide nanoparticles using Penicillium aurantiogriseum, Penicillium citrinum and Penicillium waksmanii. Digest Journal of Nanomaterials and Biostructures 7(3), 999-1005.

Hou, J., Dong, G., Ye, Y. & Chen, V. (2014). Laccase immobilization on titania nanoparticles and titania-functionalized membranes. Journal of Membrane Science, 452, 229-240.

Husain, S., Afreen, S., Yasin, D., Afzal, B. & Fatma, T. (2019). Cyanobacteria as a bioreactor for synthesis of silver nanoparticles-an effect of different reaction conditions on the size of nanoparticles and their dye decolorization ability. Journal of Microbiological Methods, 162, 77-82.

Husain, S., Sardar, M., & Fatma, T. (2015). Screening of cyanobacterial extracts for synthesis of silver nanoparticles. World Journal of Microbiology and Biotechnology, 31, 1279–1283.

Hussain, M., Ceccarelli, R., Marchisio, D.L., Fino, D., Russo, N. & Geobaldo, F. (2010). Synthesis, character(rization, and photocatalytic application of novel TiO2 nanoparticles. Chemical Engineering Journal, 157(1), 45-51.

Ibrahim, R.K., Hayyan, M., AlSaadi, M.A., Hayyan, A. and Ibrahim, S. (2016). Environmental application of nanotechnology: air, soil, and water. Environmental Science and Pollution Research, 23, 13754-13788.

Iravani, S., Thota, S. & Crans, D.C., 2017. Methods for preparation of metal nanoparticles, Vol. 63, pp. 15-31. Wiley: Weinheim, Germany.

Jacob, J.M., Ravindran, R., Narayanan, M., Samuel, S.M., Pugazhendhi, A. & Kumar, G. (2021). Microalgae: A prospective low cost green alternative for nanoparticle synthesis. Current Opinion in Environmental Science & Health, 20, p.100163.

Jain, N., Bhargava, A., Majumdar, S., Tarafdar, J.C. & Panwar, J. (2011). Extracellular biosynthesis and characterization of silver nanoparticles using Aspergillus flavus NJP08: a mechanism perspective. Nanoscale, 3(2), S635-641.

Jeffryes, C., Agathos, S.N. & Rorrer, G., (2015). Biogenic nanomaterials from photosynthetic microorganisms. Current Opinion in Biotechnology, 33, 23-31.

Jena, J., Pradhan, N., Aishvarya, V., Nayak, R.R., Dash, B.P., Sukla, L.B., Panda, P.K. & Mishra, B.K., (2015). Biological sequestration and retention of cadmium as CdS nanoparticles by the microalga Scenedesmus-24. Journal of Applied Phycology, 27, 2251-2260.

Joutey, N.T., Bahafid, W., Sayel, H. & El Ghachtouli, N. (2013). Biodegradation: involved microorganisms and genetically engineered microorganisms. Biodegradation-life of Science, 1, 289-320.

Kariim, I., Abdulkareem, A.S. & Abubakre, O.K. (2020). Development and characterization of MWCNTs from activated carbon as adsorbent for metronidazole and levofloxacin sorption from pharmaceutical wastewater: kinetics, isotherms, and thermodynamic studies. Scientific African, 7, p.e00242.

Karthikeyan, P., Mohan, D., Abishek, G. and Priya, R. (2015). Synthesis of silver nanoparticles using Phytoplankton and its characteristics. International Journal of Fisheries and Aquatic Studies, 2(6), 398-401.

Kaur, S. & Roy, A. (2021). Bioremediation of heavy metals from wastewater using nanomaterials. Environment, Development and Sustainability, 23(7), 9617-9640.

Kica, E. & Wessel, R.A. (2017). Transactional arrangements in the governance of emerging technologies: The case of nanotechnology. In Embedding New Technologies into Society (pp. 219-257). Jenny Stanford Publishing.

Kitching, M., Ramani, M. & Marsili, E. (2015). Fungal biosynthesis of gold nanoparticles: mechanism and scale up. Microbial Biotechnology, 8(6), 904-917.

Koul, B. and Taak, P., 2018. Biotechnological strategies for effective remediation of polluted soils. Springer.

Kumar, P., Govindaraju, M., Senthamilselvi, S. & Premkumar, K. (2013). Photocatalytic degradation of methyl orange dye using silver (Ag) nanoparticles synthesized from Ulva lactuca. Colloids and Surfaces B: Biointerfaces, 103, 658-661.

Kumar, V., Thakur, I.S. & Shah, M.P. (2020). Bioremediation approaches for treatment of pulp and paper industry wastewater: Recent advances and challenges. Microbial bioremediation & biodegradation, 1-48.

Kumar[1], N., Kumar, V., Panwar, R. & Ram, C., 2017. Efficacy of indigenous probiotic Lactobacillus strains to reduce cadmium bioaccessibility-an in vitro digestion model. Environmental Science and Pollution Research, 24, 1241-1250.

Kumar[2], N., Kumari, V., Ram, C., Bharath Kumar, B.S. & Verma, S., 2018. Impact of oral cadmium intoxication on levels of different essential trace elements and oxidative stress measures in mice: a response to dose. Environmental Science and Pollution Research, 25, 5401-5411.

Kumar[3], N., Kumari, V., Ram, C., Thakur, K. & Tomar, S.K. (2018). Bio-prospectus of cadmium bioadsorption by lactic acid bacteria to mitigate health and environmental impacts. Applied Microbiology and Biotechnology, 102, 1599-1615.

Kumari, B. & Singh, D.P. (2016). A review on multifaceted application of nanoparticles in the field of bioremediation of petroleum hydrocarbons. Ecological Engineering, 97, 98-105.

Kumari, P., Alam, M. & Siddiqi, W.A., 2019. Usage of nanoparticles as adsorbents for waste water treatment: An emerging trend. Sustainable Materials and Technologies, 22, p. 00128.

Lall, S., Pandey, A. & Man, J. (2022). Biosorption potential of Saraca asoca bark powder for removal of Cr (VI) ions from aqueous solution. Pollution, 8(1), 249–267.

Lead, J.R., Batley, G.E., Alvarez, P.J., Croteau, M.N., Handy, R.D., McLaughlin, M.J., Judy, J.D. & Schirmer, K., 2018. Nanomaterials in the environment: behavior, fate, bioavailability, and effects—an updated review. Environmental Toxicology and Chemistry, 37(8), 2029-2063.

Lee, B.U., Yun, S.H., Jung, J.H. & Bae, G.N. (2010). Effect of relative humidity and variation of particle number size distribution on the inactivation effectiveness of airborne silver nanoparticles against bacteria bioaerosols deposited on a filter. Journal of aerosol Science, 41(5), 447-456.

Lee, K.J., Shiratori, N., Lee, G.H., Miyawaki, J., Mochida, I., Yoon, S.H. & Jang, J. (2010). Activated carbon nanofiber produced from electrospun polyacrylonitrile nanofiber as a highly efficient formaldehyde adsorbent. Carbon, 48(15), pp.4248-4255.

Li, X., Liu, Y., Zhang, C., Wen, T., Zhuang, L., Wang, X., Song, G., Chen, D., Ai, Y., Hayat, T. and Wang, X., 2018. Porous Fe2O3 microcubes derived from metal organic frameworks for efficient elimination of organic pollutants and heavy metal ions. Chemical Engineering Journal, 336, 241-252.

Li, Z., Chen, Z., Zhu, Q., Song, J., Li, S. & Liu, X. (2020). Improved performance of immobilized laccase on Fe3O4@ C-Cu2+ nanoparticles and its application for biodegradation of dyes. Journal of Hazardous Materials, 399, p.123088.

Lin, Z., Pang, S., Zhou, Z., Wu, X., Li, J., Huang, Y., Zhang, W., Lei, Q., Bhatt, P., Mishra, S. & Chen, S. (2022). Novel pathway of acephate degradation by the microbial consortium ZQ01 and its potential for environmental bioremediation. Journal of Hazardous Materials, 426, p.127841.

Linkov, I., Satterstrom, F.K., Steevens, J., Ferguson, E. & Pleus, R.C. (2007). Multi-criteria decision analysis and environmental risk assessment for nanomaterials. Journal of Nanoparticle Research, 9, 543-554.

Lithoxoos, G.P., Labropoulos, A., Peristeras, L.D., Kanellopoulos, N., Samios, J. & Economou, I.G. (2010). Adsorption of N2, CH4, CO and CO2 gases in single walled carbon nanotubes: A combined experimental and Monte Carlo molecular simulation study. The Journal of Supercritical Fluids, 55(2), 510-523.

Mahanty, S., Chatterjee, S., Ghosh, S., Tudu, P., Gaine, T., Bakshi, M., Das, S., Das, P., Bhattacharyya, S., Bandyopadhyay, S. & Chaudhuri, P. (2020). Synergistic approach towards the sustainable management of heavy metals in wastewater using mycosynthesized iron oxide nanoparticles: biofabrication, adsorptive dynamics and chemometric modeling study. Journal of Water Process Engineering, 37, p.101426.

Manzoor, N., Ahmed, T., Noman, M., Shahid, M., Nazir, M.M., Ali, L., Alnusaire, T.S., Li, B., Schulin, R. & Wang, G. (2021). Iron oxide nanoparticles ameliorated the cadmium and salinity stresses in wheat plants, facilitating photosynthetic pigments and restricting cadmium uptake. Science of the Total Environment, 769, p.145221.

Masindi, V. & Muedi, K.L. (2018). Environmental contamination by heavy metals. Heavy Metals, 10(4), 115-133.

Merajin, M.T., Sharifnia, S., Hosseini, S.N. & Yazdanpour, N. (2013). Photocatalytic conversion of greenhouse gases (CO2 and CH4) to high value products using TiO2 nanoparticles supported on stainless steel webnet. Journal of the Taiwan Institute of Chemical Engineers, 44(2), 239-246.

Mishra, A., Kumari, M., Pandey, S., Chaudhry, V., Gupta, K.C. & Nautiyal, C.S. (2014). Biocatalytic and antimicrobial activities of gold nanoparticles synthesized by Trichoderma sp. Bioresource Technology, 166, 235-242.

Mishra, S., Lin, Z., Pang, S., Zhang, Y., Bhatt, P. & Chen, S. (2021). Biosurfactant is a powerful tool for the bioremediation of heavy metals from contaminated soils. Journal of Hazardous Materials, 418, p.126253.

Mohanpuria, P., Rana, N.K. & Yadav, S.K., (2008). Biosynthesis of nanoparticles: technological concepts and future applications. Journal of Nanoparticle Research, 10, pp.507-517.

Mourdikoudis, S., Pallares, R.M., & Thanh, N.T. (2018). Characterization techniques for nanoparticles: comparison and complementarity upon studying nanoparticle properties. Nanoscale 10, 12871–12934.

Narayanan, K.B. & Sakthivel, N. (2011). Synthesis and characterization of nano-gold composite using Cylindrocladium floridanum and its heterogeneous catalysis in the degradation of 4-nitrophenol. Journal of Hazardous Materials, 189(1-2), 519-525.

Nguyen, N.H. & Bai, H., (2015). Effect of washing pH on the properties of titanate nanotubes and its activity for photocatalytic oxidation of NO and NO2. Applied Surface Science, 355, pp.672-680.

Nogueira, H.P., Toma, S.H., Silveira, A.T., Carvalho, A.A., Fioroto, A.M. & Araki, K., 2019. Efficient Cr (VI) removal from wastewater by activated carbon superparamagnetic composites. Microchemical Journal, 149, p.104025.

Panahi, Y., Mellatyar, H., Farshbaf, M., Sabet, Z., Fattahi, T. and Akbarzadehe, A., 2018. Biotechnological applications of nanomaterials for air pollution and water/wastewater treatment. Materials Today: Proceedings, 5(7), 15550-15558.

Panwar, R., Kumar, N., Kashyap, V., Ram, C. & Kapila, R. (2019). Aflatoxin M 1 detoxification ability of probiotic Lactobacilli of Indian origin in in vitro digestion model. Probiotics and Antimicrobial Proteins, 11, 460-469.

Patel, H.K., Kalaria, R.K., & Khimani, M.R. (2020). Nanotechnology: a promising tool for bioremediation. Removal of toxic pollutants through microbiological and tertiary treatment, 515-547.

Raffa, C.M., Chiampo, F. & Shanthakumar, S. (2021). Remediation of metal/metalloid-polluted soils: A short review. Applied Sciences, 11(9), p.4134.

Raheman, F., Deshmukh, S., Ingle, A., Gade, A. & Rai, M. (2011). Silver nanoparticles: novel antimicrobial agent synthesized from an endophytic fungus Pestalotia sp. isolated from leaves of Syzygium cumini (L). Nano Biomed Eng, 3(3), 174-178.

Raj, R., Dalei, K., Chakraborty, J. & Das, S., 2016. Extracellular polymeric substances of a marine bacterium mediated synthesis of CdS nanoparticles for removal of cadmium from aqueous solution. Journal of Colloid and Interface Science, 462, 166-175.

Ramakrishna, M., Babu, D.R., Gengan, R.M., Chandra, S., & Rao, G.N. (2016). Green synthesis of gold nanoparticles using marine algae and evaluation of their catalytic activity. Journal of Nanostructure in Chemistry, 6, 1–13.

Raman, J., Reddy, G.R., Lakshmanan, H., Selvaraj, V., Gajendran, B., Nanjian, R., Chinnasamy, A. & Sabaratnam, V. (2015). Mycosynthesis and characterization of silver nanoparticles from Pleurotus djamor var. roseus and their in vitro cytotoxicity effect on PC3 cells. Process Biochemistry, 50(1), 140-147.

Rasmussen, K., González, M., Kearns, P., Sintes, J.R., Rossi, F. and Sayre, P., 2016. Review of achievements of the OECD Working Party on Manufactured Nanomaterials' Testing and Assessment Programme. From exploratory testing to test guidelines. Regulatory Toxicology and Pharmacology, 74, pp.147-160.

Ray, S., Sarkar, S. & Kundu, S. (2011). Extracellular biosynthesis of silver nanoparticles using the mycorrhizal mushroom Tricholoma crassum (Berk.) Sacc: its antimicrobial activity against pathogenic bacteria and fungus, including multidrug resistant plant and human bacteria. Digest Journal of Nanomaterials and Biostructures, 6(3), 1289-1299.

Rizwan, M.D., Singh, M., Mitra, C.K. & Morve, R.K. (2014). Ecofriendly application of nanomaterials: nanobioremediation. Journal of Nanoparticles, 2014, 1-7.

Rodriguez, J.A., Liu, P., Pérez, M., Liu, G. and Hrbek, J., (2010). Destruction of SO2 on Au and Cu nanoparticles dispersed on MgO (100) and CeO2 (111). The Journal of Physical Chemistry A, 114(11), 3802-3810.

Saleem, H. & Zaidi, S.J., (2020). Recent developments in the application of nanomaterials in agroecosystems. Nanomaterials, 10(12), p.2411.

Saleh, T.A., Shuaib, T.D., Danmaliki, G.I. & Al-Daous, M.A., (2016). Carbon-based nanomaterials for desulfurization: classification, preparation, and evaluation. In Applying nanotechnology to the desulfurization process in petroleum engineering (pp. 154-179). IGI Global.

Salvadori, M.R., Ando, R.A., Oller do Nascimento, C.A. & Corrêa, B. (2014). Intracellular biosynthesis and removal of copper nanoparticles by dead biomass of yeast isolated from the wastewater of a mine in the Brazilian Amazonia. PloS one, 9(1), p.e87968.

Sanghi, R. & Verma, P. (2009). Biomimetic synthesis and characterisation of protein capped silver nanoparticles. Bioresource Technology, 100(1), 501-504.

Sathyavathi, S., Manjula, A., Rajendhran, J. & Gunasekaran, P. (2014). Extracellular synthesis and characterization of nickel oxide nanoparticles from Microbacterium sp. MRS-1 towards bioremediation of nickel electroplating industrial effluent. Bioresource Technology, 165 270-273.

Seshadri, S., Saranya, K. & Kowshik, M. (2011). Green synthesis of lead sulfide nanoparticles by the lead resistant marine yeast, Rhodosporidium diobovatum. Biotechnology Progress, 27(5).

Shan, G., Xing, J., Zhang, H. & Liu, H. (2005). Biodesulfurization of dibenzothiophene by microbial cells coated with magnetite nanoparticles. Applied and Environmental Microbiology, 71(8), 4497-4502.

Sharma, G., Jasuja, N.D., Kumar, M. & Ali, M.I. (2015). Biological Synthesis of Silver Nanoparticles by Cell-Free Extract ofSpirulina platensis. Journal of Nanotechnology, 2015, 1-6.

Sheng, P.X., Ting, Y.P., Chen, J.P., & Hong, L. (2004). Sorption of lead, copper, cadmium, zinc, and nickel by marine algal biomass: characterization of biosorptive capacity and investigation of mechanisms. Journal of Colloidal Interface Science. 275, 131–141.

Shi, C., Zhu, N., Cao, Y. U & Wu, P. (2015). Biosynthesis of gold nanoparticles assisted by the intracellular protein extract of Pycnoporus sanguineus and its catalysis in degradation of 4-nitroaniline. Nanoscale Research Letters, 10, 1-8.

Shukla, A. & Srivastava, S. (2017). Emerging aspects of bioremediation of arsenic. Green Technologies and Environmental sustainability, 395-407.

Singh, E., Osmani, R.A.M. & Banerjee, R. (2020). Nanobioremediation: an emerging approach for a cleaner environment. Microbial bioremediation & biodegradation, 309-363.

Singh, J., Mukherjee, A., Sengupta, S.K., Im, J., Peterson, G.W. & Whitten, J.E. (2012). Sulfur dioxide and nitrogen dioxide adsorption on zinc oxide and zirconium hydroxide nanoparticles and the effect on photoluminescence. Applied Surface Science, 258(15), 5778-5785.

Slavin, Y.N., Asnis, J., Hńfeli, U.O. & Bach, H. (2017). Metal nanoparticles: understanding the mechanisms behind antibacterial activity. Journal of Nanobiotechnology, 15, 1-20.

Song, B., Zeng, G., Gong, J., Liang, J., Xu, P., Liu, Z., Zhang, Y., Zhang, C., Cheng, M., Liu, Y. and Ye, S. (2017). Evaluation methods for assessing effectiveness of in situ remediation of soil and sediment contaminated with organic pollutants and heavy metals. Environment International, 105, 43-55.

Su, F., Lu, C., Cnen, W., Bai, H. & Hwang, J.F. (2009). Capture of CO2 from flue gas via multiwalled carbon nanotubes. Science of the total environment, 407(8), 3017-3023.

Subramaniyam, V., Subashchandrabose, S.R., Thavamani, P., Megharaj, M., Chen, Z. & Naidu, R. (2015). Chlorococcum sp. MM11—a novel phyco-nanofactory for the synthesis of iron nanoparticles. Journal of Applied Phycology, 27, 1861-1869.

Sun, Y., Liang, J., Tang, L., Li, H., Zhu, Y., Jiang, D., Song, B., Chen, M. & Zeng, G., (2019). Nano-pesticides: A great challenge for biodiversity?. Nano Today, 28, p.100757.

Takmil, F., Esmaeili, H., Mousavi, S.M. & Hashemi, S.A. (2020). Nano-magnetically modified activated carbon prepared by oak shell for treatment of wastewater containing fluoride ion. Advanced Powder Technology, 31(8), 3236-3245.

Tan, W., Peralta-Videa, J.R. & Gardea-Torresdey, J.L. (2018). Interaction of titanium dioxide nanoparticles with soil components and plants: current knowledge and future research needs–a critical review. Environmental Science: Nano, 5(2), 257-278.

Thatoi, H., Behera, B.C. & Mishra, R.R. (2013). Ecological role and biotechnological potential of mangrove fungi: a review. Mycology, 4(1), 54-71.

Tsolcha, O.N., Tekerlekopoulou, A.G., Akratos, C.S., Antonopoulou, G., Aggelis, G., Genitsaris, S., Moustaka-Gouni, M. & Vayenas, D.V. (2018). A Leptolyngbya-based microbial consortium for agro-industrial wastewaters treatment and biodiesel production. Environmental Science and Pollution Research, 25, 17957-17966.

Ullah, A., Heng, S., Munis, M.F.H., Fahad, S. & Yang, X. (2015). Phytoremediation of heavy metals assisted by plant growth promoting (PGP) bacteria: a review. Environmental and Experimental Botany, 117, 28-40.

Upadhyaya, H., Panda, S.K., Bhattacharjee, M.K. & Dutta, S. (2010). Role of arbuscular mycorrhiza in heavy metal tolerance in plants: Prospects for phytoremidiation. Journal of Phytology, 2(7).

Upendar, K., Kumar, A.S.H., Lingaiah, N., Rao, K.R. & Prasad, P.S. (2012). Low-temperature CO2 adsorption on alkali metal titanate nanotubes. International Journal of Greenhouse Gas Control, 10, 191-198.

Vahabi, K., Mansoori, G.A. & Karimi, S. (2011). Biosynthesis of silver nanoparticles by fungus Trichoderma reesei (a route for large-scale production of AgNPs). Insciences J., 1(1), 65-79.

Van Dillewijn, P., Caballero, A., Paz, J. A., González-Pérez, M. M., Oliva, J. M., & Ramos, J. L. 2007. Bioremediation of 2-4, 6-trinitrotoluene under field conditions. Environmental Science & Technology, 41(4), 1378–1383.

Verma, V.C., Singh, S.K., Solanki, R. and Prakash, S. (2011). Biofabrication of anisotropic gold nanotriangles using extract of endophytic Aspergillus clavatus as a dual functional reductant and stabilizer. Nanoscale Research Letters, 6, 1-7.

Walker, C.H., Sibly, R.M., Sibly, R.M., & Peakall, D.B. (2012). Principles of Ecotoxicology (4th ed.). CRC Press.

Wang, H.Y. and Lua, A.C., 2012. Development of metallic nickel nanoparticle catalyst for the decomposition of methane into hydrogen and carbon nanofibers. The Journal of Physical Chemistry C, 116(51), 26765-26775.

Wang, L., Hu, C. & Shao, L. (2017). The antimicrobial activity of nanoparticles: present situation and prospects for the future. International Journal of Nanomedicine, pp.1227-1249.

Wang, L., Zhang, C., Gao, F., Mailhot, G. & Pan, G., (2017). Algae decorated TiO2/Ag hybrid nanofiber membrane with enhanced photocatalytic activity for Cr (VI) removal under visible light. Chemical Engineering Journal, 314, 622-630.

Wei, Y., Fang, Z., Zheng, L. & Tsang, E.P. (2017). Biosynthesized iron nanoparticles in aqueous extracts of Eichhornia crassipes and its mechanism in the hexavalent chromium removal. Applied Surface Science, 399, 322-329.

Wu, C.M., Baltrusaitis, J., Gillan, E.G. and Grassian, V.H. (2011). Sulfur dioxide adsorption on ZnO nanoparticles and nanorods. The Journal of Physical Chemistry C, 115(20), 10164-10172.

Xu, S., Luo, X., Huang, Q. & Chen, W. (2021). Calcium-crosslinked alginate-encapsulated bacteria for remediating of cadmium-polluted water and production of CdS nanoparticles. Applied Microbiology and Biotechnology, 105, 2171-2179.

Yadav, A., Kon, K., Kratosova, G., Duran, N., Ingle, A.P. & Rai, M., 2015. Fungi as an efficient mycosystem for the synthesis of metal nanoparticles: progress and key aspects of research. Biotechnology Letters, 37, 2099-2120.

Yang, J., Hou, B., Wang, J., Tian, B., Bi, J., Wang, N., Li, X. & Huang, X. (2019). Nanomaterials for the removal of heavy metals from wastewater. Nanomaterials, 9(3), p.424.

Yang, X., Shen, Z., Zhang, B., Yang, J., Hong, W.X., Zhuang, Z. & Liu, J. (2013). Silica nanoparticles capture atmospheric lead: implications in the treatment of environmental heavy metal pollution. Chemosphere, 90(2), 653-656.

Yogalakshmi, K.N., Das, A., Rani, G., Jaswal, V., & Randhawa, J.S. (2020). Nanobioremediation: a new age technology for the treatment of dyes in textile effluents. Bioremediation of Industrial Waste for Environmental Safety, 313–347.

Yu, G., Wang, X., Liu, J., Jiang, P., You, S., Ding, N., Guo, Q. & Lin, F., 2021. Applications of nanomaterials for heavy metal removal from water and soil: a review. Sustainability 13: 713.

Zare, B., Babaie, S., Setayesh, N. & Shahverdi, A.R. (2013). Isolation and characterization of a fungus for extracellular synthesis of small selenium nanoparticles. Nanomedicine Journal, 1(1), pp.13-19.

Zhang, C., Li, M., Xu, X., & Liu, N. (2015). Effects of carbon nanotubes on atrazine biodegradation by Arthrobacter sp. Journal of Hazardous Materials., 287, 1–6.

Zhang, M., Atkinson, K.R. & Baughman, R.H. (2004). Multifunctional carbon nanotube yarns by downsizing an ancient technology. Science, 306(5700), 1358-1361.

Zhang, M., Yi, K., Zhang, X., Han, P., Liu, W. & Tong, M. (2020). Modification of zero valent iron nanoparticles by sodium alginate and bentonite: enhanced transport, effective hexavalent chromium removal and reduced bacterial toxicity. Journal of Hazardous Materials, 388, p.121822.

Zhou, Z., Wu, X., Lin, Z., Pang, S., Mishra, S., & Chen, S. (2021). Biodegradation of fipronil: current state of mechanisms of biodegradation and future perspectives. Applied Microbiology Biotechnology, 105, 7695–7708

Zou, Y., Wang, X., Khan, A., Wang, P., Liu, Y., Alsaedi, A., Hayat, T., & Wang, X. (2016). Environmental remediation and application of nanoscale zero-valent iron and its composites for the removal of heavy metal ions: a review. Environmental Science Technolology, 50, 7290-7304.

14

Cleaner and Greener Nanotechnology Based Approaches for Wastewater Remediation

Riddhi M. Kathrotiya[1], Bhargav Raval[1,2], Shivani Yagnik Raval[1] and Vikram H. Raval[1*]

[1]*Department of Microbiology and Biotechnology, University School of Sciences Gujarat University, Ahmedabad-380009, Gujarat, India*

[2]*Gujarat Pollution Control Board, Gandhinagar-382010, Gujarat, India*

Abstract

One of the biggest issues the world is facing today is the lack of clean and pure water. Due to rising water demands, rigid health regulations, and xenobiotic pollutants, traditional wastewater treatment technologies cannot provide safe water. The present era necessitates innovative approaches to wastewater treatment technologies and improved water resource conservation and management materials. By utilizing innovative nanomaterial, nanotechnology guarantees a continuous provision of potable water and contributes to the preservation of water sources. Nanomaterials expedite rapid processes for the purification, conservation, and reuse of wastewater, complementing other enduring industrial methods. Recent developments in nanotechnology show that the numerous issues surrounding water quality can be easily solved by using nanomaterials as catalysts, adsorbents, and catalytic membranes for improved filtration. Countries with emerging economies stand to gain significant advantages by implementing nanotechnology in essential areas such as low-cost water disinfection, bioremediation, and separations. This chapter not only covers the recent advancements in nanoparticle technology, and the mechanism of remediation but also a sincere effort to comprehend the advantages, limitations, and prospects.

Keywords: *Bioremediation, Membrane, Nano-adsorbents, Nanotechnology, Nanotubes, Photocatalysis*

* *Corresponding author e-mail: vhraval@gujaratuniversity.ac.in; vikramhraval@gmail.com*

1. Introduction

Water, a ubiquitous molecule (H_2O), is the essence of life on Earth (Morrison *et al.*, 2020). Its unique physicochemical properties, notably its exceptional solvent power, make it indispensable for all biological processes (Rubilar *et al.*, 2020). However, the availability of clean freshwater, vital for human and ecosystem health, is increasingly threatened by anthropogenic (human-caused) contamination (Gitis and Hankins, 2018; Hodges *et al.*, 2018). Several human activities contribute to this global challenge. Inadequate wastewater treatment facilities allow untreated sewage to enter waterways, introducing harmful pathogens and nutrients that disrupt aquatic ecosystems. Industrial discharges often contain hazardous substances, heavy metals, and persistent organic pollutants, further compromising water quality (Umar *et al.*, 2019a; Elbeltagi *et al.*, 2021). Unsustainable agricultural practices, including excessive fertilizer use and pesticide discharge, also contribute to water pollution. Additionally, the dumping of waste materials, including plastics and radioactive elements, into oceans and other water bodies poses a significant long-term hazard (Jiang *et al.*, 2019; Edokpayi *et al.*, 2020). The consequences of water pollution are far-reaching. Aquatic ecosystems are directly impacted, with reduced biodiversity, algal outbreaks, and dead zones devoid of oxygen. Water pollution can also exacerbate air pollution through processes like volatilization (transformation of liquids to gases) of hazardous chemicals (Wu *et al.*, 2019). This contaminated air and water pose significant risks to human health, leading to waterborne diseases, respiratory diseases, and even cancers. Furthermore, economic development can be hampered by water scarcity and the high costs associated with water treatment and remediation (Liu *et al.*, 2020; Shah *et al.*, 2020). Social inequalities are often exacerbated, with vulnerable communities often facing the effects of polluted water sources. A recent report by the United Nations highlights the growing challenge of securing clean, accessible freshwater for all in the 21st century (Wu *et al.*, 2020). As the global population continues to rise, ensuring sustainable water management practices is critical for safeguarding the health of our planet and its inhabitants (Deshpande *et al.*, 2020; Wong *et al.*, 2019). The term "contaminated water" describes the presence of unwanted materials in reservoirs or bodies of water, making the water unfit for human use or other uses. Various chemical, physical, and mechanical techniques have been used to solve this growing problem. Fortunately, researchers are actively exploring innovative and cost-effective methods to enhance water purification (Sekoai *et al.*, 2019; Umar *et al.*, 2019b). One particularly promising avenue is the rapidly growing field of nanotechnology as shown in figure 1. Nanomaterials, with their unique properties at the atomic and molecular level, offer exciting alternatives for water treatment. With its low costs, high efficiency in removing pollutants, and reusability, the new discipline of nanotechnology presents an attractive option for water purification (Chhabra and Kumar, 2020). Nanomaterials have already shown success in a variety of sectors, including medicine and enzymology. Due to their unique qualities, which include robust mechanical characteristics, porosity, hydrophilic nature, dispersibility, hydrophobicity, high reactivity, large surface area, and extremally small size, nanomaterials have become a superior option for wastewater treatment as the world challenges with serious drinking water problem (Cheriyamundath and Vavilala, 2021). Numerous nanomaterials demonstrate potential in eliminating hazardous microorganisms, chemical, and inorganic pollutants, including heavy metals like Pb and Mo (Wang *et al.*, 2021). The World Health Organization (WHO) estimates that water pollution causes approximately 1.7 million deaths worldwide each year. The water crisis extends far beyond fatalities. A staggering 4 billion cases of waterborne infections are reported each year, highlighting the enormous impact on world health (Karthigadevi *et al.*, 2021). Table 1 indicates different types of water pollutants with sources.

In recent years, significant developments have been made in nanomaterials, including nano photocatalysts, nanomotors, nanomembranes, and nano-absorbents (notably valued for their high sorption capacity, making them versatile for water treatment applications). Additionally, certain imprinted polymers have been effective in treating polluted water. Research into nanomaterial uses in water purification is a ray of optimism with the potential to revolutionize water treatment and enhance global wellness. (Liu *et al.*, 2021). This book chapter tries to summarize the importance of nanomaterials in addressing water crises through wastewater treatment. Nanomaterials offer promising and adaptable water treatment approaches, yet certain imperfections require further attention, as outlined in this article.

Table 1: Different types of water pollutants with sources.

Source of wastewater	Sub-Source	Name of contaminant	Conventional wastewater treatment	Advanced waste-water treatment	References
Industrial waste	Mining and quarrying	Solid rock, slag, phosphogypsum, mud, tailings	Activated Sludge process, trickling filters, etc.	Membranefiltra-tion, Advanced Oxidation Process, Electro-chemical process-es, Microbial fuel cells	(Millati *et al.*, 2019)
	Energy production units	Coal combustion residuals like fly ash and bottom ash, industrial byproducts such as boiler slag and particulates, used oils and sludge			
	Manufacturing	Discarded materials like plastics and water, reactive materials, alkalis, acids, various solvents, spent catalysts, used oils, pigments, peroxide, organic stabilizer, heavy metals			
	Construction	Building materials like blocks (cinder and masonry), asphalt, roofing (wood shingles and slate), interior finishes (plaster), gypsum, concrete			
	Waste/water services	Spent adsorbent, sludge			
Agricultural waste	Field waste	Corncob, Rice husk, Rice straw, Sugarcane bagasse, Wheat straw	Composting, vermicomposting, Anaerobic digestion, Open Burning	Enzyme-assisted technologies, Hydrothermal conversion, Plasma Gasification	(Ahmad *et al.*, 2016)
	Forest waste	Logging residues, Wood processing residues			

Municipal solid waste	Residential/household	Food waste, plastics, paper, yard waste, cardboard, wood, metals, glass, etc.	Open dumping, Landfilling, Incineration,	Plasma Gasification, Anaerobic digestion, Waste-to-energy conversion, mechanical biological treatment, bioconversion technology	(Millati *et al.*, 2019)
	Commercial and trade	Glass, paper, plastics, cardboard, wood, food waste, metals, electronic waste, etc			
	Institutional/office	printer paper, cardboard, envelopes, Plastic bottles, packaging materials, Aluminum cans, foil, and other metal items, Glass bottles and jars, Food waste, E-waste			
	Medical waste	Infectious wastes, hazardous wastes radioactive waste from cancer therapies, pharmaceutical waste			
	Municipal services	general wastes from parks, beaches, and other recreational areas, Street sweepings, landscape and tree trimmings,			

Hospital wastewater	In-Patient Department	40-60 Liter water/patient/ intervention	Wastewater treatment plant (primary 20-50%, secondary 30-70%, tertiary >90%), chlorination	Ozonation (>93%), membrane bioreactor, Reverse Osmosis, Electrocoagulation and photo-catalytic methods (99%), ion exchange, nanofiltration, nano adsorption	(Khan *et al.*, 2021)
	Operation and Labor and delivery (L&D) room	100 Liter water /patient/ intervention			
	Out-Patient Department	5 Liter water /patient/ intervention			
	Community-based management of acute malnutrition (CMAM) program	5 Liter water /patient/ intervention			
	Therapeutic feeding unit (TFU)	30-50 Liter water/patient/ intervention			
	Cholera Treatment Centre	60 Liter water /patient/ intervention			
	Severe acute respiratory syndrome center	100 Liter water/patient/ intervention			
	Viral Hemorrhagic Fever treatment	300-400 Liter water /patient/ intervention			
Pharmaceutical Wastes	Waste materials containing excess drugs	syringes, IV bags, tubing, vials,	coagulation, reverse osmosis, filtration, evaporation, and sedimentation	Nanosorbents, nano-photocatalysts, and nano-sensors	(Jaseem *et al.*, 2017; Saroa *et al.*, 2023)
	acute hazardous waste drugs (discarded expired drugs)	Nicotine & salts, Cetamide, Trinitrate, 4- Aminopyridine			
	chemotherapy drug residues	Paclitaxel, Taxol, Uracil Mustard, Vepesid, Carmustine, Cytoxan, Leukeran			

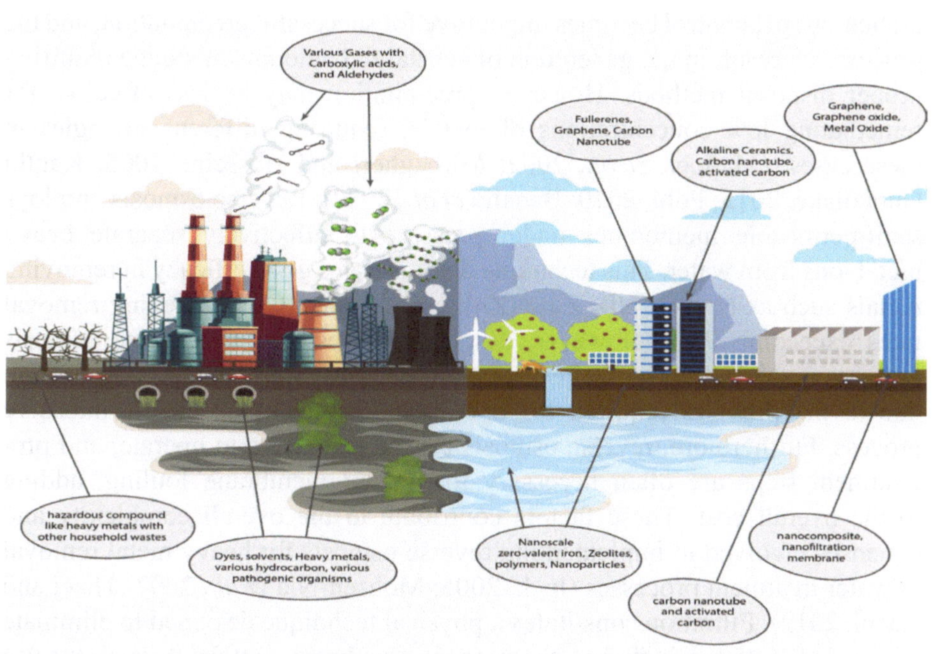

Fig. 1: The use of various Nano technological strategies to remove various contaminants

2. Conventional Wastewater Treatment Process

To address the public health risks associated with heavy metal contamination, developing efficient water treatment technologies is essential. Various conventional methods for heavy metal remediation have been employed for decades, each with its set of advantages and limitations. Ion exchange is an effective method frequently used to remove hazardous heavy metals from polluted water, particularly chromium, lead, nickel, cadmium, and copper. It operates by facilitating the interchange of metal ions present in the water with ions residing on an ion exchange resin. Despite its efficacy, ion exchange is constrained by several limitations. These include the considerable operational costs associated with its implementation, the limited capacity to effectively remove certain metals, and the necessity for resin regeneration. The regeneration process not only generates secondary waste but also demands meticulous disposal procedures to mitigate the environmental impact (Vaaramaa and Lehto, 2003; Dąbrowski *et al.*, 2004; Foroutan and Esmaeili., 2015). In precipitation, precipitating agents are introduced into contaminated water, leading to the formation of insoluble metal compounds that can be subsequently separated. This technique exhibits efficiency in treating metals such as Copper (Cu), Cadmium (Cd), and Chromium (Cr). Nevertheless, meticulous pH control becomes imperative for successful precipitation, and the process can result in the generation of substantial amounts of sludge requiring proper disposal methods. However, precipitation may be less effective for remediating low concentrations of metals, requiring different strategies in these cases. (Matlock *et al.*, 2002; Mirbagheri and Hosseini, 2005; Kanthi Gunatilake, 2015; Pohl, 2020; Benalia *et al.*, 2022). Reverse osmosis employs semi-permeable membranes under pressure to effectively separate heavy metal ions from water. This technique demonstrates good efficacy in removing metals such as Nickel (Ni), Zinc (Zn), and Copper (Cu), achieving removal rates exceeding 95%. However, despite its effectiveness, reverse osmosis presents notable drawbacks. It is characterized by high energy consumption due to the pressure required for operation, making it an energy-intensive process. Furthermore, reverse osmosis can be expensive to operate, and pre-treatment steps are often necessary to prevent membrane fouling, adding to the overall cost. These factors contribute to the overall complexity and expenses involved in implementing reverse osmosis for heavy metal removal in water treatment processes (Ipek, 2005; Mohsen-Nia *et al.*, 2007; Thaçi and Gashi, 2019). Filtration constitutes a physical technique designed to eliminate suspended particles, including those containing heavy metals, through the use of a barrier. While filtration excels at removing larger particles, it may require pre-treatment for finer ones and struggles with dissolved metal ions, limiting its effectiveness in certain situations (Barakat, 2011; Kanthi Gunatilake, 2015; Calabrò *et al.*, 2021). Coagulation provides another way to heavy metal removal in water treatment. This method involves adding coagulants to destabilize suspended particles, collecting heavy metals in the resultant flocs. This action causes these particles to coagulate, resulting in larger clumps. Following coagulation, flocculation is employed, where flocculants are added to further agglomerate the coagulated particles into larger flocs. This process facilitates their removal through settling or filtration methods. It demonstrates effectiveness, particularly with metals like Copper (Cu), Nickel (Ni), and Cobalt (Co). However, achieving optimal pH conditions, maintaining suitable temperature levels, and ensuring precise dosing of coagulants and flocculants are essential for achieving effective performance. It produces a large amount of sludge, which must be handled and disposed of cautiously. Furthermore, the process relies on chemicals, raising worries regarding their use and potentially transferring hazardous compounds into the solid phase. Furthermore, certain heavy metals may not readily coagulate or flocculate, requiring additional treatment steps or alternative methods for their successful removal (Assaad *et al.*, 2007; Teh *et al.*, 2016). Electrochemical treatment uses electricity to remove heavy metals from polluted water. This approach employs electrical currents to remove these harmful contaminants. While effective, electrochemical treatment is energy-intensive and complex, requiring skilled operation and maintenance (Yousuf *et al.*, 2001; Aziz *et al.*, 2008).

Conventional heavy metal remediation methods can be limited by high costs, the creation of secondary waste streams, incomplete contaminant removal, and potential environmental drawbacks. In contrast, nanotechnology-based approaches present a range of advantages that position them as superior and more environmentally sustainable options. Nano-bioremediation, for instance, harnesses nanomaterials like nanoparticles or nanocomposites to address heavy metal contamination in various environments. Nanomaterials have an impressive advantage in heavy metal removal due to their extremely high surface area-to-volume ratio. This results in a significantly greater surface area for contact with pollutants, which improves their ability to absorb and bind heavy metals. Nano-bioremediation processes exhibit high efficiency, necessitating lesser quantities of materials and energy compared to traditional methods (Vidu *et al.*, 2020). Moreover, they can target specific contaminants, enhancing selectivity and efficacy in heavy metal removal from the environment. Nanotechnology-based approaches are also adaptable to specific challenges associated with conventional methods. For example, nanomaterials can be designed to target particular heavy metals, function across diverse environmental conditions, and facilitate the recovery and recycling of metal contaminants, thereby minimizing waste generation. Furthermore, nano-remediation processes can be implemented in situ, reducing the need for transporting contaminated materials. This not only reduces remediation expenses but also mitigates disruptions to surrounding ecosystems (Zhang *et al.*, 2019). In summary, nano-bioremediation and other nanotechnology-based strategies present promising solutions for heavy metal remediation by overcoming the limitations of conventional methods and offering more efficient, selective, and environmentally friendly remediation alternatives. Given the limitations of traditional methods highlighted earlier, exploring the potential of nanotechnology for metal remediation offers a promising avenue for achieving efficient and environmentally friendly remediation strategies.

3. A Novel Approach: Applications of Nanotechnology in Wastewater Treatment

In the field of nanotechnology for purifying water, a particularly exciting and promising avenue involves the exploration of innovative technologies utilizing nanomaterials for various purposes (Lu *et al.*, 2016). A groundbreaking area of research is emerging that explores the potential of specifically engineered materials at the nanoscale. These innovative materials are the key to unlocking long-term and efficient solutions to today's critical water concerns. Their unique features have the potential to improve water treatment processes dramatically (Sadegh *et al.*, 2017). Many nanotechnology-based wastewater treatment methods have been developed (table 2). figure 2 depicts a graphical representation of various applications and uses of nanoparticles across different fields (Prachi *et al.*, 2013, Madannejad *et al.*, 2019); the most common methods for wastewater can be broadly classified into the following categories:

a) Utilization of Adsorption-Based Technology
b) Photocatalytic wastewater treatment using nanomaterials
c) Disinfection of wastewater with nanomaterials
d) Membrane-based technology implementation

Table 2: The Advantages and Limitations of Nanotechnology in Materials

Types/ mechanism	Nanoparticle	Contaminant	Size (nm)	Contact time	Advancement	Limitation in use	Reference
Nano-absorption	Iron Oxide nanoparticles	Heavy metals (e.g., Pd, Cu, Hg), Dyes, Organic Pollutant	10-100	Depends on the process (batch/ continuous)	High adsorption capacity for various contaminants, Can be functionalized for specific contaminants, Regenerable	Aggregation, Sludge formation during regeneration, Potential human and environmental health risks	(Aragaw *et al.*, 2021; Mostafapour *et al.*, 2023)
Nanofiltration	Silver nanoparticles	Bacteria	1-100	Continuous (in filtered media)	Enhanced antimicrobial properties in membranes	Potential for silver nanoparticle release	(Palani *et al.*, 2023)
Photocatalyst	Metal-Organic Frameworks (MOFs)	Organic pollutants (e.g., dyes, pharmaceuticals)	2-10	It depends on light source and catalyst	Engineering of MOFs with tunable pore structures for efficient pollutant capture and degradation	Limited stability of some MOFs in water - Scalability for large-scale treatment can be challenging	(Bai *et al.*, 2020)
Nano-enabled Membranes	Carbon nanotube (CNT) embedded membranes	Various contaminants (depending on pore size and functionality)	1-100 (CNTs)	Various depending on filtration methods	Development of CNT membranes with improved mechanical strength and antifouling properties	High cost of CNT production - Potential environmental and health concerns associated with CNTs	(Akther *et al.*, 2015; Mulyanti and Susanto, 2018; J. Wang *et al.*, 2020)

Biogenic Nanoparticles	Biosynthesized nanoparticles (e.g., from bacteria, plants)	Heavy metals	1-50	Depends on specific process parameters	Exploration of sustainable and eco-friendly methods for nanoparticle synthesis	Limited understanding of the long-term environmental impact of biogenic nanoparticles	(Ahmad *et al.*, 2016; Kuppusamy *et al.*, 2016; R. Wang *et al.*, 2020)
Antimicrobial nanomaterials (Antimicrobial coating, encapsulation)	Chitosan nanoparticles, microbial essential oil nanoparticles, antimicrobial peptide nanoparticles	Fungi, broad-spectrum microbes	10-1000	Hours to week	Biocompatible, Slow-release antimicrobials, Targeted therapy, reduce immunogenicity	Lower potency compared to free antimicrobials, Can be expensive, High production costs, Limited stability	(Mohammed *et al.*, 2017; Yang *et al.*, 2021)

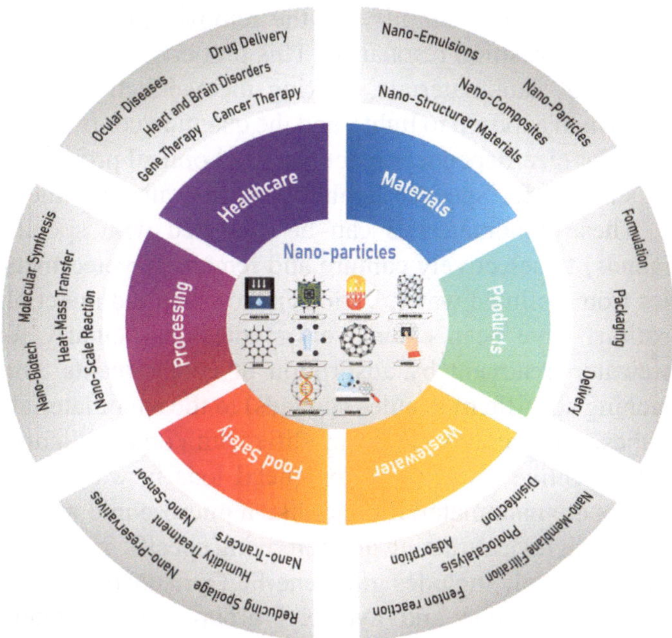

Fig. 2: A schematic representation of nanoparticle applications across diverse fields

3.1. Utilization of Adsorption-Based Technology

Adsorption is the process by which atoms, molecules, or ions from a gas, liquid, or dissolved material adhere to the surface of a solid. Adsorption is directly related to surface energy, similar to surface tension in liquids. Inside a solid material, atoms are tightly packed, surrounded by neighbors on all sides. However, atoms on the surface have less neighbors, resulting in a type of energy imbalance. This imbalance draws ions, atoms, or molecules from the surrounding liquids or gasses, causing them to adhere to the surface. During adsorption, the substance being attracted (adsorbate) sticks to the surface of the attracting material (adsorbent). This attraction can be strong or weak, depending on the specific molecules involved. There are two main types of adsorption: physical adsorption, where weak forces hold the adsorbate in place, and chemical adsorption, where a stronger chemical bond forms between the adsorbent and adsorbate. Certain breakthroughs leverage the gradually scalable size-dependent attributes of nanomaterials, which contribute to their large specific surface areas, allowing for rapid dissolution, increased reactivity, and strong absorption. Additionally, other advancements capitalize on the unique, non-continuous properties of nanomaterials, such as superparamagnetic (This property enables the easy separation of magnetic nanomaterials from water, simplifying the removal process after adsorption), localized surface plasmon resonance (This property can enhance the photocatalytic activity of nanomaterials, making them effective in the degradation of pollutants when exposed to light), and the quantum confinement effect (This effect allows for size-dependent electronic and optical properties, making nanomaterials suitable for specific applications in water treatment) (Prachi *et al.*, 2013). These nanomaterials can be modified with specific molecules (organic ligands) to selectively capture and remove targeted metal ions or other impurities from polluted water. Various nanostructured materials with surface modifications have been extensively studied, particularly for their efficacy in wastewater treatment by adsorption of contaminants. One approach involves attaching specific molecules (ligands) to the nanomaterial's surface. Carbon nanotubes, for example, can be modified with organic ligands and can become highly selective towards certain metal ions in water. This targeted attachment allows the nanomaterial to act like a microscopic magnet (Fe_3O_4/C core-shell nanoparticles), attracting and removing specific pollutants through adsorption. Other example includes graphene-Fe_3O_4 nanocomposites, silica gel functionalized with amino and hydroxyl groups, and graphene oxide. The efficacy of these materials in eliminating heavy metals and dye molecules from water has been explored. When contrasting these innovations with conventional materials, nanomaterial adsorbents demonstrate superior performance, exhibiting much greater effectiveness and quicker rates of pollution removal from wastewater. This increased efficiency is crucial in addressing the challenges of water pollution and ensuring a sustainable water supply for growing populations. This increased efficiency is crucial in addressing the challenges of water pollution and ensuring a sustainable water supply for growing populations (Kadam *et al.*, 2019).

3.1.1. Utilization of Carbon-Based Materials for the treatment of wastewater

Carbon-based materials are emerging as powerful tools for water purification. Studies have shown them to be highly effective adsorbents for a wide range of pollutants, including both organic and inorganic contaminants (Kalaitzidou *et al.*, 2020; Wang *et al.*, 2020). Nanostructured carbon materials, including carbon nanotubes, activated carbon, fullerenes, and graphene, showcase exceptional properties such as large surface area – ideal for capturing pollutants – along with high porosity, which allows for effective circulation of water. Additionally, their excellent conductivity, ability to control pore size, and robust chemical stability across various pH levels (acidic or alkaline) make them ideal for diverse water treatment applications (Lam and Luong, 2014). Additionally, carbon materials derived from metal-organic frameworks

(MOFs) encompass a diverse spectrum of structures, incorporating carbon skeletons that encapsulate metal or metal oxide nanoparticles. Carbon-based metallic-organic frameworks, encompassing non-metal-doped hybrids, metal-free carbons, and diverse composites (Hao *et al.*, 2021), are attracting significant interest. These structures, formed through the self-assembly of metal clusters and organic linkers, feature the metal clusters as connecting nodes and organic ligands providing structural support. Notably, their vast surface area, numerous pores, changeable shape, and overall versatility make them a focus of intense study for water treatment applications. Their use extends to varied sectors such as storing gases, monitoring, energy preservation, adsorption, and catalysts, as proven by research (Li *et al.*, 2018; Fan *et al.*, 2019; Man *et al.*, 2019). The fundamental versatility of MOFs generates an environment for rigorous modulation of the chemical and physical characteristics of materials (Solis *et al.*, 2020). The range of material types is greatly expanded by changes in the composition of organic ligands, functional monomers, and the process of carbonization, which open up a multitude of options for hetero Atom doping as well as performance improvements (Ren *et al.*, 2018; Fan *et al.*, 2019; Pandi *et al.*, 2020). The ability of strong carbon-based adsorbents, particularly graphitic carbons, to remove heavy metals such as uranium from wastewater demonstrates their potential for real-world applications (Tang and Wang, 2020). Developing new carbon-based nanomaterials with high efficiency and selectivity for extracting uranium (U(VI)) from nuclear waste and seawater is a key research objective (Qiu *et al.*, 2021). While these materials show promise as potent adsorbents, their high cost remains a substantial barrier to broad use in wastewater treatment. Therefore, there is an increased concentration on producing affordable adsorbents (Hu *et al.*, 2020).

3.1.1.1 Graphene-based nanomaterial

For widespread application in contaminant removal, graphene, like any other nanomaterial, needs to be evaluated based on three major factors: production cost, efficiency in pollutant elimination, and potential environmental impact. The cost-effectiveness of synthesizing graphene oxide (GO) for environmental applications puts it as a potential alternative to pristine graphene (Wadhawan *et al.*, 2020). Graphene stands out as a superior material, functioning as a substitute for carbon nanotubes (CNTs) in the treatment of wastewater. Graphene-based adsorbents offer distinct advantages over carbon nanotubes (CNTs). Unlike CNTs, which have limited adsorption sites within their inner walls, graphene's single atomic layer provides two accessible surfaces for pollutant absorption (Yang *et al.*, 2020). Furthermore, graphene oxide (GO) and reduced graphene oxide (RGO) can be easily produced through chemical exfoliation of graphite, eliminating the need for expensive metallic catalysts or complex equipment. This simplified manufacturing process makes graphene-based materials more appealing for large-scale applications. Graphene is a useful candidate for water treatment applications due to its simple production, lack of catalyst residue, and low post-processing requirements. This interest has extended beyond the area of chemistry, with graphene becoming the primary focus in environmental water quality management. Its unique characteristics, such as numerous π-bonds, oxygen-rich functional groups, and remarkable surface area, enhance its efficacy in pollution removal (Lu *et al.*, 2018). In the adsorption processes of graphene-based nanomaterials, five distinct reactions are possible: hydrophobic effects, hydrogen bonds, electrostatic forces, covalent bonds, and π-π bonds. From a physicochemical aspect, both graphene and its oxides enhance hydrophilicity and the adsorption potential of particular contaminants. The oxidation process is critical in reinforcing the physical and chemical stability of graphene, making it an attractive option for contaminant removal in wastewater treatment. (Cai *et al.*, 2019; Yin *et al.*, 2020).

3.1.1.2. Carbon nanotubes

There are three main types of carbon nanotubes: single-walled, multi-walled, and composites. (Fenati *et al.*, 2021). Due to their unique one-dimensional structure, CNTs display exceptional physical properties, including electrical, heating, and mechanical features (Mudasir and Naqash, 2020). The dimensional parameters of CNTs play a crucial role in determining their attributes. Studies reveal that improving the surface modification of CNTs leads to an enhancement of metal removal effectiveness. To enhance their effectiveness, carbon nanotubes can be engineered with specific functional groups on their surface. These tailored groups improve the interaction between the nanotubes and various contaminants in water. Numerous experiments have demonstrated the high efficacy of carbon nanotubes as potent nanomaterials for adsorbing heavy metal ions, including Lead, Copper, Zinc, and Cadmium (Huang *et al.*, 2021). Surface modification approaches for CNTs encompass chemical grafting, and oxidation, whereas physical changes such as coating. The outside wall, intramolecular gaps, interior cavities, and CNT grooves constitute four possible adsorption sites. Adsorption in carbon nanotubes, like graphene, includes interactions between the contaminant and the material's surface. However, the distinctive, hollow, and multi-layered structure of CNTs introduces an additional factor: porosity. This porous nature plays a significant role in capturing contaminants within the internal channels of the nanotubes, increasing their overall adsorption capacity compared to flat graphene sheets. Carbon nanotubes have historically been used as compact solids, therefore it has been difficult to separate them from water. To solve this, new granular composites of Al_2O_3 and carbon nanotubes have been created. These composites have shown significant sorption efficacy, durability, thermal resistance, and hydrophilic properties for two drugs: diclofenac sodium and carbamazepine. Notably, during the regeneration phase, the adsorbed drugs can deteriorate. The hybridization technique has affected the production of granular composites utilizing CNTs. Despite the advancements, some problems exist in the environmental use of CNTs, underlining the importance of further research into innovative CNT adsorbents, such as novel forms of CNT composites (Abukhadra *et al.*, 2019).

3.1.2. Metal nanoparticles

Nano-zerovalent iron, gold, and silver stand out as some of the most frequently employed metal nanoparticles for addressing heavy metal contamination. Their remarkable reactivity, capacity for degradation, magnetic characteristics, and capacity to reduce catalytic activity using oxidation and agglomeration make them unique (Tan *et al.*, 2021). When they are incorporated, core-shell and composite nanostructures are formed, which improves the selectivity and efficiency of other adsorbents including silica, alumina, and activated carbon. Particularly, nano zerovalent iron has attracted a lot of interest as a cutting-edge nano adsorbent for several heavy metal groups (Zhang *et al.*, 2019). It is composed of an outer layer of mixed iron oxide and an inside core of zero-oxidation state iron. It acts as the center of adsorption processes for pollutants through electrostatic interactions and surface complexation processes. However, because of its high reactivity and toxicity, nano zerovalent iron is still difficult to handle (Rambabu *et al.*, 2021).

3.1.2.1. Metal Oxide as Nanomaterial

Since metal oxide-based nanomaterials have specific physicochemical characteristics, they are frequently used for the remediation of hazardous metals in polluted wastewater. The most prominent is iron oxide, but there are also important alternatives such as zinc oxide, copper oxide, nickel oxide, cesium oxide, titanium oxide, manganese oxide, and tin oxide (Pang *et al.*, 2019). The effectiveness of these substances in removing lead, arsenic, cadmium, and mercury ions has been demonstrated. The effect of polyacrylic acid-stabilized iron oxide nanoparticles (Fe_3O_4) on the elimination of cadmium from contaminated soil was examined in a study by Mohamadiun *et al.* (2018) (Mohamadiun *et al.*, 2018). The research examined several factors, such as dosage, pH, concentration, and contact time. According to reports, the metal had its highest removal rate (100%) at 500 ppm concentration,

24-hour contact duration, and pH 6.5. A methyltrimethoxysilane/iron oxide nanocomposite was produced and studied for its capacity to remove lead ions from wastewater in a study by Nodeh *et al.* (2020) (Nodeh *et al.*, 2020). Several characterization techniques were used to determine the shape, elemental composition, and crystal structure of the nanomaterials. These include Scanning Electron Microscopy (SEM) provides high-resolution images, revealing the size, shape, and surface properties of nanomaterials. Energy-dispersive X-ray spectroscopy (EDX) offered a quantitative analysis of the elemental composition, confirming the presence and relative abundance of the desired components within the nanostructure. Finally, X-ray Diffraction (XRD) revealed important information regarding the crystal structure and phase purity of the produced materials. To assess the adsorption process, numerous isotherm and kinetic models were used. The data indicate effectiveness with the Freundlich isotherm, indicating a multi-layered adsorption mechanism, and pseudo-second-order kinetics, showing a chemical interaction between the adsorbent and the pollutants (Ma *et al.*, 2018). It is interesting that at 105.5 mg/g, the rate of adsorption for Pb ions was found to be substantially higher. A novel and successful nano adsorbent was developed by Hossein *et al.* (2017) using the sol-gel method by generating an iron oxide magnetic core encased in mesoporous silica (Hossein *et al.*, 2017). The compound was then modified using bis(3-triethoxysiliclpropyl) tetrasulfide before being thoroughly characterized (Mahmoud and Abdelwahab, 2021). At its most optimal state, the metal oxide adsorbent showed outstanding effectiveness in removing metals by absorption rates over 270 mg/g for Pb, 256 mg/g for Pd, and 98 mg/g for Hg ions, Based on research. In addition, the evaluation of versatility revealed that the nano adsorbent could be reused successfully for a maximum of five cycles without experiencing a substantial decrease in its performance. In their 2016 study, Dargahi and colleagues explored the efficacy of MgO nanoparticles in eliminating Cr^{+6} and Pb^{+2} from polluted sources (Dargahi *et al.*, 2016). To achieve the optimal performance in removing metal ions, the study identified key parameters: an initial metal concentration of 10 milligrams per liter, a contact duration of 280 minutes, an adsorbent quantity of 0.8 grams per liter, and a stirring rate of 250 rotations per minute at a pH of 9 (Ebrahimpour *et al.*, 2021). Their findings revealed the involvement of reduction processes, and the adsorption model followed Langmuir isotherms (Singh *et al.*, 2021).

3.1.3. Hybrid nanomaterials

These nanomaterials are synthesized by combining two or more previously defined techniques. Essentially, polymeric nanoparticles serve as a foundational support for a spectrum of inorganic as well as organic nanomaterials in this context. The production of hybrid nano adsorbents has been accomplished via procedures such as impregnation, polymerization, solvent casting, and encapsulation. According to Jeevanandam *et al.* (2018), the efficiency of these nano adsorbents in the elimination of heavy metals has been greatly increased by the addition of nano hybridization methods (Jeevanandam *et al.*, 2018).

3.1.4. Polymer-based nanomaterials

In the last ten years, polymeric nano adsorbents have developed as a novel category of materials in the area of water treatment plant (Yin and Deng, 2015). Polymeric materials coated with metals, characterized by their particular surface area, porous structure, and various groups of functions on the surface, are recognized for their ability to effectively capture organic dyes and metal such as zinc, lead, and cadmium from water. Although several adsorbents in metal ion or dye removal from wastewater have been discovered, difficulties such as decreased adsorption capacity, insufficient selectivity, and low recyclability prevent the general availability of a variety of these nanomaterials. hybrid polymers, which have greater adsorption capacity, increased thermal stability, and enhanced recyclable materials (Lofrano *et al.*, 2016). Polymer-based nanomaterials for water treatment can be designed using various methods, including sol-gel synthesis, surfactant-assisted co-precipitation, and hierarchical/interpenetrating networks. The sol-gel method, for example, has been effectively used to produce high surface area nano-adsorbents such as mesoporous silica functionalized with organic groups. These materials remove hazardous metal ions from water with remarkable effectiveness (Li *et al.*, 2014). Another promising strategy is to modify magnetic nanoparticles with specific molecules, as exemplified by the work of Fei *et al.* (2012), who used 3-aminopropyltriethoxysilane and acrylic acid/ crotonic acid copolymers on Fe3O4 nanoparticles (Ge *et al.*, 2012). These diverse synthesis techniques provide a path towards developing specialized polymeric materials for effective water purification. The magnetic polymer nano-composites that were produced were utilized to extract heavy metals like Zinc, Cadmium, Copper, and Lead from wastewater across a range of pH values. The ability for metal ion uptake in relation to contact time and concentration of metal ions was the subject of the investigation. It is worth mentioning that these nano-composites of magnetic polymers and metal ions showed remarkable efficiency in adsorption, even when exposed to acidic pH (Agarwal *et al.*, 2016).

3.2. Photocatalytic wastewater treatment using nanomaterials

In recent times, photocatalytic degradation has become an attractive option for the remediation of a variety of harmful. Extensive research has been undertaken on the photodegradation of dyes utilizing nanoparticles of titanium dioxide (TiO_2) under visible or UV light. The phenomena of photocatalysis are surface-oriented, and the mechanism that underlies it consists of five basic phases (Zou *et al.*, 2011; Bai *et al.*, 2020).

a) Transport of reactants and contaminants to the photocatalyst surface
b) Reactants and/or contaminants adhering to the surface of the photocatalyst
c) Adsorption of reactants/pollutants and their reactions
d) Desorption from the surface of the resultant products
e) Products are removed or diffused from the interface

Traditional biological wastewater treatment methods often struggle to eliminate persistent contaminants (Zhang *et al.*, 2011; Lee *et al.*, 2012). Photocatalysis provides a promising alternative. Several mechanisms for dye degradation by photocatalysts have been postulated, including adsorption onto the catalyst surface, reactions involving the formation of surface electron-hole pairs (e-/h+), and the formation of hydroxyl radicals (OH•) that lead to the formation of end products (Lázaro-Navas *et al.*, 2015). An alternative mechanism suggests the combined role of adsorption, reactions initiated by energized electron-hole pairs or OH• radicals, and subsequent end-product production. Variations in photodegradation kinetics can be attributed to the particular type of reaction which depends on solution-phase species as well as surface adsorption. Photocatalytic oxidation of volatile organic compounds (VOCs) in the state of gas using visible light typically employs one of two strategies: 1) utilizing semiconductors with a smaller bandgap as catalysts or sensitizers, and 2) introducing titanium dioxide (TiO_2) to produce intermediary electronic states that promote the transportation of electrons to the conduction band. Semiconductor nanoparticles have shown remarkable efficacy in removing a wide range of organic pollutants, making them helpful tools for wastewater treatment. Among the well-studied semiconductors, ZnO stands out as a particularly promising material for photocatalysis. Noteworthy properties of ZnO include its robust oxidative capability, water insolubility, cost-effectiveness, and photocatalytic capacity in the near-UV spectroscopic range (Zhang *et al.*, 2014). While ZnO has exhibited strong photocatalytic characteristics, there is a need for improvement, particularly in terms of easy recovery from aquatic conditions. Coating a magnetic substrate with ZnO is one fascinating prospect for future research. $ZnFe_2O_4$, a ternary oxide, appears as a desirable option due to its magnetic properties due to its magnetic attributes, coupled with photocatalytic and antibacterial characteristics. The magnetic characteristics of this ternary oxide remain unchanged even when exposed to high temperatures (Maynez-Navarro *et al.*, 2020). Commercially available visible light photocatalysts are limited despite their efficacy because of structural instability or a need for specific, intense light sources. Numerous studies have delved into the exploration of novel materials showing both highly stable structures as well as high degradation capabilities for organic contaminants under visible light. Chen, Xiao *et al.* (2012) highlighted the significance of such materials as active photocatalysts (Chen *et al.*, 2012). This study investigated the photocatalytic properties of zinc cadmium sulfide (ZnCdS) nanoparticles covered on sheets of reduced graphene oxide for removing organic dyes in a two-dimensional (2D) platform. The nanoparticles were initially produced and then reduced graphene oxide-coated before evaluating their performance in dye removal.

3.3 Disinfection of wastewater with nanomaterials

The presently known disinfection techniques prove effective in removing microorganisms from drinking water (Kidd *et al.*, 2020). Recent research efforts in water treatment have strived to bridge the gap between effective disinfection and minimizing the formation of harmful disinfection byproducts (DBPs). In water treatment plants, disinfectants like chlorine, ozone, and chloramines are frequently used (Mazhar *et al.*, 2020; Srivastav *et al.*, 2020; Tang *et al.*, 2020). However, these disinfectants can react with other compounds, leading to the formation of hazardous DBPs, many of which are carcinogenic (Verma and Balomajumder, 2020).

3.3.1. Titanium Oxide nanoparticles as disinfectants

The antimicrobial properties of TiO_2 nanoparticles have been extensively studied and recognized for their efficacy in inhibiting the growth of various microorganisms. Titanium dioxide (TiO_2) nanoparticles show antimicrobial action mainly through the generation of reactive oxygen species (ROS) upon exposure to ultraviolet (UV-A) light. The photocatalytic activity of TiO_2, induced by UV irradiation, leads to the production of ROS such as hydroxyl radicals and superoxide ions. These ROS have a destructive effect on microbial cells by causing oxidative damage to cell membranes, proteins, and DNA (Gao *et al.*, 2019). A damaged cell membrane, unable to maintain its barrier function, leaks cellular contents, eventually resulting in cell death. Furthermore, the oxidative stress caused by TiO_2 nanoparticles can interfere with vital cellular functions, enhancing the antimicrobial effect. The ability of TiO2 nanoparticles to provide sustainable and long-lasting antimicrobial effects makes them a promising option in combating the spread of hazardous microorganisms (Wang *et al.*, 2020).

3.3.2. Silver nanoparticles as disinfectants

Silver has been known to have antimicrobial properties since ancient times. The application of silver nanoparticles covers an array of manufacturing applications, including healthcare and external medicine. In recent years, silver nanoparticles have gained popularity as efficient antifungal nanomaterials. These particles have generated significant interest in water decontamination due to their broad-spectrum and potent antibacterial properties, which contribute to a healthier and more secure water supply. When silver is ionized in water, it forms a link with hydrogen-SH groups, which is what deactivates vital enzymes using means of nickel-silver particle absorption (Albukhari *et al.*, 2019; Kumar *et al.*, 2020; Metreveli *et al.*, 2020). The release rate of silver ions, a key factor in their toxicity, is regulated by several features of the nanoparticles themselves. These characteristics include shape, size, surface coating (padding), and the arrangement of atoms on the surface (crystallographic facets). Significantly, the presence of ubiquitous ligands raises toxicity and bioavailability. Silver nanoparticles are a useful tool for bacterial defense in the context of ceramic micro-filters used in underdeveloped nations (Najafpoor *et al.*, 2020).

3.3.3. Zinc oxide nanoparticles as disinfectants

Zinc oxide nanoparticles have a unique set of characteristics, making them a valuable material for a variety of applications. Their great ability to absorb ultraviolet (UV) radiation makes them ideal for sunscreens and protective coatings. Furthermore, their transparency to visible light allows them to be incorporated into a variety of products without altering their appearance, making them valuable for cosmetics and paints. These nanoparticles exhibit excellent antibacterial efficacy across a diverse range of bacteria. The proposed mechanism for their antimicrobial activity involves the photocatalytic production of H_2O_2 (Hee *et al.*, 2019). While zinc oxide nanoparticles and Zn^{2+} ions display antibacterial activity, there is a notable susceptibility of water species to dissolved zinc. Despite their benefits, the limited solubility and potential instability of zinc oxide nanoparticles make them difficult to use directly in drinking water treatment (Shkir *et al.*, 2020).

3.4 Nanomembranes and other filtration techniques

Membranes are very thin sheets of semi-permeable material that allow certain substances to pass through based on their morphology. Membrane-based methods have become widely used in modern wastewater treatment. In membrane water purification, the membrane acts as a selective barrier. It allows clean water molecules to pass through its tiny pores while blocking larger contaminants, resulting in purified water on the other side. Membrane technology is now widely used in many different applications, from desalination to antimicrobial activity. Advancements in membrane technology have led to the development of highly efficient filters capable of removing a wide spectrum of pollutants from water. These include microbes, micro pollutants, particulate materials, and even organic compounds that affect color, taste, and odor. This leads to significant improvement in water quality. Membranes can be classed into numerous categories based on their permeability, structure, and mode of action. These categories include aquaporin-based, nanocomposite, thin-film composite, self-assembling, nanofiber, nanofiltration, and biological membranes. Additionally, membranes have been developed specifically for forward and reverse osmosis processes (Mulyanti and Susanto, 2018). Nanofiltration (NF) stands as a membrane-based process utilizing pressure to separate pollutants from water streams while preventing suspended solids from entering the purified water stream. This method demonstrates exceptional versatility in eliminating a broad spectrum of contaminants from water, including microorganisms, suspended particulates, dissolved organic matter, viruses, large versatile ions, pesticides, herbicides, and many other impurities. Compared to microfilters and ultrafilters, nanofiltration is not only more efficient but also utilizes a smaller amount of electricity. Additionally, the running costs of nanofiltration are lower. Furthermore, the wastewater treatment plant requires less area and is simpler to set up and operate than previous treatment methods. In membrane filtration, only a portion of the initial water (feed) passes through a semi-permeable membrane. This procedure separates the entering stream into two parts: permeate and retentate. The permeate, or filtered water, emerges from the other side of the membrane, while the retentate, which contains concentrated pollutants, remains behind. Nanofiltration, in particular, gained attention for its capacity to soften water and remove both synthetic and naturally occurring organic matter. In an innovative approach highlighted by I. S. Al-Husaini electrospun nanofibrous membranes with iron oxide nanoparticles were employed successfully for oil spill remediation in water (Al-Husaini *et al.*, 2019). Another study referenced by F. M. Gunawan addressed the purification of dye-containing wastewater using a cross-linked polyetherimide-based nanofiltration membrane containing m-phenylenediamine (Gunawan *et al.*, 2019). Furthermore, M. J. Lodo and L. J. Diaz studied combined cyanide with a Fe-modified montmorillonite nanomembrane as an adsorbent material to purify wastewater containing mercury in aqueous solutions (Lodo and Diaz, 2019).

4. Limitations of Nanoparticles in Wastewater Treatment Application

While nanoparticles have enormous promise for wastewater treatment, recent studies highlight some significant challenges (Sultana *et al.*, 2015, Dimapilis *et al.*, 2018). These materials offer diverse functionalities due to their unique properties and varied particle morphologies. The advancements in the commercial application of nanomaterials are quick, resulting in a worldwide expansion in nanomaterial production (He *et al.*, 2014). Despite this, nanoparticles demonstrate the potential for purifying water through several approaches, including photocatalysis, adsorption, and the utilization of nano-absorbents (Mir *et al.*, 2013). Current limitations hinder the optimal application of nanomaterials in wastewater treatment. While existing methods like photocatalysis and adsorption hold promise, they require further optimization for improved effectiveness. A crucial challenge lies in the knowledge gap concerning the potential risks of nanotechnology (Bushra *et al.*, 2014). A crucial aspect of utilizing nanomaterials for water treatment involves understanding their potential environmental impact and long-term health effects. Research is ongoing to assess the release pathways of these materials into the environment, their behavior in water, and any potential human health risks. Some studies have indicated possibilities for health problems, but further investigation is necessary for a comprehensive understanding. Drawing definitive conclusions about the safety of nanomaterials in wastewater treatment remains a challenge. Continuous research is crucial to bridge this knowledge gap. This ongoing effort necessitates conducting experiments at ecologically useful dosages, mimicking real-world exposure scenarios. Furthermore, in-depth toxicological studies are important to comprehensively understand the potential health risks associated with nanomaterial use. Membrane filtration, another effective method in wastewater treatment, faces limitations due to a lack of extensive study. A major problem lies in membrane filtration, where pores become clogged after filtration, reducing efficiency. While nano-absorbents show promise for absorbing pollutants via adsorption, their reusability remains an obstacle to practical application. Developing easily recoverable nanosorbents that effectively remove pollutants from wastewater is critical for their widespread use. The US Environmental Protection Agency (USEPA) has found challenges associated with removing nanoparticles during water treatment processes (Mir *et al.*, 2012; Saritha *et al.*, 2022). Moreover, water is naturally polluted, and after being treated using the methods described earlier, pollutants either decompose or degrade. It is necessary to plan experiments to identify the byproducts and/or intermediates that are produced during these procedures. In addition, developing cost-effective methodologies for the characterization of nanomaterials is important for their broader application in wastewater treatment. Such methods should be capable of accurately evaluating complex nanomaterials to ensure their efficacy and safety (Lu and Astruc, 2020).

5. Conclusion

Nanotechnology is revolutionizing water treatment. Nanomaterials' unique properties allow them to effectively remove contaminants like organic compounds, microorganisms, and heavy metals. Carbon-based nanoparticles, including activated carbons and carbon nanotubes, are particularly optimistic for wastewater treatment. However, creating new, modified nanomaterials and hybrid nano-based frameworks is crucial to optimize wastewater treatment. Some obstacles still exist, such as high costs, limitations in technology, and potential risks to human and environmental health. Extensive collaboration between governments, research institutions, industries, and investors is needed to address these concerns. Overcoming these limitations will ensure environmentally friendly and sustainable water treatment solutions. Continuous development of nanotechnology-based methods will provide robust solutions for future water treatment needs. Nanomaterials hold immense promise for environmental remediation and wastewater treatment.

References

Abukhadra, M.R., Bakry, B.M., Adlii, A., Yakout, S.M., El-Zaidy, M.A. (2019). Facile conversion of kaolinite into clay nanotubes (KNTs) of enhanced adsorption properties for toxic heavy metals (Zn^{2+}, Cd^{2+}, Pb^{2+}, and Cr^{6+}) from water. Journal of Hazardous Material,s 374, 296–308. https://doi.org/10.1016/j.jhazmat.2019.04.047

Agarwal, S., Sadegh, H., Monajjemi, M., Hamdy, A.S., Ali, G.A.M., Memar, A.O.H., Shahryari-Ghoshekandi, R., Tyagi, I., Gupta, V.K., (2016). Efficient removal of toxic bromothymol blue and methylene blue from wastewater by polyvinyl alcohol. Journal of Molecular Liquid, 218, 191–197. https://doi.org/10.1016/j.molliq.2016.02.060

Ahmad, H.R., Aziz, T., Zia-ur-Rehman, M., Sabir, M., Khalid, H., (2016). Sources and composition of waste water: Threats to plants and soil health, in: Soil Science: Agricultural and Environmental Prospectives. Springer International Publishing, pp. 349–370. https://doi.org/10.1007/978-3-319-34451-5_16

Akther, N., Sodiq, A., Giwa, A., Daer, S., Arafat, H.A., Hasan, S.W., (2015). Recent advancements in forward osmosis desalination: A review. Chemical Engineering Journal, 281, 502–522. https://doi.org/10.1016/j.cej.2015.05.080

Albukhari, S.M., Ismail, M., Akhtar, K., Danish, E.Y., (2019). Catalytic reduction of nitrophenols and dyes using silver nanoparticles @ cellulose polymer paper for the resolution of waste water treatment challenges. Colloids and Surfaces A: Physicochemical and Engineering Aspects, 577, 548–561. https://doi.org/10.1016/j.colsurfa.2019.05.058

Al-Husaini, I.S., Yusoff, A.R.M., Lau, W.J., Ismail, A.F., Al-Abri, M.Z., Wirzal, M.D.H., (2019). Iron oxide nanoparticles incorporated polyethersulfone electrospun nanofibrous membranes for effective oil removal. Chemical Engineering Research and Design, 148, 142–154. https://doi.org/10.1016/j.cherd.2019.06.006

Aragaw, T.A., Bogale, F.M., Aragaw, B.A., (2021). Iron-based nanoparticles in wastewater treatment: A review on synthesis methods, applications, and removal mechanisms. Journal of Saudi Chemical Society. https://doi.org/10.1016/j.jscs.2021.101280

Assaad, E., Azzouz, A., Nistor, D., Ursu, A. V., Sajin, T., Miron, D.N., Monette, F., Niquette, P., Hausler, R., (2007). Metal removal through synergic coagulation-flocculation using an optimized chitosan-montmorillonite system. Applied Clay Science 37, 258–274. https://doi.org/10.1016/j.clay.2007.02.007

Aziz, H.A., Adlan, M.N., Ariffin, K.S., (2008). Heavy metals (Cd, Pb, Zn, Ni, Cu and Cr(III)) removal from water in Malaysia: Post treatment by high quality limestone. Bioresource Technology, 99, 1578–1583. https://doi.org/10.1016/j.biortech.2007.04.007

Bai, L., Wei, M., Hong, E., Shan, D., Liu, L., Yang, W., Tang, X., Wang, B., (2020). Study on the controlled synthesis of Zr/TiO2/SBA-15 nanophotocatalyst and its photocatalytic performance for industrial dye reactive red X–3B. Materials Chemistry and Physics, 246. https://doi.org/10.1016/j.matchemphys.2020.122825

Barakat, M.A., (2011). New trends in removing heavy metals from industrial wastewater. Arabian Journal of Chemistry. https://doi.org/10.1016/j.arabjc.2010.07.019

Benalia, M.C., Youcef, L., Bouaziz, M.G., Achour, S., Menasra, H., 2022. Removal of Heavy Metals from Industrial Wastewater by Chemical Precipitation: Mechanisms and Sludge Characterization. The Arabian Journal for Science and Engineering 47, 5587–5599. https://doi.org/10.1007/s13369-021-05525-7

Bushra, R., Shahadat, M., Ahmad, A., Nabi, S.A., Umar, K., Oves, M., Raeissi, A.S., Muneer, M., 2014. Synthesis, characterization, antimicrobial activity and applications of polyanilineTi (IV) arsenophosphate adsorbent for the analysis of organic and inorganic pollutants. Journal of Hazardous Materials 264, 481–489. https://doi.org/10.1016/j.jhazmat.2013.09.044

Cai, Y., Jiang, Y., Feng, L., Hua, Y., Liu, H., Fan, C., Yin, M., Li, S., Lv, X., Wang, H., 2019. Q-graphene-scaffolded covalent organic frameworks as fluorescent probes and sorbents for the fluorimetry and removal of copper ions. Analytica Chimica Acta 1057, 88–97. https://doi.org/10.1016/j.aca.2018.12.054

Calabrò, P.S., Bilardi, S., Moraci, N., (2021). Advancements in the use of filtration materials for the removal of heavy metals from multicontaminated solutions. Current Opinion in Environmental Science & Health. https://doi.org/10.1016/j.coesh.2021.100241

Chen, P., Xiao, T.Y., Li, H.H., Yang, J.J., Wang, Z., Yao, H. Bin, Yu, S.H., (2012). Nitrogen-doped graphene/ZnSe nanocomposites: Hydrothermal synthesis and their enhanced electrochemical and photocatalytic activities. ACS Nano 6, 712–719. https://doi.org/10.1021/nn204191x

Cheriyamundath, S., Vavilala, S.L., (2021). Nanotechnology-based wastewater treatment. Water and Environment Journal. https://doi.org/10.1111/wej.12610

Chhabra, H., Kumar, M., (2020). Size and shape dependent equation of state for nanomaterials with application to bulk materials. Journal of Physics and Chemistry of Solids, 139. https://doi.org/10.1016/j.jpcs.2019.109308

Dąbrowski, A., Hubicki, Z., Podkościelny, P., Robens, E., 2004. Selective removal of the heavy metal ions from waters and industrial wastewaters by ion-exchange method. Chemosphere. https://doi.org/10.1016/j.chemosphere.2004.03.006

Dargahi, A., Golestanifar, H., Darvishi, P., Karami, A., Hasan, S.H., Poormohammadi, A., Behzadnia, A., (2016). An investigation and comparison of removing heavy metals (Lead and chromium) from aqueous solutions using magnesium oxide nanoparticles. Polish Journal of Environmental Studies 25, 557–562. https://doi.org/10.15244/pjoes/60281

Deshpande, B.D., Agrawal, P.S., Yenkie, M.K.N., Dhoble, S.J., (2020). Prospective of nanotechnology in degradation of waste water: A new challenges. Nano-Structures and Nano-Objects. https://doi.org/10.1016/j.nanoso.2020.100442

Dimapilis, E.A.S., Hsu, C.S., Mendoza, R.M.O., Lu, M.C., (2018). Zinc oxide nanoparticles for water disinfection. Sustainable Environment Research. https://doi.org/10.1016/j.serj.2017.10.001

Ebrahimpour, Z., Pliekhova, O., Cabrera, H., Abdelhamid, M., Korte, D., Segbéya Gadedjisso-Tossou, K., Niemela, J., Lavrencic Stangar, U., Franko, M., (2021). Photodegradation mechanisms of reactive blue 19 dye under UV and simulated solar light irradiation. Spectrochimica Acta, Part A: Molecular and Biomolecular Spectroscopy 252. https://doi.org/10.1016/j.saa.2021.119481

Edokpayi, J.N., Makungo, R., Mathivha, F., Rivers, N., Volenzo, T., Odiyo, J.O., (2020). Influence of global climate change on water resources in South Africa: toward an adaptive management approach, in: Water Conservation and Wastewater Treatment in BRICS Nations: Technologies, Challenges, Strategies and Policies. Elsevier, pp. 83–115. https://doi.org/10.1016/B978-0-12-818339-7.00005-9

Elbeltagi, A., Azad, N., Arshad, A., Mohammed, S., Mokhtar, A., Pande, C., Etedali, H.R., Bhat, S.A., Islam, A.R.M.T., Deng, J. (2021). Applications of Gaussian process regression for predicting blue water footprint: Case study in Ad Daqahliyah, Egypt. Agricultural water management 255. https://doi.org/10.1016/j.agwat.2021.107052

Fan, W., Wang, Xia, Liu, X., Xu, B., Zhang, X., Wang, W., Wang, Xiaokang, Wang, Y., Dai, F., Yuan, D., Sun, D., (2019). Regulating C 2 H 2 and CO 2 Storage and Separation through Pore Environment Modification in a Microporous Ni-MOF. ACS Sustainable Chemistry & Engineering 7, 2134–2140. https://doi.org/10.1021/acssuschemeng.8b04783

Fenati, R.A., Sherrell, P.C., Khodakov, D.A., Shapter, J.G., Ellis, A. V., (2021). Spatially isolated redox processes enabled by ambipolar charge transport in multi-walled carbon nanotube mats. Materials Horizons 8, 1304–1313. https://doi.org/10.1039/d0mh01967d

Foroutan, R., Esmaeili, H., (2015). Investigation into ion exchange and adsorption methods for removing heavy metals from aqueous solutions. International Journal of Biology, Pharmacy and Allied Sciences 4, 620-629.

Gao, D., Zheng, S., Wang, L., Wang, C., Zhang, H., Wang, Q., (2019). SILAR preparation of visible-light-driven TiO2 NTs/Ag2WO4-AgI photoelectrodes for waste water treatment and photoelectric conversion. Separation and Purification Technology 224, 308–314. https://doi.org/10.1016/j.seppur.2019.05.050

Ge, F., Li, M.M., Ye, H., Zhao, B.X., 2012. Effective removal of heavy metal ions Cd 2+, Zn 2+, Pb 2+, Cu 2+ from aqueous solution by polymer-modified magnetic nanoparticles. Journal of Hazardous Materials 211–212, 366–372. https://doi.org/10.1016/j.jhazmat.2011.12.013

Gitis, V., Hankins, N., (2018). Water treatment chemicals: Trends and challenges. Journal of Water Process Engineering 25, 34–38. https://doi.org/10.1016/j.jwpe.2018.06.003

Gunawan, F.M., Mangindaan, D., Khoiruddin, K., Wenten, I.G., (2019). Nanofiltration membrane cross-linked by m-phenylenediamine for dye removal from textile wastewater. Polymers for Advanced Technologies 30, 360–367. https://doi.org/10.1002/pat.4473

Hao, M., Qiu, M., Yang, H., Hu, B., Wang, X., 2021. Recent advances on preparation and environmental applications of MOF-derived carbons in catalysis. Science of the Total Environment. https://doi.org/10.1016/j.scitotenv.2020.143333

He, D., Sun, Y., Xin, L., Feng, J., 2014. Aqueous tetracycline degradation by non-thermal plasma combined with nano-TiO_2. Chemical Engineering Journal 258, 18–25. https://doi.org/10.1016/j.cej.2014.07.089

Hodges, B.C., Cates, E.L., Kim, J.H., 2018. Challenges and prospects of advanced oxidation water treatment processes using catalytic nanomaterials. Nature Nanotechnology, 13, 642–650. https://doi.org/10.1038/s41565-018-0216-x

Hu, B., Ai, Y., Jin, J., Hayat, T., Alsaedi, A., Zhuang, L., Wang, X., 2020. Efficient elimination of organic and inorganic pollutants by biochar and biochar-based materials. Biochar. https://doi.org/10.1007/s42773-020-00044-4

Huang, F., Singer, G., Zhou, Y., Sha, Z., Chen, J., Han, Z., Brown, S.A., Zhang, J., Wang, C.H., (2021). Creating ionic pathways in solid-state polymer electrolyte by using PVA-coated carbon nanofibers. Composites Science and Technology 207. https://doi.org/10.1016/j.compscitech.2021.108710

Ipek, U., (2005). DESALINATION Removal of Ni(II) and Zn(II) from an aqueous solution by reverse osmosis. Desalination 174, 161–169. https://doi.org/10.1016/j

Jaseem, M., Kumar, P., Mariam John, R., Muhammed Jaseem, C., (2017). An overview of waste management in pharmaceutical industry. The Pharma Innovation Journal 6, 158–161.

Jeevanandam, J., Barhoum, A., Chan, Y.S., Dufresne, A., Danquah, M.K., (2018). Review on nanoparticles and nanostructured materials: History, sources, toxicity and regulations. Beilstein Journal of Nanotechnology. https://doi.org/10.3762/bjnano.9.98

Jiang, Y., Carrijo, D., Huang, S., Chen, J., Balaine, N., Zhang, W., van Groenigen, K.J., Linquist, B., (2019). Water management to mitigate the global warming potential of rice systems: A global meta-analysis. Field Crops Research 234, 47–54. https://doi.org/10.1016/j.fcr.2019.02.010

Kadam, A., Saratale, R.G., Shinde, S., Yang, J., Hwang, K., Mistry, B., Saratale, G.D., Lone, S., Kim, D.Y., Sung, J.S., Ghodake, G., (2019). Adsorptive remediation of cobalt oxide nanoparticles by magnetized A-cellulose fibers from waste paper biomass. Bioresource Technology 273, 386–393. https://doi.org/10.1016/j.biortech.2018.11.041

Kalaitzidou, K., Zouboulis, A., Mitrakas, M., (2020). Cost evaluation for Se(IV) removal, by applying common drinking water treatment processes: Coagulation/precipitation or adsorption. The Journal of Environmental Chemical Engineering 8. https://doi.org/10.1016/j.jece.2020.104209

Kanthi Gunatilake, S., (2015). Methods of Removing Heavy Metals from Industrial Wastewater. Journal of multidisciplinary Engineering Science Studies.

Karthigadevi, G., Manikandan, S., Karmegam, N., Subbaiya, R., Chozhavendhan, S., Ravindran, B., Chang, S.W., Awasthi, M.K., 2021. Chemico-nanotreatment methods for the removal of persistent organic pollutants and xenobiotics in water – A review. Bioresource Technology. https://doi.org/10.1016/j.biortech.2021.124678

Khan, M.T., Shah, I.A., Ihsanullah, I., Naushad, M., Ali, S., Shah, S.H.A., Mohammad, A.W., (2021). Hospital wastewater as a source of environmental contamination: An overview of management practices, environmental risks, and treatment processes. Journal of Water Process Engineering. https://doi.org/10.1016/j.jwpe.2021.101990

Kidd, J., Westerhoff, P., Maynard, A., (2020). Public Perceptions for the Use of Nanomaterials for In-Home Drinking Water Purification 1 Devices 2.

Kumar, A., Mishra, B., Tripathi, B.P., (2020). Polydopamine assisted synthesis of ultrafine silver nanoparticles for heterogeneous catalysis and water remediation. Nano-Structures and Nano-Objects 23. https://doi.org/10.1016/j.nanoso.2020.100489

Kuppusamy, P., Yusoff, M.M., Maniam, G.P., Govindan, N., (2016). Biosynthesis of metallic nanoparticles using plant derivatives and their new avenues in pharmacological applications – An updated report. Saudi Pharmaceutical Journal. https://doi.org/10.1016/j.jsps.2014.11.013

Lam, E., Luong, J.H.T., (2014). Carbon materials as catalyst supports and catalysts in the transformation of biomass to fuels and chemicals. ACS Catalysis. https://doi.org/10.1021/cs5008393

Lázaro-Navas, S., Prashar, S., Fajardo, M., Gómez-Ruiz, S., (2015). Visible light-driven photocatalytic degradation of the organic pollutant methylene blue with hybrid palladium–fluorine-doped titanium oxide nanoparticles. Journal of Nanoparticle Research 17, 1–14. https://doi.org/10.1007/s11051-015-2902-z

Lee, E., Hong, J.Y., Kang, H., Jang, J., 2012. Synthesis of TiO_2 nanorod-decorated graphene sheets and their highly efficient photocatalytic activities under visible-light irradiation. Journal of Hazardous Materials 219–220, 13–18. https://doi.org/10.1016/j.jhazmat.2011.12.033

Li, J., Wang, Xiangxue, Zhao, G., Chen, C., Chai, Z., Alsaedi, A., Hayat, T., Wang, Xiangke, 2018. Metal-organic framework-based materials: Superior adsorbents for the capture of toxic and radioactive metal ions. Chemical Society Reviews. https://doi.org/10.1039/c7cs00543a

Li, M., Li, M.Y., Feng, C.G., Zeng, Q.X., 2014. Preparation and characterization of multi-carboxyl-functionalized silica gel for removal of Cu (II), Cd (II), Ni (II) and Zn (II) from aqueous solution. Applied Surface Science 314, 1063–1069. https://doi.org/10.1016/j.apsusc.2014.06.038

Liu, F., Zhu, S., Li, D., Chen, G., Ho, S.H., 2020. Detecting Ferric Iron by Microalgal Residue-Derived Fluorescent Nanosensor with an Advanced Kinetic Model. Science 23. https://doi.org/10.1016/j.isci.2020.101174

Liu, Xiaolu, Pang, H., Liu, Xuewei, Li, Q., Zhang, N., Mao, L., Qiu, M., Hu, B., Yang, H., Wang, X., (2021). Orderly Porous Covalent Organic Frameworks-based Materials: Superior Adsorbents for Pollutants Removal from Aqueous Solutions. Innovation. https://doi.org/10.1016/j.xinn.2021.100076

Lodo, M.J., Diaz, L.J., 2019. Reusability of Fe-modified MMT nanomembranes and the retrieval of the adsorbed mercury metal, in: IOP Conference Series: Earth and Environmental Science. Institute of Physics Publishing. https://doi.org/10.1088/1755-1315/345/1/012011

Lofrano, G., Carotenuto, M., Libralato, G., Domingos, R.F., Markus, A., Dini, L., Gautam, R.K., Baldantoni, D., Rossi, M., Sharma, S.K., Chattopadhyaya, M.C., Giugni, M., Meric, S., 2016. Polymer functionalized nanocomposites for metals removal from water and wastewater: An overview. Water Resource. https://doi.org/10.1016/j.watres.2016.01.033

Lu, F., Astruc, D., 2020. Nanocatalysts and other nanomaterials for water remediation from organic pollutants. Coordination Chemistry Reviews. https://doi.org/10.1016/j.ccr.2020.213180

Lu, H., Wang, J., Stoller, M., Wang, T., Bao, Y., Hao, H., 2016. An Overview of Nanomaterials for Water and Wastewater Treatment. Advances in Materials Science and Engineering. https://doi.org/10.1155/2016/4964828

Lu, S.Y., Jin, M., Zhang, Y., Niu, Y.B., Gao, J.C., Li, C.M., 2018. Chemically Exfoliating Biomass into a Graphene-like Porous Active Carbon with Rational Pore Structure, Good Conductivity, and Large Surface Area for High-Performance Supercapacitors. Advanced Energy Materials 8. https://doi.org/10.1002/aenm.201702545

Ma, Q., Cui, L., Zhou, S., Li, Y., Shi, W., Ai, S., (2018). Iron nanoparticles in situ encapsulated in lignin-derived hydrochar as an effective catalyst for phenol removal. Environmental Science and Pollution Research 25, 20833–20840. https://doi.org/10.1007/s11356-018-2285-7

Madannejad, R., Shoaie, N., Jahanpeyma, F., Darvishi, M.H., Azimzadeh, M., Javadi, H., (2019). Toxicity of carbon-based nanomaterials: Reviewing recent reports in medical and biological systems. Chemico-Biological Interactions. https://doi.org/10.1016/j.cbi.2019.04.036

Mahmoud, M.E., Abdelwahab, M.S., (2021). One-step synthesis of zero-valent Sn nanoparticles and potential microwave remediation of lead from water. Materials Research Bulletin 134. https://doi.org/10.1016/j.materresbull.2020.111090

Man, P., He, B., Zhang, Q., Zhou, Z., Li, C., Li, Q., Wei, L., Yao, Y., (2019). A one-dimensional channel self-standing MOF cathode for ultrahigh-energy-density flexible Ni-Zn batteries. Journal of Materials Chemistry 7, 27217–27224. https://doi.org/10.1039/c9ta11759h

Matlock, M.M., Howerton, B.S., Atwood, D.A., (2002). Chemical precipitation of heavy metals from acid mine drainage, Water Research.

Maynez-Navarro, O.D., Mendez-Rojas, M.A., Flores-Cervantes, D.X., Sanchez-Salas, J.L., 2020. Hydroxyl Radical Generation by Recyclable Photocatalytic Fe3O4/ZnO Nanoparticles for Water Disinfection. Air, Soil and Water Research 13. https://doi.org/10.1177/1178622120970954

Mazhar, M.A., Khan, N.A., Ahmed, S., Khan, A.H., Hussain, A., Rahisuddin, Changani, F., Yousefi, M., Ahmadi, S., Vambol, V., (2020). Chlorination disinfection by-products in municipal drinking water – A review. The Journal of Cleaner Production. https://doi.org/10.1016/j.jclepro.2020.123159

Metreveli, G., David, J., Schneider, R., Kurtz, S., Schaumann, G.E., (2020. Morphology, structure, and composition of sulfidized silver nanoparticles and their aggregation dynamics in river water. Science of the Total Environment 739. https://doi.org/10.1016/j.scitotenv.2020.139989

Millati, R., Cahyono, R.B., Ariyanto, T., Azzahrani, I.N., Putri, R.U., Taherzadeh, M.J., (2019). Agricultural, industrial, municipal, and forest wastes: An overview, in: Sustainable Resource Recovery and Zero Waste Approaches. Elsevier, pp. 1–22. https://doi.org/10.1016/B978-0-444-64200-4.00001-3

Mir, N.A., Haque, M.M., Khan, A., Muneer, M., Boxall, C., (2012). Photoassisted degradation of a herbicide derivative, dinoseb, in aqueous suspension of titania. The Scientific World Journal 2012. https://doi.org/10.1100/2012/251527

Mir, NiyazA., Khan, A., Umar, K., Muneer, M., (2013). Photocatalytic Study of a Xanthene Dye Derivative, Phloxine B in Aqueous Suspension of TiO2 : Adsorption Isotherm and Decolourization Kinetics . Energy and Environment Focus 2, 208–216. https://doi.org/10.1166/eef.2013.1052

Mirbagheri, S.A., Hosseini, S.N., (2005). Pilot plant investigation on petrochemical wastewater treatment for the removal of copper and chromium with the objective of reuse. Desalination 171, 85–93. https://doi.org/10.1016/j.desal.2004.03.022

Mohamadiun, M., Dahrazma, B., Saghravani, S.F., Khodadadi Darban, A., (2018). Removal of cadmium from contaminated soil using iron (III) oxide nanoparticles stabilized with polyacrylic acid. Journal of Environmental Engineering and Landscape Management 26, 98–106. https://doi.org/10.3846/16486897.2017.1364645

Mohammed, M.A., Syeda, J.T.M., Wasan, K.M., Wasan, E.K., (2017). An overview of chitosan nanoparticles and its application in non-parenteral drug delivery. Pharmaceutics. https://doi.org/10.3390/pharmaceutics9040053

Mohsen-Nia, M., Montazeri, P., Modarress, H., (2007). Removal of Cu2+ and Ni2+ from wastewater with a chelating agent and reverse osmosis processes. Desalination 217, 276–281. https://doi.org/10.1016/j.desal.2006.01.043

Morrison, E.S., Shields, M.R., Bianchi, T.S., Liu, Y., Newman, S., Tolic, N., Chu, R.K., (2020). Multiple biomarkers highlight the importance of water column processes in treatment wetland organic matter cycling. Water Resource 168. https://doi.org/10.1016/j.watres.2019.115153

Mostafapour, F.K., Miri, A., Khatibi, A., Balarak, D., Kyzas, G.Z., (2023). Survey of Fe3O4 Magnetic Nanoparticles Modified with Sodium Dodecyl Sulfate for Removal P-Cresol and Pyrocatechol from Aqueous Solutions. Biointerface Research in Applied Chemistry, 13. https://doi.org/10.33263/BRIAC136.554

Mudasir, P., Naqash, J.A., (2020). The effect of water cement ratio on the characteristics of multi-walled carbon nanotube reinforced concrete, in: Materials Today: Proceedings. Elsevier Ltd, pp. 3852–3855. https://doi.org/10.1016/j.matpr.2020.12.1176

Mulyanti, R., Susanto, H., (2018). Wastewater treatment by nanofiltration membranes, in: IOP Conference Series: Earth and Environmental Science. Institute of Physics Publishing. https://doi.org/10.1088/1755-1315/142/1/012017

Najafpoor, A., Norouzian-Ostad, R., Alidadi, H., Rohani-Bastami, T., Davoudi, M., Barjasteh-Askari, F., Zanganeh, J., (2020). Effect of magnetic nanoparticles and silver-loaded magnetic nanoparticles on advanced wastewater treatment and disinfection. Journal of Molecular Liquids 303. https://doi.org/10.1016/j.molliq.2020.112640

Nodeh, H.R., Shakiba, M., Gabris, M.A., Bidhendi, M.E., Shahabuddin, S., Khanam, R., (2020). Spherical iron oxide methyltrimethoxysilane nanocomposite for the efficient removal of lead(II) ions from wastewater: Kinetic and equilibrium studies. Desalination and Water Treatment 192, 297–305. https://doi.org/10.5004/dwt.2020.25767

O, K.H., Rupa, E.J., Anandapadmanaban, G., Chokkalingam, M., Li, J.F., Markus, J., Soshnikova, V., Perez, Z.E.J., Yang, D.C., (2019). Cationic and anionic dye degradation activity of Zinc oxide nanoparticles from Hippophae rhamnoides leaves as potential water treatment resource. Optik (Stuttg) 181, 1091–1098. https://doi.org/10.1016/j.ijleo.2018.12.152

Palani, G., Trilaksana, H., Sujatha, R.M., Kannan, K., Rajendran, S., Korniejenko, K., Nykiel, M., Uthayakumar, M., (2023). Silver Nanoparticles for Waste Water Management. Molecules. https://doi.org/10.3390/molecules28083520

Pandi, K., Prabhu, S.M., Ahn, Y., Park, C.M., Choi, J., (2020). Design and synthesis of biopolymer-derived porous graphitic carbon covered iron-organic frameworks for depollution of arsenic from waters. Chemosphere 254. https://doi.org/10.1016/j.chemosphere.2020.126769

Pang, Y., Ruan, Y., Feng, Y., Diao, Z., Shih, K., Hou, L., Chen, D., Kong, L., (2019). Ultrasound assisted zero valent iron corrosion for peroxymonosulfate activation for Rhodamine-B degradation. Chemosphere 228, 412–417. https://doi.org/10.1016/j.chemosphere.2019.04.164

Pohl, A., (2020). Removal of Heavy Metal Ions from Water and Wastewaters by Sulfur-Containing Precipitation Agents. Water, Air, & Soil Pollution. https://doi.org/10.1007/s11270-020-04863-w

Prachi, Gautam, P., Madathil, D., Brijesh Nair, A.N., 2013. Nanotechnology in Waste Water Treatment: A Review, International Journal of ChemTech Research.

Qiu, M., Liu, Z., Wang, S., Hu, B., (2021). The photocatalytic reduction of U(VI) into U(IV) by ZIF-8/g-C3N4 composites at visible light. Environmental Research 196. https://doi.org/10.1016/j.envres.2020.110349

Rambabu, K., Bharath, G., Banat, F., Show, P.L., (2021. Green synthesis of zinc oxide nanoparticles using Phoenix dactylifera waste as bioreductant for effective dye degradation and antibacterial performance in wastewater treatment. Journal of Hazardous Materials 402. https://doi.org/10.1016/j.jhazmat.2020.123560

Ren, S., Tao, J., Tan, F., Cui, Y., Li, X., Chen, J., He, X., Wang, Y., (2018). Diffusive gradients in thin films based on MOF-derived porous carbon binding gel for in-situ measurement of antibiotics in waters. Science of the Total Environment 645, 482–490. https://doi.org/10.1016/j.scitotenv.2018.07.013

Rubilar, R., Hubbard, R., Emhart, V., Mardones, O., Quiroga, J.J., Medina, A., Valenzuela, H., Espinoza, J., Burgos, Y., Bozo, D., (2020). Climate and water availability impacts on early growth and growth efficiency of Eucalyptus genotypes: The importance of GxE interactions. Forest Ecology and Management 458. https://doi.org/10.1016/j.foreco.2019.117763

Sadegh, H., Ali, G.A.M., Gupta, V.K., Makhlouf, A.S.H., Shahryari-ghoshekandi, R., Nadagouda, M.N., Sillanpää, M., Megiel, E., (2017). The role of nanomaterials as effective adsorbents and their applications in wastewater treatment. Journal of Nanostructure in Chemistry. https://doi.org/10.1007/s40097-017-0219-4

Saritha, G.N.G., Anju, T., Kumar, A., (2022). Nanotechnology - Big impact: How nanotechnology is changing the future of agriculture? The Journal of Agriculture and Food Research 10. https://doi.org/10.1016/j.jafr.2022.100457

Saroa, A., Singh, A., Jindal, N., Kumar, R., Singh, K., Guleria, P., Boopathy, R., Kumar, V., (2023). Nanotechnology-assisted treatment of pharmaceuticals contaminated water. Bioengineered. https://doi.org/10.1080/21655979.2023.2260919

Sekoai, P.T., Ouma, C.N.M., du Preez, S.P., Modisha, P., Engelbrecht, N., Bessarabov, D.G., Ghimire, A., (2019). Application of nanoparticles in biofuels: An overview. Fuel. https://doi.org/10.1016/j.fuel.2018.10.030

Shah, A.I., Din Dar, M.U., Bhat, R.A., Singh, J.P., Singh, K., Bhat, S.A., (2020). Prospectives and challenges of wastewater treatment technologies to combat contaminants of emerging concerns. Ecological engineering. https://doi.org/10.1016/j.ecoleng.2020.105882

Shkir, M., Al-Shehri, B.M., Pachamuthu, M.P., Khan, A., Chandekar, K. V., AlFaify, S., Hamdy, M.S., (2020). A remarkable improvement in photocatalytic activity of ZnO nanoparticles through Sr doping synthesized by one pot flash combustion technique for water treatments. Colloids and Surfaces A: Physicochemical and Engineering Aspects 587. https://doi.org/10.1016/j.colsurfa.2019.124340

Singh, J., Juneja, S., Soni, R.K., Bhattacharya, J., (2021). Sunlight mediated enhanced photocatalytic activity of TiO2 nanoparticles functionalized CuO-Cu_2O nanorods for removal of methylene blue and oxytetracycline hydrochloride. The Journal of Colloid and Interface Science 590, 60–71. https://doi.org/10.1016/j.jcis.2021.01.022

Solis, K.L.B., Kwon, Y.H., Kim, M.H., An, H.R., Jeon, C., Hong, Y., (2020). Metal organic framework UiO-66 and activated carbon composite sorbent for the concurrent adsorption of cationic and anionic metals. Chemosphere 238. https://doi.org/10.1016/j.chemosphere.2019.124656

Srivastav, A.L., Patel, N., Chaudhary, V.K., (2020). Disinfection by-products in drinking water: Occurrence, toxicity and abatement. Environmental Pollution. https://doi.org/10.1016/j.envpol.2020.115474

Sultana, S., Rafiuddin, Khan, M.Z., Umar, K., Ahmed, A.S., Shahadat, M., (2015). SnO_2-SrO based nanocomposites and their photocatalytic activity for the treatment of organic pollutants. The Journal of Molecular Structure 1098, 393–399. https://doi.org/10.1016/j.molstruc.2015.06.032

Tan, W., Ruan, Y., Diao, Z., Song, G., Su, M., Hou, L., Chen, D., Kong, L., Deng, H., (2021). Removal of levofloxacin through adsorption and peroxymonosulfate activation using carbothermal reduction synthesized nZVI/carbon fiber. Chemosphere 280. https://doi.org/10.1016/j.chemosphere.2021.130626

Tang, J., Wang, J., (2020). Iron-copper bimetallic metal-organic frameworks for efficient Fenton-like degradation of sulfamethoxazole under mild conditions. Chemosphere 241. https://doi.org/10.1016/j.chemosphere.2019.125002

Tang, Y., Long, X., Wu, M., Yang, S., Gao, N., Xu, B., Dutta, S., (2020). Bibliometric review of research trends on disinfection by-products in drinking water during 1975–2018. Separation and Purification Technology. https://doi.org/10.1016/j.seppur.2020.116741

Teh, C.Y., Budiman, P.M., Shak, K.P.Y., Wu, T.Y., (2016). Recent Advancement of Coagulation-Flocculation and Its Application in Wastewater Treatment. Industrial & Engineering Chemistry Research. https://doi.org/10.1021/acs.iecr.5b04703

Thaçi, B.S., Gashi, S.T., (2019). Reverse osmosis removal of heavy metals from wastewater effluents using biowaste materials pretreatment. Polish Journal of Environmental Studies 28, 337–341. https://doi.org/10.15244/pjoes/81268

Umar, K., Ibrahim, M.N.M., Ahmad, A., Rafatullah, M., (2019a). Synthesis of Mn-doped TiO_2 by novel route and photocatalytic mineralization/intermediate studies of organic pollutants. Research on Chemical Intermediates 45, 2927–2945. https://doi.org/10.1007/s11164-019-03771-x

Umar, K., Parveen, T., Khan, M.A., Ibrahim, M.N.M., Ahmad, A., Rafatullah, M., (2019b). Degradation of organic pollutants using metal-doped TiO_2 photocatalysts under visible light: A comparative study. Desalination and Water Treatment 161, 275–282. https://doi.org/10.5004/dwt.2019.24298

Vaaramaa, K., Lehto, J., (2003). Removal of metals and anions from drinking water by ion exchange.

Verma, B., Balomajumder, C., (2020). Surface modification of one-dimensional Carbon Nanotubes: A review for the management of heavy metals in wastewater. Environmental Technology & Innovation. https://doi.org/10.1016/j.eti.2019.100596

Vidu, R., Matei, E., Predescu, A.M., Alhalaili, B., Pantilimon, C., Tarcea, C., Predescu, C., (2020). Removal of heavy metals from wastewaters: A challenge from current treatment methods to nanotechnology applications. Toxics. https://doi.org/10.3390/toxics8040101

Vojoudi, H., Badiei, A., Bahar, S., Mohammadi Ziarani, G., Faridbod, F., Ganjali, M.R., (2017). A new nano-sorbent for fast and efficient removal of heavy metals from aqueous solutions based on modification of magnetic mesoporous silica nanospheres. The Journal of Magnetism and Magnetic Materials 441, 193–203. https://doi.org/10.1016/j.jmmm.2017.05.065

Wadhawan, S., Jain, A., Nayyar, J., Mehta, S.K., (2020). Role of nanomaterials as adsorbents in heavy metal ion removal from waste water: A review. Journal of Water Process Engineering. https://doi.org/10.1016/j.jwpe.2019.101038

Wang, H., Zhang, Q., Qiu, M., Hu, B., (2021). Synthesis and application of perovskite-based photocatalysts in environmental remediation: A review. Journal of Molecular Liquids. https://doi.org/10.1016/j.molliq.2021.116029

Wang, J., Tang, X., Xu, Y., Cheng, X., Li, G., Liang, H., (2020). Hybrid UF/NF process treating secondary effluent of wastewater treatment plants for potable water reuse: Adsorption vs. coagulation for removal improvements and membrane fouling alleviation. Environmental Resources 188. https://doi.org/10.1016/j.envres.2020.109833

Wang, R., Chen, D., Wang, Q., Ying, Y., Gao, W., Xie, L., (2020). Recent advances in applications of carbon nanotubes for desalination: A review. Nanomaterials. https://doi.org/10.3390/nano10061203

Wang, W., Lin, F., Yan, B., Cheng, Z., Chen, G., Kuang, M., Yang, C., Hou, L., (2020). The role of seashell wastes in TiO_2/Seashell composites: Photocatalytic degradation of methylene blue dye under sunlight. Environmental Resource 188. https://doi.org/10.1016/j.envres.2020.109831

Wong, J.K.H., Tan, H.K., Lau, S.Y., Yap, P.S., Danquah, M.K., (2019). Potential and challenges of enzyme incorporated nanotechnology in dye wastewater treatment: A review. The Journal of Environmental Chemical Engineering. https://doi.org/10.1016/j.jece.2019.103261

Wu, W., Shi, Y., Liu, G., Fan, X., Yu, Y., (2020). Recent development of graphene oxide based forward osmosis membrane for water treatment: A critical review. Desalination. https://doi.org/10.1016/j.desal.2020.114452

Wu, Y., Pang, H., Liu, Y., Wang, Xiangxue, Yu, S., Fu, D., Chen, J., Wang, Xiangke, (2019). Environmental remediation of heavy metal ions by novel-nanomaterials: A review. Environmental Pollution. https://doi.org/10.1016/j.envpol.2018.12.076

Yang, X., Guo, N., Yu, Y., Li, H., Xia, H., Yu, H., (2020). Synthesis of magnetic graphene oxide-titanate composites for efficient removal of Pb(II) from wastewater: Performance and mechanism. The Journal of Environmental Management 256. https://doi.org/10.1016/j.jenvman.2019.109943

Yang, Z., He, S., Wu, H., Yin, T., Wang, L., Shan, A., (2021). Nanostructured Antimicrobial Peptides: Crucial Steps of Overcoming the Bottleneck for Clinics. Frontiers in Microbiology. https://doi.org/10.3389/fmicb.2021.710199

Yin, J., Deng, B., (2015). Polymer-matrix nanocomposite membranes for water treatment. The Journal of Membrane Science. https://doi.org/10.1016/j.memsci.2014.11.019

Yin, Z., Cui, C., Chen, H., Duoni, Yu, X., Qian, W., (2020). The Application of Carbon Nanotube/Graphene-Based Nanomaterials in Wastewater Treatment. Small. https://doi.org/10.1002/smll.201902301

Yousuf, M., Mollah, A., Schennach, R., Parga, J.R., Cocke, D.L., (2001). Electrocoagulation (EC)-science and applications, Journal of Hazardous Materials.

Zhang, H., Ruan, Y., Liang, A., Shih, K., Diao, Z., Su, M., Hou, L., Chen, D., Lu, H., Kong, L., (2019). Carbothermal reduction for preparing nZVI/BC to extract uranium: Insight into the iron species dependent uranium adsorption behavior. The Journal of Cleaner Production 239. https://doi.org/10.1016/j.jclepro.2019.117873

Zhang, T., Lowry, G. V., Capiro, N.L., Chen, J., Chen, W., Chen, Y., Dionysiou, D.D., Elliott, D.W., Ghoshal, S., Hofmann, T., Hsu-Kim, H. *et al.*, (2019). In situ remediation of subsurface contamination: Opportunities and challenges for nanotechnology and advanced materials. Environmental Science: Nano. https://doi.org/10.1039/c9en00143c

Zhang, Y., Chi, H., Zhang, W., Sun, Y., Liang, Q., Gu, Y., Jing, R., (2014). Highly Efficient Adsorption of Copper Ions by a PVP-Reduced Graphene Oxide Based on a New Adsorptions Mechanism. Nano-Micro Letters 6. https://doi.org/10.5101/nml.v6i1.p80-87

Zou, R., Zhang, Z., Yu, L., Tian, Q., Chen, Z., Hu, J., (2011). A general approach for the growth of metal oxide nanorod arrays on graphene sheets and their applications. Chemistry - A European Journal 17, 13912–13917. https://doi.org/10.1002/chem.201101981t

15

Zinc Oxide Nanoparticles: Pioneering Sustainable Green Synthesis & Diverse Environmental Applications

Ayushi Bhatnagar, Rama Kanwar Khangarot and Gangotri Pemawat*

Department of Chemistry, University College of Science, Mohanlal Sukhadia University, Udaipur-313001, Rajasthan, India

Abstract

Nanotechnology holds immense promise in bolstering environmental protection and sustainability by revolutionizing approaches to monitoring, addressing, and amending environmental pollutants. This paradigm shift extends to resource conservation and energy efficiency through the innovation of more adept technologies. Yet, as we harness the potential of nanotechnology, it is of prime importance to meticulously evaluate its environmental and health implications, instating stringent measures to mitigate risks. Responsible and sustainable development, coupled with effective regulation, is imperative. Nanomaterials, particularly ZnO stand as a pivotal player in this arena. ZnO's exceptional physical and chemical properties position it as a cost-effective substance in diverse applications, ranging from the cosmetic industry to nano-fertilizers and electrical devices. Its utility encompasses to bioimaging, targeted drug delivery, gene delivery, and environmental sensing for pollutant detection and amendment. ZnO's functionality expands to real-time pollutant detection sensors and its application in nanomaterial-based solar cells, further solidifying its significance in advancing clean energy solutions. This chapter emphasizes on various methods of ZnO nanoparticles synthesis encircling biological routes which spindles focus towards eco-friendly approaches, as well as offer reduced hazards. ZnO's photocatalytic properties aid in breaking down pollutants in air and water, and also assist in soil remediation. Ensuring responsible use is crucial, considering environmental and health implications, to ensure justifiable pollution management.

Keywords: *Nanotechnology, ZnO, Nanoparticles, Synthesis Environmental, Remediation, Applications.*

* *Correspondence author: drgpemawat@mlsu.ac.in*

1. Introduction

Zinc oxide nanoparticles (ZnO NPs) have emerged as highly promising entities in the realm of nanotechnology, owing to their distinctive properties and versatile applications. Nanoparticles (NPs), characterized by their small size and expansive surface area-to-volume ratio, exhibit exceptional efficiency across a spectrum of processes (Bandeira, Giovanela, Roesch-Ely, Devine, & da Silva Crespo, 2020). Their significance lies in their potential to revolutionize industries spanning electronics, healthcare, and environmental improvement, offering solutions that are both effective and environmentally friendly (Jiang *et al.*, 2023). One of the paramount advantages of ZnO NPs is their efficacy and versatility in numerous domains (Islam *et al.*, 2022). With their elevated reactivity and ample surface area, these NPs markedly reduce reaction time in catalytic processes, thereby augmenting productivity (Droepenu, Wee, Chin, Kok, & Maligan, 2022). In catalysis, they function as proficient catalysts in diverse chemical reactions, such as oxidation and reduction processes (Shaba, Jacob, Tijani, & Suleiman, 2021). The applications of ZnO NPs are manifold and extensive. Their potent antimicrobial properties render them indispensable in medical settings, where they can combat pathogens with efficacy (Mendes *et al.*, 2022) and also find utility in biomedical fields including wound healing, drug delivery systems, and antibacterial coatings (Mendes *et al.*, 2022). In electronics, these NPs are essential in the fabrication of sensors, photovoltaic devices, and lithium-ion batteries, because of their excellent electrical conductivity and optical properties.(Alexandrov, Zvaigzne Lypenko, Nabiev, & Samokhvalov, 2020) Additionally, in opto-electronic applications, these NPs demonstrate outstanding performance, contributing to the advancement of efficient energy conversion systems (Luque *et al.*, 2020). Moreover, ZnO NPs play a crucial role in environmental remediation endeavors, such as wastewater treatment and pollutant abatement, underlining their potential for sustainable solutions (Faisal *et al.*, 2021). Utilizing a variety of plant extracts, biologically synthesized NPs exhibit a spectrum of distinctive traits, surrounding potent antimicrobial and many other properties. These eco-friendly NPs offer an acceptable approach to synthesis, with wide-ranging applications spanning biomedicine, agriculture, and environmental rectification. (Figure 1) Leveraging the rich array of phytochemicals found in plant extracts, these NPs hold promise for targeted drug delivery, biosensing, and innovative nanotechnology applications.

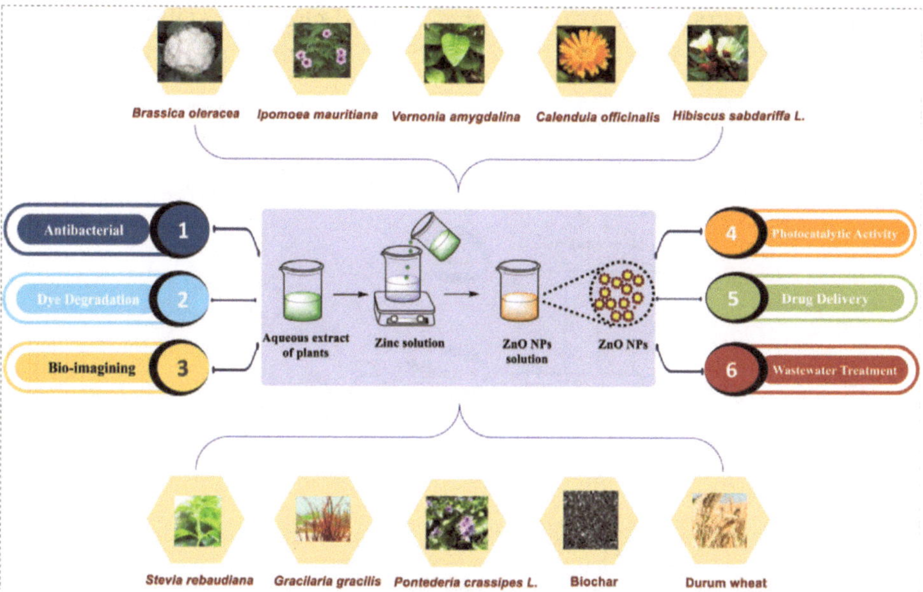

Fig. 1. Schematic presentation for the synthesis of ZnO NPs and their applications

2. Production of ZnO NPs Consuming Plant Extracts and Their Applications:

ZnO NPs epitomize a significant leap forward in nanotechnology, delivering multifaceted advantages across various sectors. Their ability to enhance efficiency, expedite processes, and offer forward-looking solutions to pressing challenges is noteworthy (Droepenu *et al.*, 2022). With applications spanning catalysis, and beyond, these NPs continue to propel innovation and pave the way for a more sustainable future (Czyżowska & Barbasz, 2022). The increasing alarm regarding pollutants like bisphenol A (BPA) present in water sources has catalyzed the exploration of advanced treatment methods (Velumani, Rajamohan, Pandey, Pham, & Hoang, 2024). In a study led by Mahmood, biochar, an economically viable and porous carbon material sourced from *Pontederia crassipes* L. leaves, was used as a carrier for ZnO in the photocatalytic degradation of rhodamine B (RB) dye in aqueous suspension under solar irradiation. (Figure 2) Additionally, unmodified ZnO and biochar samples were prepared for comparative analysis. The resulting hybrid showcased enhanced properties, including a reduced band gap energy and an augmented surface area of the photocatalyst. Upon exposure to solar light irradiation for 120 minutes in lieu of H_2O_2, the nanohybrid exhibited remarkable photocatalytic efficiency, achieving a degradation rate of RB of 99.67%. Furthermore, a study on the recyclability of the ZnO/biochar catalyst demonstrated its sustained physical and chemical stability over five consecutive cycles, with an 82.24% retention of its initial activity. (Figure 2)

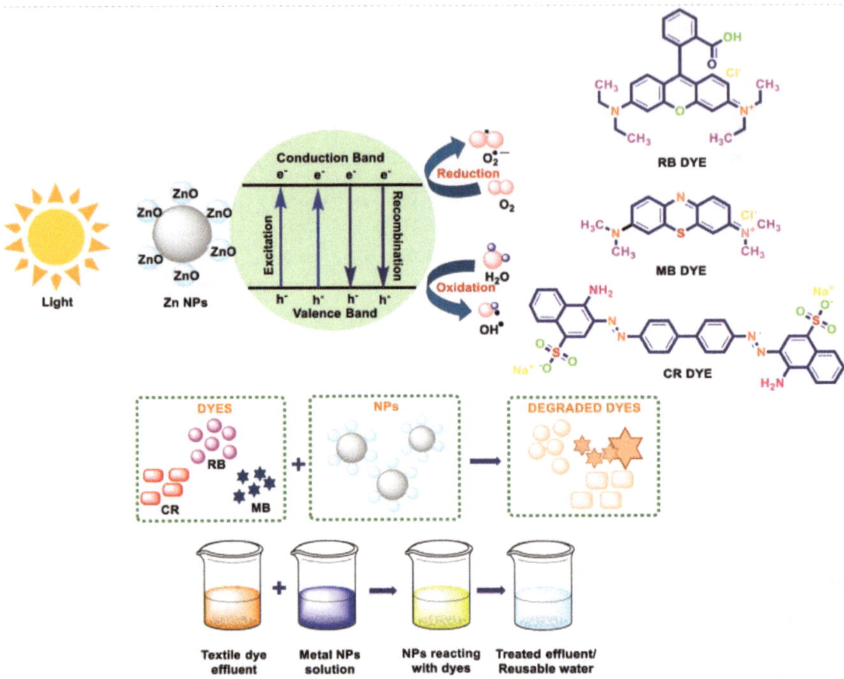

Fig. 2: Degradation of various dyes using ZnO NPs

Chemingui *et al.* delved into the environment friendly production of ZnO NPs utilizing an aqueous solution of *Hibiscus sabdariffa* L. flower extract, acting as both a reducing and capping agent. The identification of flavonoids and phenols within the plant extract exhibited their function in reducing $Zn(NO_3)_2$ salt. Testing the antibacterial effectiveness through agar dilutions displayed strong activity against both gram-positive and negative microorganisms. Moreover, these NPs showcased antioxidative characteristics, exhibiting an IC_{50} value of 38 µg/mL as determined by the 2,2-diphenyl-1-picrylhydrazyl (DPPH) radical scavenging assay. Furthermore, they demonstrated considerable catalytic ability in the degradation of methyl orange (MO) dye under UV irradiation. Overall, these findings underscored the promising potential in various applications, including biomedical and wastewater treatment (Chemingui, Moulahi, Missaoui, Al-Marri, & Hafiane, 2024).

Orie *et al.* explored the efficacy of plant extracts and ZnO NPs against specific clinical pathogens. Their study focused on examining the influence of varied ZnO concentrations on the bioactivity of NPs derived from *Vernonia amygdalina.* The synthesis involved employing different concentrations alongside a consistent volume of the leaf extract. The evaluation of bioactive potential against clinical pathogens revealed that NPs synthesized with higher ZnO concentrations displayed larger zones of inhibitions as compared to gentamycin. It was suggested that the sensitivity of clinical pathogens enhanced with increased concentrations (Orie & Okocha, 2024).

Acar *et al.* conducted research to fabricate ZnO NPs utilizing *Calendula officinalis* flower extract via green synthesis and assessed their antioxidant and wound healing properties. These NPs showed no evident cytotoxicity up to a concentration of 10 μg/mL, with an IC_{50} value of 25.96 μg/mL. Furthermore they exhibited partial cell migration and wound closure. Antioxidant activities were evaluated using 2,2-azino-bis-3-ethylbenzothiazoline-6-sulphonic acid (ABTS) and DPPH assays, revealing radical scavenging ability. These results indicated that the extract served as an efficient reducing agent for green synthesis, showcasing significant activities as mentioned above (Aydin Acar, Gencer, Pehlivanoglu, Yesilot, & Donmez, 2024).

The synthesis of ZnO NPs using algae extracts presents a promising and eco-friendly approach, as demonstrated by Alparol *et al.* Seaweeds and marine microalgae served as abundant biomass sources, with seaweed extracts acting as cost-effective reducing and stabilizing agents. This green synthesis offered various advantages, including the ability to produce NPs of different sizes and shapes using non-toxic bioactive compounds. These NPs exhibited practical applications in biomedicine, wastewater treatment, and food packaging. They played a crucial role in the photocatalytic degradation of pollutants, where light irradiation led to the separation of hole and electron, followed by the scattering of charge carriers on the NP's surface. Subsequently, light-driven catalytic oxidation and reduction reactions occurred on the active sites of the NPs. In their study, Francavilla *et al.* introduced an alternative method for synthesizing ZnO NPs, employing a reactive milling process utilizing agar derived from red seaweed *Gracilaria gracilis* as a sacrificial template material along with the $Zn(NO_3)_2$ precursor. These porous NPs displayed promising photocatalytic properties and were demonstrated to degrade aqueous solutions of phenol after calcination at 600 °C (Francavilla *et al.*, 2014).

Missaoui *et al.* emphasized the importance of optimizing crop output, quality, and reduced fertilizer usage. The study displayed the effects of NPs, bio-stimulants, and their combination on plant cell metabolism across various cellular compartments. Results showed that the combined application of NPs and bio-stimulants yielded superior benefits for plant health indicators compared to individual treatments. Notably, the combination of folcare biostimulant with ZnO NPs enhanced the pea crop growth, attributed to increased antioxidant activities during plant growth phases (Missaoui, Boughdiri, Chemingui, Al Sobeai, & Smiri, 2023).

Natraj *et al.* successfully synthesized ZnO NPs using *Ipomoea mauritiana* aqueous leaf extract, highlighting its eco-friendly nature. The radical scavenging activity (RSA) was evaluated using DPPH and metal chelation (MC) assays, demonstrating half inhibition at concentrations of 1.66 and 1.79 mg/mL, respectively. Moreover, cytotoxic assays conducted on human colorectal adenocarcinoma cell line (HT-29) indicated significant inhibition of cell division at the G0/G1 (Quiescent/Growth 1) phase. Notably, 73.78% of cells showed early signs of apoptosis 24 hours post-treatment, with results comparable to conventional medications. These findings suggested that they hold potential as a viable alternative to synthetic medications for cytotoxicity treatment (Nataraj, Murali, Satish, & Amruthesh 2023).

Manojkumar *et al.* prepared ZnO NPs using *Brassica oleracea* var. botrytis leaf extract through co-precipitation and assessed their photocatalytic and antibacterial properties. The photocatalytic activity was evaluated for methylene blue (MB) dye degradation under direct sunlight, achieving an 80% degradation rate within 75 minutes at pH 8. The antimicrobial study demonstrated the formation of inhibition zones against various bacterial pathogens. Additionally, the larvicidal activity against *C. quinquefasciatus* mosquito larvae was evaluated. Results are presented in Table 1, which proved that they had excellent potential for diverse environmental and antipathogenic applications (Manojkumar *et al.*, 2023).

Table 1: Antimicrobial and larvicidal activity of ZnO NPs synthesized from Brassica oleracea leaf extract

S. No.	Bacteria/Pathogens	Observations
1.	*B. subtilis*	16 mm (zone inhibition)
2.	*S. aureus* and *K. pneumonia*	13 mm (zone inhibition)
3.	*C. quinquefasciatus* mosquito larvae	LC_{50}- 76.03 ppm LC_{90}-190.03 ppm

Guirguis *et al.* developed the idea for the synthesis of ZnO NPs utilizing *Stevia rebaudiana* as a green reducing agent, aiming to develop eco-friendly adsorbents for eliminating antibiotics from aqueous solutions. They exhibited considerable antioxidant efficacy and antibacterial characteristics against *E. coli* and *S. aureus*. Optimal adsorption efficiency was attained at particular dosages and pH levels, yielding substantial elimination rates and adsorption capacities for both ciprofloxacin (CIP) and tetracycline (TET). Moreover, in a binary system of the two contaminants in tap water, these NPs exhibited considerable removal efficiency. Notably, these NPs were reusable over multiple cycles, highlighting their potential as cost-effective and sustainable adsorbents for wastewater treatment (Guirguis *et al.* 2024).

Hasaan *et al.* conducted a study examining the production of ZnO NPs from durum wheat extract, employing both chemical and green synthesis methods. Experimental evaluations were conducted to analyse the anaerobic digestion (AD) conversion efficiency with and without NP treatment on biogas production yield. Various parameters such as total solid (TS), volatile solid (VS), ash content (AC), and elemental composition (C, H, and O) of the biomass were assessed. The outcomes revealed that durum wheat displayed the highest biogas yield compared to the untreated biomass. Furthermore, the treatment of durum wheat with both chemical and green NPs led to a significant increase in specific biogas production. These findings suggested that NP treatment enhanced biogas production efficiency in AD processes (Hassaan *et al.*, 2021).

Sachin *et al.* employed a green chemistry approach to synthesize ZnO NPs, utilizing lychee peel extract. Their study aimed to evaluate the NPs efficacy in capturing congo red (CR) dye from wastewater and explored their antimicrobial properties. The research showcased remarkable efficiency in CR dye removal, achieving over 98% removal within 120 minutes across a broad pH range from 2 to 10. The primary mechanism of dye removal was attributed to the electrostatic interaction between the ZnO adsorbent and CR dye molecules. Additionally, the antimicrobial activity revealed significant inhibition rates against various pathogenic bacteria. Overall, the study demonstrated promising antimicrobial performance and highlighted the potential of ZnO NPs as efficient adsorbents for CR dye removal from wastewater (Singh, Singh, Shah, & Pramanik, 2023).

In this study, Gaur *et al.* examined the photocatalytic degradation of CR dye using ZnO NPs prepared by the reduction of a zinc salt with *Carica papaya* leaf extract. They possessed remarkable photocatalytic efficiency compared to other systems, with a minimum energy consumption. Overall, the findings showed that they represented a promising option for the effective degradation of industrial dyes (Gaur *et al.*, 2023).

Similarly, Maijan *et al.* successfully synthesized ZnO NPs by utilizing agricultural waste materials: para rubber leaf extract as a capping agent and para wood ash extract as a natural alkaline source, employing a precipitation method. These NPs displayed significant antimicrobial activity against various pathogens. Furthermore, they showcased efficient photodegradation of MB dye under ultraviolet irradiation for 5 hours, with remarkable photocatalytic performance observed under sunlight. Scavenger tests unveiled hydroxyl radicals as the primary reactive species responsible for the photocatalytic degradation process. Additionally, these NPs exhibited good stability and recyclability over five iterations, demonstrating great potential for degrading organic pollutants and controlling microorganisms (Maijan, Waen-ngoen, Suwanboon, Chantarak, & Voravuthikunchai, 2024). Different characteristics such as used plant extract, size, and shape of all biologically synthesized ZnO NPs and their applications are summarized in Table 2.

Table 2: Summary of biologically synthesized ZnO NPs from various plant extracts

S. No.	Plant (family)	Common name (Part taken)	Size (nm)	Shape	Applications	Reference
1.	*Hibiscus sabdariffa L.*	Indian sorell (Flower)	15 (XRD)	Spherical, Wurtzite	- Antibacterial activity - Antioxidant activity - Catalytic degradation of MO dye	Chemingui *et al.*, 2024
2.	*Vernonia amygdalin*	Bitter leaf (Leaves)	-	Spherical, hexagonal wurtzite	- Antibacterial activity	Orie *et al.*, 2024
3.	*Calendula officinalis*	Pot marigold (Flower)	17.66 (XRD)	Spherical with irregularity	- Antioxidant activity - Wound healing	Acar *et al.*, 2024
4.	*Gracilaria gracilis*	Slender wart weed (Whole plant)	18-50 (XRD)	Hexagonal	- Photocatalytic activity - Degradation of phenol	Francavilla *et al.*, 2014
5.	*Ipomoea mauritiana*	Morning glory (Flower)	10.83 (XRD)	Crystalline	- Antioxidant activity - Cytotoxic activity	Natraj *et al.*, 2023
6.	*Brassica oleracea* var. botrytis	Cauliflower (Leaves)	52 (XRD)	Hexagonal wurtzite	- Antibacterial activity - Degradation of MB dye	Manojkumar *et al.*, 2023
7.	*Stevia rebaudiana*	Candy leaf or sugar leaf (Leaves)	117.01 ± 2.08 and 373.38 ± 1.18	Non-uniform spherical	- Antibacterial activity	Guirguis *et al.*, 2024
8.	*Triticum durum*	Durum wheat (Seeds)	5-28 (XRD)	Semi-flat with rod-like particles	- Biogas production	Hassan *et al.*, 2021
9.	*Litchi chinensis*	Lychee/litchee (Peel)	2.18 (XRD)	Hexagonal	- Antibacterial activity	Singh *et al.*, 2023
10.	*Carica papaya*	Papaya (Leaves)	21	Spherical, semi-spherical, hexagonal, and rod-like	- Degradation of industrial dye	Gaur *et al.*, 2023
11.	*Hevea brasiliensis*	Para rubber tree (Leaves)	91.60	Spherical	- Degradation of MB dye	Maijan *et al.*, 2024

3. Conclusion

In conclusion, the remarkable adaptability and multifaceted usefulness of ZnO NPs have proven their crucial role in modern research and technological amelioration. These NPs have epitomized the amalgamation of scientific creativity and societal development and have provided transformative solutions across diverse fields. From fundamental contributions in addressing environmental challenges to extensive applications in biomedical engineering, renewable energy, consumer goods, materials science, and sensing technology, they are serving as a cornerstone for progress. Their distinct characteristics and sustainable approaches have not only tackled various environmental concerns but have also ensured a cleaner and healthier environment for present and future generations. These NPs have demonstrated their significance not only in environmental applications but also in healthcare by serving as vital components in drug delivery systems, biosensors, medical imaging agents, and tissue engineering scaffolds. Their biocompatibility and minimal toxicity offer promising prospects for advancements in diagnostics, therapeutics, and regenerative medicine. In the renewable energy sector also, they have made significant contributions to advancing solar energy conversion and energy storage technologies, driving the shift towards a sustainable energy future. From formulations in sunscreen to antibacterial coatings and electronic components, these NPs augment various facets of daily life, enhancing living standards and fostering technological progress.

In essence, the thorough exploration of ZnO NPs presented here underscores their innovative potential across numerous disciplines. As researchers continue to explore synthesis techniques, characterization methods, and application strategies, the potential of these NPs in driving technological invention that remains limitless. Through collaborative efforts across disciplines and unwavering commitment to sustainability, it can be fully employed to its potential to create a brighter, cleaner, and more resilient future for humanity.

References

Alexandrov, A., Zvaigzne, M., Lypenko, D., Nabiev, I., Samokhvalov, P. (2020). Al-, Ga-, Mg-, or Li-doped zinc oxide nanoparticles as electron transport layers for quantum dot light-emitting diodes. Scientific Reports, 10(1), 7496.

Aydin Acar, C., Gencer, M. A., Pehlivanoglu, S., Yesilot, S., Donmez, S. (2024). Green and eco□friendly biosynthesis of zinc oxide nanoparticles using Calendula officinalis flower extract: Wound healing potential and antioxidant activity. International Wound Journal, 21(1), e14413.

Bandeira, M., Giovanela, M., Roesch-Ely, M., Devine, D. M., da Silva Crespo, J. (2020). Green synthesis of zinc oxide nanoparticles: A review of the synthesis methodology and mechanism of formation. Sustainable Chemistry and Pharmacy, 15, 100223.

Chemingui, H., Moulahi, A., Missaoui, T., Al-Marri, A. H., Hafiane, A. (2024). A novel green preparation of zinc oxide nanoparticles with Hibiscus sabdariffa L.: photocatalytic performance, evaluation of antioxidant and antibacterial activity. Environmental Technology, 45(5), 926-944.

Czyżowska, A., Barbasz, A. (2022). A review: zinc oxide nanoparticles–friends or enemies? International Journal of Environmental Health Research, 32(4), 885-901.

Droepenu, E. K., Wee, B. S., Chin, S. F., Kok, K. Y., Maligan, M. F. (2022). Zinc oxide nanoparticles synthesis methods and its effect on morphology: A review. Biointerface Res. Appl. Chem, 12, 4261-4292.

Faisal, S., Jan, H., Shah, S. A., Shah, S., Khan, A., Akbar, M. T., Akhtar, N. (2021). Green synthesis of zinc oxide (ZnO) nanoparticles using aqueous fruit extracts of Myristica fragrans: their characterizations and biological and environmental applications. ACS Omega, 6(14), 9709-9722.

Francavilla, M., Pineda, A., Romero, A. A., Colmenares, J. C., Vargas, C., Monteleone, M., Luque, R. (2014). Efficient and simple reactive milling preparation of photocatalytically active porous ZnO nanostructures using biomass derived polysaccharides. Green Chemistry, 16(5), 2876-2885.

Gaur, J., Vikrant, K., Kim, K. H., Kumar, S., Pal, M., Badru, R., Momoh, J. (2023). Photocatalytic degradation of Congo red dye using zinc oxide nanoparticles prepared using Carica papaya leaf extract. Materials Today Sustainability, 22, 100339.

Guirguis, H. A., Youssef, N., William, M., Abdel-Dayem, D., El-Sayed, M. M. (2024). Bioinspired Stevia rebaudiana Green Zinc Oxide Nanoparticles for the Adsorptive Removal of Antibiotics from Water. ACS Omega.

Hassaan, M. A., Pantaleo, A., Tedone, L., Elkatory, M. R., Ali, R. M., Nemr, A. E., Mastro, G. D. (2021). Enhancement of biogas production via green ZnO nanoparticles: Experimental results of selected herbaceous crops. Chemical Engineering Communications, 208(2), 242-255.

Islam, F., Shohag, S., Uddin, M. J., Islam, M. R., Nafady, M. H., Akter, A., Cavalu, S. (2022). Exploring the journey of zinc oxide nanoparticles (ZnO-NPs) toward biomedical applications. Materials, 15(6), 2160.

Jiang, Z., Liu, B., Yu, L., Tong, Y., Yan, M., Zhang, R., Zhang, S. (2023). Research progresses in preparation methods and applications of zinc oxide nanoparticles. Journal of Alloys and Compounds, 170316.

Luque, P. A., Nava, O., Romo-Cardenas, G., Nieto-Hipolito, J. I., Vilchis-Nestor, A. R., Valdez, K., Murrieta-Rico, F. N. (2020). Facile zinc oxide nanoparticle green synthesis using citrus reticulata extract for use in optoelectronic sensors. IEEE Sensors Journal, 21(10), 11275-11282.

Maijan, P., Waen-ngoen, T., Suwanboon, S., Chantarak, S., Voravuthikunchai, S. P. (2024). Sustainable environmental-based synthesis of zinc oxide nanoparticles using para rubber leaf extract for photocatalytic degradation of organic pollutants and microbial control in wastewater treatment. Inorganic Chemistry Communications, 162, 112086.

Manojkumar, U., Kaliannan, D., Srinivasan, V., Balasubramanian, B., Kamyab, H., Mussa, Z. H., Palaninaicker, S. (2023). Green synthesis of zinc oxide nanoparticles using Brassica oleracea var. botrytis leaf extract: Photocatalytic, antimicrobial and larvicidal activity. Chemosphere, 323, 138263.

Mendes, C. R., Dilarri, G., Forsan, C. F., Sapata, V. d. M. R., Lopes, P. R. M., de Moraes, P. B., Bidoia, E. D. (2022). Antibacterial action and target mechanisms of zinc oxide nanoparticles against bacterial pathogens. Scientific Reports, 12(1), 2658.

Missaoui, T., Boughdiri, N., Chemingui, H., Al Sobeai, S. M., Smiri, M. (2023). Growth and metabolism of pea (Pisum sativum) via biostimulants based on greener ZnO nanoparticles. International Journal of Phytoremediation, 1-9.

Nataraj, K., Murali, M., Satish, A., Amruthesh, K. (2023). Structural characterization and biological properties of novel biosynthesized zinc oxide nanoparticles from the giant potato plant Ipomoea mauritiana Jacq. Biomass Conversion and Biorefinery, 1-10.

Orie, K. J., Okocha, B. I. (2024). Biosynthesis and antimicrobial studies of zinc oxide nanoparticles of Vernonia Amygdalina Leaf with varying concentration of zinc oxide. Inorganic Chemistry Communications, 112178.

Shaba, E. Y., Jacob, J. O., Tijani, J. O., Suleiman, M. A. T. (2021). A critical review of synthesis parameters affecting the properties of zinc oxide nanoparticle and its application in wastewater treatment. Applied Water Science, 11(2), 48.

Singh, N., Singh, R., Shah, K., Pramanik, B. K. (2023). Green synthesis of zinc oxide nanoparticles using lychee peel and its application in anti-bacterial properties and CR dye removal from wastewater. Chemosphere, 327, 138497.

Velumani, M., Rajamohan, S., Pandey, A., Pham, N. D. K., Hoang, A. T. (2024). Nanocomposite from tannery sludge-derived biochar and Zinc oxide nanoparticles for photocatalytic degradation of Bisphenol A toward dual environmental benefits. Science of The Total Environment, 907, 167896.

16

Nanotechnology: A Step Towards Sustainable Agriculture

***Kavita*[1*] *and Yogesh Kumar Verma*[2]**

[1]*Department of Zoology, MBD Govt. College, Kushalgarh, Banswara, Rajasthan India*

[2]*Department of Chemistry, MBD Govt. College, Kushalgarh, Banswara Rajasthan India*

Abstract

Agriculture serve as a backbone in improving economic growth of a country as well as it provides a base in meeting various needs of people in direct and indirect ways. The recent advancement in nanotechnology has shown a path to develop traditional agriculture into a sustainable practice. Several nanochemicals emerged as promising agents for better plant growth, used as fertilizer, pesticides, control weeds and also used as antimicrobial agents in food packaging. Nanotechnology may act as sensor for checking soil health of agriculture fields. Nanofertilizers and nanopesticides improve crop productivity without imposing negative impacts on soil and water health. Present chapter focuses on the application of nanotechnology for sustainable agriculture to minimize hazardous effects on environment on one hand and on the other hand to enhance crop production with improving soil quality.

Keywords: *Nanotechnology; nanochemicals; sustainable agriculture; nano-pesticides*

1. Introduction

1.1 Importance of agriculture for human/nation

Food is the basic need of any species;thus, the agriculture has been always played the most vital role in the survival of any civilization from ancient time to the present day. Agriculture is also a significant factor in the development of a country. The tradition of practicing agriculture has changed in drastic way to fulfill the needs of expanding human population. Most of the farmers and agricultural lands belong to rural areas which are shrinking year by year because of urbanization. Size of the farm lands are decreasing year by year.

* *Corresponding author e-mail: kavitabansiwal21@gmail.com*

On the other hand, the world population is growing beyond the carrying capacity. Studies predict that the demand of food might increase to 1.5 to 2-fold by the year 2050 as the world population would reach 9 billion mark (Duro *et al.*, 2020). Before this ongoing situation turns into a panic call, it is high time to incorporate science and technology to mold the conventional agricultural ways into a sustainable agricultural ecosystem to combat worldwide food insecurities.

1.2 Essentials of good agriculture practice

Development of science in past few decades lead to inclusion of humongous amount of chemicals such as fertilizers, weedicides, pesticides and ripening chemicals in agricultural processes. These chemicals are now integrated into the system to such an extent that they become unavoidable in all types of farming. A modern agriculture practice has several essentials in it, which includes soil quality, (such as P^H, chemical compositions, microbial community),viability of seed, irrigation, tilling, fertilizers, pesticides, insecticides, monitoring, and production of crop, packaging, and processing of food products. To obtain adequate amount of food products with high nutritional quality, every component of the agriculture process needs to be taken care of equally. Efforts are always been made in quest of achieving more amount of quantity and quality of food with appropriate labor.

1.3 Need of sustainable agriculture

Despite the use of so many traditional chemicals and scientific tools, there are various challenges in meeting the agricultural production par with the increasing demands. Therefore, it is an urgent need towards sustainable agriculture because of the factors which are discussed below.

1.3.1 Excessive use of chemicals

Almost all countries had green revolutions during last century to enhance the production of crops that are essential for human use. That lead to the incorporation of chemical entities in agriculture ecosystem in form of fertilizers, pesticides, weedicides and in other ways. The excessive use of chemicals has deeply impacted the quality of food and depleted the soil fertility. To maintain the same amount of production every single time, the farming has to be dependent on those chemicals forever.

1.3.2 Invasion of novel weeds species

In past decades, the chemical overuse encouraged weeds to evolve in more robust and resistant forms. Along with that, occurrence of several invasive species is spread due to global trade and exchange from one part of the world to other. Due to evolved weeds there is a need of more amount of weedkillers otherwise the food production will hamper drastically.

1.3.3 Declining Quality of food

the quality of food in terms of nutritional and productive value is gone down because of the repetitive farming and over chemical use.

1.3.3. Climate factors

There are many climatic factors such as delayed seasonal rain, storms, tsunami, floods, cyclones which affects either the decline in food production or devastation of entire vegetation of a region. The climate change, deforestation and decent of ground water level are also a challenge to the farmers in recent years.

All the above factors are rising recently and reports reveal that these factors will retain as new normal situation in coming years too. To combat all those agricultural inhibitions, it is high time where we need to incorporate advance scientific achievements/technology to focus sustainable agriculture practice so that the high food demand could be fulfilled.

2. What is Nanoscience and Nanotechnology?

Nanoscience is basically the study of phenomenaand manipulations of matter at nanoscale i.e. 1 nm to 100 nm in size. The properties such as surface behavior, magnetic, optical, electrical properties of matter significantly differ at nanosizefrom large scale. Quantum effects become very significant at this size of materials. Nanotechnology manifests those properties by adequate amount of fabrications and manipulations in desired materials in the betterment of mankind. In the recent years, nanotechnology has broaden itself in diverse areas which is not just limited to semiconductor science, energy storage, space technology, digital technologies, computer, electronics, and communication engineering along with food and agriculture. This chapter exclusively deals with the broad application of nanotechnology in agriculture and food industry.

3. Applications of Nanotechnology in Agriculture

3.1 Nanotechnology and soil quality

Agricultural soil quality basically comprises the chemical composition, pH of soil, moisture content, microbial community and salinity of the soil. Uses of nanotechnology enableus to sense pollution, contamination of drinking water & ground water and agricultural lands.On the other hand, Nanosensors are designed to detect moisture content, chemical contaminations and several microbes even at minute quantity.

Several nano photo catalyst such as metal oxides can convert organic pesticides and industrial pollutants into environmentally harmless components (Baruah and Dutta, 2009), thus helps in treating polluted agricultural lands and water bodies. Nanoscale iron particles could be used in detoxification of variety of common contaminants, such as chlorinated solvents, chlorinated pesticides, and polychlorinated biphenyls (Zhang, 2003). Lanthanum nanoparticles can be a useful method to absorb phosphates in aqueous systems, thus inhibiting algal growth (Joseph and Morrison, 2006).

Microbial community present in soil plays significant role in mineralization of organic matter, recycling of minerals and maintaining fertility of the agricultural land. Some bacteria develop symbiotic relationship in legume plants and fix atmospheric nitrogen. Population of soil microbes depend on physicochemical environment such as pH, salinity, moisture, oxygen content and thus serve as indicators to soil health. Extensive applications of nanomaterials affect the diversity of microbes through accumulation in land and water. Nano Silver particles show broad antimicrobial activity against various plant pathogens. Smaller size increase surface area of silver particles which increase reactivity against the microbes (Choi *et al.* 2008). Fullerence C_{60}, have been found growth inhibitory for common soil and water bacteria (Fortner *et al.* 2005; Oberdorster *et al.* 2004).

3.2 Nano fertilizers

Nano-size and high volume to surface area make Nanofertilizers (NFs) more efficient in comparison to normal fertilizers. Recently various NFs have been developed such as iron, titanium, silver, zinc, silica, carbon-nanotubes CNTS and molybdenum which are successfully applied on different crop systems (Table 1). Reports have been published that nanomaterials could impact on root elongation, shoot elongation, plant biomass, chlorophyll content and seed germination at certain concentrations. Mochizuki, *et al.* (2009) showed that NFs move faster into the cell membrane of plants as compared to conventional fertilizers because of more entropy due to colloid suspension state. Kottegoda *et al.*, (2017) demonstrated that controlled delivery of urea up to 1 week could be achieved by using urea-hydroxyapatite nanohybrids.Hydroxyapatite isrich source of phosphorus and bio-compatible in nature, which make them a sustainable smart carrier of fertilizers. They proved that weakly bound urea to hydroxyapatite NPs slowly release urea into the soil enhancing plant growth. Abdel *et al.*, (2016) have reported synthesis of nano-chitosan- NPK fertilizers and explored their application in the growth of wheat plant. These NFs enter into wheat plants through stomata and were transported via phloem tissue thus cause a significant reduction in the life cycle of wheat plants by 23.5%. The nano-NPK improved photosynthesis via enhancing water absorption (Ma, *et al.*, 2009; Wu, 2013) and enhanced plasma membrane permeability (Du *et al.*, 2011; Wang *et al.*, 2013).

ZnO NPs showed increased photosynthesis and biomass in lettuce (Xu, *et al.*, 2018). Zinc oxide nanoparticles also found to enhance the CO_2 supply at the carboxylation site in chloroplasts by improvement of carbonic anhydrase (DiMario *et al.*, 2017).

TiO_2 nanoparticles have also been reported to increase chlorophyll, spike length, numbers of grain and grain mass in barley crop (Janmohammadi *et al.*, 2016).

Rai *et al.* (2015) observed a synergistic effect of bio fertilizers with mycorrhizal fungi *Piriformosporaindica* & *Glomusmosseae*separately in combination of calcium phosphate nanoparticles. They applied them on *Zea maize* and found increased chlorophyll content and root proliferation significantly. Several renowned companies have manifested the potential of NFs and developed useful NF based commercial products (table 1) for farmers to enhance their crop production.

Table 1: Some commercial products of Nano fertilizer (Source-Prasad *et al.*, 2017)

S. No.	Commercial Product	Content	Company
1	Nano-GroTM	Plant growth regulator and immunity enhancer	Agro Nanotechnology Corp., FL, United States
2	Nano Green	Extracts of corn, grain, soybeans, potatoes, coconut, and palm	Nano Green Sciences, Inc.,India
3	Nano-Ag Answer R	Microorganism, sea kelp, and mineral electrolyte	Urth Agriculture, CA, United States
4	Biozar Nano-Fertilizer	Combination of organic materials, micronutrients, and macromolecules	Fanavar Nano-Pazhoohesh Markazi Company, Iran
5	Nano Max NPK Fertilizer	Multiple organic acids chelated with major nutrients, amino acids, organic carbon, organic micro nutrients/trace elements, vitamins, and probiotic	JU Agri Sciences Pvt. Ltd, Janakpuri, New Delhi, India
6	Master Nano Chitosan Organic Fertilizer	Water soluble liquid chitosan, organic acid and salicylic acids, phenolic compounds	PannarajIntertrade, Thailand

7	TAG NANO (NPK, PhoS, Zinc, Cal, etc.) fertilizers	Proteino-lacto-gluconate chelated with micronutrients, vitamins, probiotics, seaweed extracts, humic acid	Tropical Agrosystem India (P) Ltd, India

3.3 Detection and controlling plant pathogens

Plant pathogens and crop pests are one of the major problems in agriculture that always has been a matter of serious concern. Various pathogens like bacteria, fungi, viruses, insects, protozoans etc. have been reported to impose a negative impact on crop yield.Chemical fertilizers which are used to combat these pathogens, also affects non-targeted organisms present in soil environment that indirectly leads to ecological imbalance. Various Nano formulation have been introduced to control these pathogens in a smart way that produces less harmful impacts on nontarget organisms as well as on environment. These formulation include essential oils (such as Neem oil, garlic essential LippaSidoides oil) and plant extracts (such as Capsaicin from Chilli Pepper and Lansiumamide B extract), which are used to enhance self-life of food items (Anjali *et al.*, 2012; Bohu and Zizoung, 2011;Lai *et al.*, 2006; Xu *et al.*, 2010; Yin *et al.*, 2012).

It has been reported that silver nanoparticles reduces growth of many fungal pathogens such as *Magnaporthegrissa, fusariumoxysporum, colletotrichum gloeosporoides*. (Rabab and El-Shafey, 2013; Angiular *et al.*, 2011;Jo Kim and Jung, 2009; Kasprowicz and Gorczyca, 2010; Min *et al.*, 2009; Musarrat *et al.*, 2010). Ag-Silica nanoparticles were found to have antifungal activity against powdery mildew infection in Pumpkin leaves (Kim *et al.*, 2006). ZnO nanoparticles also found to be suitable as antifungal agents against *Penicilliumexpansum, Botritiscinerea, Aspergillus flavus and Aspergillusniger* (He *et al.*, 2011; Jayaseelan *et al.*, 2012).

Some nanoparticles also have been reported as antibacterial agents against bacterial pathogens such as copper nanoparticles effectively control the *Xanthomonasoriyzae and Xanthomonas campetris* which causes Rice blast and leaf spot disease respectively (Mondal and Mani, 2012). Silver nanoparticles too have shown antibacterial properties against phyto-pathogens (*Citrobacterfreundii and Erwiniacacticida*) as well as sun root tuber extract based silver nanoparticles also have antibacterial effects on *Xanthomonas exonopodis* and *Ralstonia solanacearum* (Paulkumar *et al.*, 2014; Arvinthuan *et al.*, 2015). Some nano-gel based formulations such as Silica nano particle are reported to have insecticidal properties that has been used as effective tool to control insect pests (Barik and Gowswami, 2012).

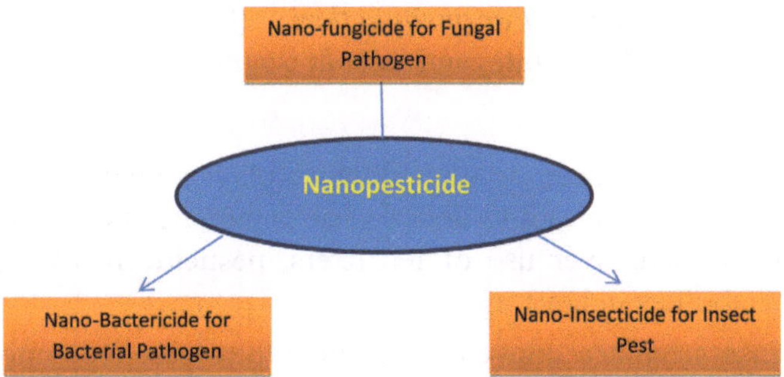

Fig. 1: Various form of Nano pesticide

3.4 Impact on plant growth

Treatment of specific plants with specific Nanoparticles could influence seed germination and plant growth. The results are affected with many factors such as type of nanoparticle, their concentration and the species of plant on which growth is sought. These nanoparticles enter plant cells along with the carrier proteins via ion channels or endocytosis (Rico *et al.* 2011). NMs can also enter a seed by creating new pores thereby enhancing water uptake too which is beneficial in germination process. Khodakovskaya *et al.* (2009) showed that several CNTs could enter the seed coat which is tougher to traverse than the normal plant cell and thus influence germination process and growth.

The beneficial effects of CNTs on plant growth were observed in the plants such as tomato, rape (*Brassica napus*), lettuce (*Lactuca sativa*), cucumber (*Cucumis sativus*), radish (*Raphanus sativus*), rye grass (*Lolium perenne*), corn (Zea mays) and (Lin and Xing 2007). CNT treatment is very helpful to improve germination in Rice seeds (Stamphoulis *et al.* 2009). Similarly, the treatment of soyabean (*Glycine max*) with SiO_2 and TiO_2 nanoparticles result in improved plant growth and seed germination. Lu *et al.* (2002) showed that those nanoparticles at low concentration enhance fertilizer uptake, water uptake, nitrate reductase activity and better antioxidant activity.

In spinach plant nano-TiO_2 is able to improve nitrogen metabolism and photosynthesis rate. Similarly, SiO_2 nanoparticles showed a positive effect on growth of Changbai larch (*Larix olgensis*) too.

Apart from enhancing growth there are certain studies that show inhibitory effect over seed germination, root and plant growth. ZnO inhibits germination in Rye grass and corn (Lin and Xing 2007), inhibits root elongation in radish, rye grass and *Allium cepa* (Ghodake *et al.* 2011). Nano particles of ZnO were found highly toxic in *Arabidopsis thaliana* (Lee *et al.* 2010) and cause inhibition in desert plants such as *Parkinsonia florida* (blue palo verde), *Prosopis juliflora-velutina* (velvet mesquite) and *Salsola tragus* (tumbleweed).

3.5 Monitoring

In recent years, Nanosensors have been developed which plays a vital role in agriculture by acting as an effective tool to detect various parameters related with growth, sprouting of seed, over use of fertilizers, pesticide residues, humidity in soil, and metabolites in beverages etc.

Application of Nanosensors replace conventional agriculture practices into smart agriculture practices which favor less effort in making a grip upon sustainable agriculture. Various Nanosensors have been used in agriculture to detect pesticide residue which includes gold nanoparticles (Au-NPs), Carbon Nanotubes (CNT), Quantum dots (QD) etc. (Cesarino, *et al.*, 2012; Liu *et al.*, 2012; Talerico *et al.*, 2016; Zheng *et al.*, 2011).

3.6 Nanotechnology in various industrial applications

After the ripening of crop in the field, there are many essential steps involved in the following process from collecting grain/fruit/vegetable from the field to reaching it to the consumers. These steps include food security/safety, food processing, food packaging, food preservation and labeling etc. Use of nanotechnology has been reported for facilitating and improving each of the above steps significantly, some of them are discussed below.

3.6.1 Nanotechnology for food security

Nanotechnology has also shown its potential in food safety and in the food processing industry. Nano materials have been used as food packaging materials to preserve and enhance shelf-life of packaged food materials.

Nanotechnology plays an important role in improving physicochemical values and nutrient stability as well as their bio availability. Some food includes nanosized elements that are different from synthetically produced NMs. The most important synthetic NMs are –

Liposomes- 100 nm to 400 nm (made up of lipid bilayer)

Micelles – 10 nm to 100 nm (self-assembling amphiphilic particles)

Nanocapsules – 10 nm-1000 nm- encapsulate significant quantities of medicines and nucleic acid

Nano conjugates- polymers covalently bonded with medicinal molecules

Dendrimers – nano-disperse molecules that encapsulate medicines.

Nano capsules of vitamins and taste enhancers prevent degradation and significantly enhance shelf life (Katouzian & Jafari, 2016). Nanoencapsulation of anthocyanin has been reported to increase pigment stability (Zhang *et al.*, 2014), whereas ferritin nano capsules inhance solubility of rutinflavinoid (Yang *et al.*, 2015).

All the above NMs are built from food grade materials and have a great impact on food processing, preservation, packaging and storage.

3.6.2 Nanotechnology for food packaging and preservation

Food packaging plays an important role to protect food from various hazards that can arise due to microbial infections, temperature, humid environment etc. Commonly used packaging materials are impermeable so that the stored material could be protected from UV rays and environmental factors. Modern packaging materials have been used after incorporation with Nanosensors known as smart packaging. These nanosensors are used for the detection of chemicals, oxidation process, gases and pathogens in food efficiently. Nanocomposites have been found to enhance mechanical and thermal barrier of packaging and also act as antimicrobial agents. ZnO NPs are reported to exhibit antimicrobial activity against *Staphylococcus aureus* (Liu *et al.*, 2009), Silver NPs shows inhibition against *Escherichia coli, Aeromonas hydrophila,* and *Klebsiella pneumoniae* (Aziz *et al.*, 2016). The antimicrobial mechanism of NPs is typically considered because of cell damage, oxidative stress, metal ion release, or other non-oxidative mechanisms (Wang *et al.*, 2017). Silver nanoparticles are most commonly used against a variety of microorganism (Berekaa, 2015). These Ag NPs might change the surface electric charge on cell membrane of microbes leading to change in permeability and eventually to apoptosis.

NP based Nanosensors are designed to indicate colour change in food materials. For smart food packaging ZnO, Cu, Cu_2O, CuO, TiO_2 and Ag based Nano filers are utilized (Lee, 2010). TiO_2 and SiO_2 were also used as colour, and flavour enhancement agents in the food industry (Dekkers *et al.*, 2011).

Plastic polymers are the most widely used material in packaging industries. Reports have shown that Ag NPs coated into plastic polymers are now being used to make storage bins. The arrangement enhances mechanical strength of material and keeps the bins disinfected (Bumbudsanpharoke and Ko, 2015).

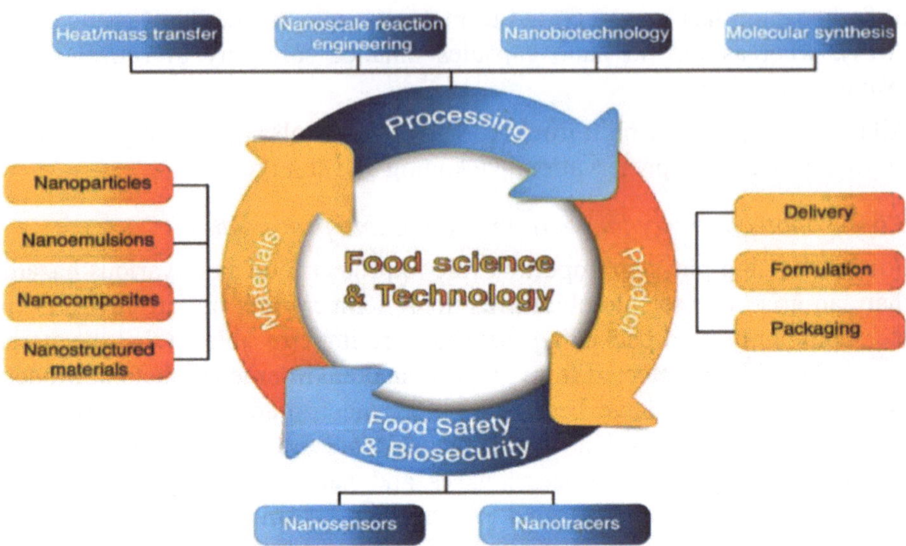

Fig. 2: Application of nanotechnology in food science (Source-Roohinejad and Greiner, 2017)

4. Conclusion

Nanotechnology has been successfully incorporated in contemporary agricultural practices and act as a strong tool to overcome hurdles in the path of achieving worldwide food demands. Throughout the chapter we have mentioned the positive outcomes of nanotechnology in agriculture. Though the journey is successful so far, there are still several challenges and precautions need to be taken care of.

- The technology must not cause a threat to disturb ecological balance between different biotic components of ecosystem.
- The negative effects of excess use of NPs must be taken into account that might harm the soil microflora.
- There should be no harmful serious side effects to the consumers by intake of NPs through the direct consumption or through the food chain.
- The properties of nanomaterials such as size, dose, exposure time, surface chemistry, structures, immune response,accumulation, retention time, etc., and other effects should be accessed carefully.
- Authorities or regulatory bodies should provide clear guidelines and roadmaps for reducing risks of the use of nanotechnological products.
- Life-cycle analysis of nanomaterials/nanofoods should be done.
- New communication channels and debates should be opened with participation of various stake holders such as consumers, researchers,

authorities, industrial sectors, etc. to discuss impacts of any novel technology on human life, nature, economy, and science.

References

Abdel-Aziz, H.M.M., Hassaneen, M.N.A. and Omer, A.M. (2016) Nano Chitosan-NPK Fertilizer Enhances the Growth and Productivity of Wheat Plants Grown in Sandy Soil. Spanish Journal of Agricultural Research, 1, 1-10.

Aguilar-Mendez, M. A., San Martı´n-Martı´nez, E., Ortega-Arroyo, L., Cobia´n-Portillo, G., &Sa´nchez-Espı´ndola, E. (2011). Synthesis and characterization of silver nanoparticles: Effect on phytopathogenColletotrichumgloesporioides. Journal of Nanoparticle Research, 13(6), 2525–2532.

Anjali, C. H., Sharma, Y.,Mukherjee, A., &Chandrasekaran, N. (2012). Neem oil (Azadirachtaindica) nanoemulsion as potent larvicidal agent against Culexquinquefasciatus. Pest Management Science, 68(2), 158–163.

Aravinthan, A., Govarthanan, M., Selvam, K., Praburaman, L., Selvankumar, T., Balamurugan, R., *et al.*, (2015). Sunroot mediated synthesis and characterization of silver nanoparticles and evaluation of its antibacterial and rat splenocyte cytotoxic effects. International Journal of Nanomedicine, 10, 1977.

Aziz, N., Pandey, R., Barman, I., & Prasad, R. (2016). Leveraging the Attributes ofMucorhiemalis-Derived Silver Nanoparticles for a Synergistic Broad-Spectrum Antimicrobial Platform. Frontiers in Microbiology, 7, 1984. https://doi.org/10.3389/fmicb.2016.01984

Barik, T. K., Kamaraju, R., &Gowswami, A. (2012). Silica nanoparticle: A potential new insecticide for mosquito vector control. Parasitology Research, 111(3), 1075–1083.

Baruah, S., & Dutta, J. (2009). Nanotechnology applications in pollution sensing and degradation in agriculture: a review. Environmental Chemistry Letters, 7, 191-204.

Berekaa, M. M. (2015). Nanotechnology in food industry; advances in food processing, packaging and food Safety. Int. J. Curr. Microbiol. App. Sci. 4, 345–357.

Bohua, F., & Ziyong, Z. (2011). Carboxymethy chitosan grafted ricinoleic acid group for nanopesticide carriers. Advanced Materials Research, 236–238, 1783–1788.

Bumbudsanpharoke, N., & Ko, S. (2015). Nano-food packaging: an overview of market, migration research, and safety regulations. Journal of food science, 80(5), R910–R923. https://doi.org/10.1111/1750-3841.12861

Cesarino, I., Moraes, F. C., Lanza, M. R., & Machado, S. A. (2012). Electrochemical detection of carbamate pesticides in fruit and vegetables with a biosensor based on acetylcholinesterase immobilised on a composite of polyaniline–carbon nanotubes. Food Chemistry, 135(3), 873–879.

Choi, O., Deng, K. K., Kim, N. J., Ross Jr, L., Surampalli, R. Y., & Hu, Z. (2008). The inhibitory effects of silver nanoparticles, silver ions, and silver chloride colloids on microbial growth. Water research, 42(12), 3066-3074.

Dekkers, S., Krystek, P., Peters, R. J., Lankveld, D. P., Bokkers, B. G., van Hoeven-Arentzen, P. H., Bouwmeester, H., &Oomen, A. G. (2011). Presence and risks of nanosilica in food products. Nanotoxicology, 5(3), 393–405. https://doi.org/10.3109/17435390.2010.519836

DiMario, R. J., Clayton, H., Mukherjee, A., Ludwig, M., &Moroney, J. V. (2017). Plant carbonic anhydrases: Structures, locations, evolution, and physiological roles. Molecular Plant, 10(1), 30–46.

Du, W., Sun, Y., Ji, R., Zhu, J., Wu, J., &Guo, H. (2011). TiO2 and ZnO nanoparticles negatively affect wheat growth and soil enzyme activities in agricultural soil. Journal of Environmental Monitoring, 13(4), 822–828.

Duro, J. A., Lauk, C., Kastner, T., Erb, K. H., &Haberl, H. (2020). Global inequalities in food consumption, cropland demand and land-use efficiency: A decomposition analysis. Global Environmental Change, 64, 102124.

Fortner, J. D., Lyon, D. Y., Sayes, C. M., Boyd, A. M., Falkner, J. C., Hotze, E. M., Alemany, L. B., Tao, Y. J., Guo, W., Ausman, K. D., Colvin, V. L., & Hughes, J. B. (2005). C60 in water: nanocrystal formation and microbial response. Environmental science & Technology, 39(11), 4307–4316. https://doi.org/10.1021/es048099n

Ghodake, G., Seo, Y. D., Park, D., & Lee, D. S. (2010). Phytotoxicity of carbon nanotubes assessed by Brassica juncea and Phaseolusmungo. Journal of Nanoelectronics and Optoelectronics, 5(2), 157–160.

He, L., Liu, Y., Mustapha, A., & Lin, M. (2011). Antifungal activity of zinc oxide nanoparticles against Botrytis cinerea and Penicilliumexpansum. Microbiological Research, 166(3), 207–215.

Janmohammadi, M., Amanzadeh, T., Sabaghnia, N., &Dashti, S. (2016). Impact of foliar application of nano micronutrient fertilizers and titanium dioxide nanoparticles on the growth and yield components of barley under supplemental irrigation. Acta Agriculturae Slovenica, 107(2), 265–276.

Jayaseelan, C., Rahuman, A. A., Kirthi, A. V., Marimuthu, S., Santhoshkumar, T., Bagavan, A., *et al.*, (2012). Novel microbial route to synthesize ZnO nanoparticles using Aeromonashydrophila and their activity against pathogenic bacteria and fungi. SpectrochimicaActa Part A: Molecular and Biomolecular Spectroscopy, 90, 78–84.

Jo, Y. K., Kim, B. H.,&Jung, G. (2009). Antifungal activity of silver ions and nanoparticles on phytopathogenic fungi. Plant Disease, 93(10), 1037–1043.

Joseph, T. and M. Morrison, 2006. Nanotechnology in Agriculture and Food. A Nanoforum report, Institute of Nanotechnology May 2006, www.nanoforum.org.

Kasprowicz, M. J., Kozioł, M., & Gorczyca, A. (2010). The effect of silver nanoparticles on phytopathogenic spores of Fusariumculmorum. Canadian Journal of Microbiology, 56(3), 247–253.

Katouzian, I., &Jafari, S. M. (2016). Nano-encapsulation as a promising approach for targeted delivery and controlled release of vitamins. Trends in Food Science & Technology, 53, 34–48.

Khodakovskaya, M., Dervishi, E., Mahmood, M., Xu, Y., Li, Z., Watanabe, F., &Biris, A. S. (2009). Carbon nanotubes are able to penetrate plant seed coat and dramatically affect seed germination and plant growth. ACS nano, 3(10), 3221–3227. https://doi.org/10.1021/nn900887m

Kottegoda, N., Sandaruwan, C., Priyadarshana, G., Siriwardhana, A., Rathnayake, U. A., BerugodaArachchige, D. M., Kumarasinghe, A. R., Dahanayake, D., Karunaratne, V., &Amaratunga, G. A. (2017). Urea-Hydroxyapatite Nanohybrids for Slow Release of Nitrogen. ACS nano, 11(2), 1214–1221. https://doi.org/10.1021/acsnano.6b0778

Lai, F., Wissing, S. A., M€uller, R. H., &Fadda, A. M. (2006). Artemisia arborescens L essential oil-loaded solid lipid nanoparticles for potential agricultural application: Preparation and characterization. AAPS PharmSciTech, 7(1), E10.

Lee, C. W., Mahendra, S., Zodrow, K., Li, D., Tsai, Y. C., Braam, J., & Alvarez, P. J. (2010). Developmental phytotoxicity of metal oxide nanoparticles to Arabidopsis thaliana. Environmental toxicology and chemistry, 29(3), 669–675. https://doi.org/10.1002/etc.58

Lin, D., & Xing, B. (2007). Phytotoxicity of nanoparticles: inhibition of seed germination and root growth. Environmental pollution (Barking, Essex: 1987), 150(2), 243–250. https://doi.org/10.1016/j.envpol.2007.01.016

Liu, D., Chen, W., Wei, J., Li, X., Wang, Z., & Jiang, X. (2012). A highly sensitive, dualreadout assay based on gold nanoparticles for organophosphorus and carbamate pesticides. Analytical Chemistry, 84(9), 4185–4191.

Liu, Y., He, L., Mustapha, A., Li, H., Hu, Z. Q., & Lin, M. (2009). Antibacterial activities of zinc oxide nanoparticles against Escherichia coli O157:H7. Journal of applied microbiology, 107(4), 1193–1201. https://doi.org/10.1111/j.1365-2672.2009.04303.x

Lu, C., Zhang, C., Wen, J., Wu, G., & Tao, M. (2002). Research of the effect of nanometer materials on germination and growth enhancement of glycine max and its mechanism. Soybean Science, 21(3), 168–171.

Ma, J., Liu, J.,&Zhang, Z. M. (2009). Application study of carbon nano-fertilizer on growth of winter wheat. Humic Acid, 2, 14–20.

Min, J. S., Kim, K. S., Kim, S. W., Jung, J. H., Lamsal, K., Kim, S. B., *et al.*, (2009). Effects of colloidal silver nanoparticles on sclerotium-forming phytopathogenic fungi. The Plant Pathology Journal, 25(4), 376–380.

Mochizuki, H., Gautam, P. K., Sinha, S., & Kumar, S. (2009). Increasing fertilizer and pesticide use efficiency by nanotechnology in desert afforestation, arid agriculture. Journal of Arid Land Studies, 19(1), 129-132.

Mondal, K. K., & Mani, C. (2012). Investigation of the antibacterial properties of nanocopper against Xanthomonasaxonopodispv. punicae, the incitant of pomegranate bacterial blight. Annals of Microbiology, 62(2), 889–893.

Musarrat, J., Dwivedi, S., Singh, B. R., Al-Khedhairy, A. A., Azam, A., &Naqvi, A. (2010). Production of antimicrobial silver nanoparticles in water extracts of the fungus Amylomycesrouxii strain KSU-09. Bioresource Technology, 101(22), 8772–8776.

Oberdörster E. (2004). Manufactured nanomaterials (fullerenes, C60) induce oxidative stress in the brain of juvenile largemouth bass. Environmental health perspectives, 112(10), 1058–1062. https://doi.org/10.1289/ehp.7021

Park, H. J., Kim, S. H., Kim, H. J., & Choi, S. H. (2006). A new composition of nanosized silica-silver for control of various plant diseases. The plant pathology journal, 22(3), 295–302.

Paulkumar, K., Gnanajobitha, G., Vanaja, M., Rajeshkumar, S., Malarkodi, C., Pandian, K., *et al.*, (2014). Piper nigrum leaf and stem assisted green synthesis of silver nanoparticles and evaluation of its antibacterial activity against agricultural plant pathogens. The Scientific World Journal, 2014, 1–9.

Prasad, R., Bhattacharyya, A., & Nguyen, Q. D. (2017). Nanotechnology in sustainable agriculture: recent developments, challenges, and perspectives. Frontiers in microbiology, 8, 1014.

Rabab, M. A. E.,&El-Shafey, R. A. S. (2013). Inhibition effects of silver nanoparticles against rice blast disease caused by Magnaporthegrisea. The Egyptian Journal of Agricultural Research, 91, 1271–1283.

Rane, M., Bawskar, M., Rathod, D., Nagaonkar, D., &Rai, M. (2015). Influence of calcium phosphate nanoparticles, piriformosporaindica and Glomusmosseae on growth of Zea mays. Advances in Natural Sciences: Nanoscience and Nanotechnology, 6(4), 045014.

Rico, C. M., Majumdar, S., Duarte-Gardea, M., Peralta-Videa, J. R., &Gardea-Torresdey, J. L. (2011). Interaction of nanoparticles with edible plants and their possible implications in the food chain. Journal of agricultural and food chemistry, 59(8), 3485–3498. https://doi.org/10.1021/jf104517j

Roohinejad S., Greiner R. In: Nanotechnology in Agriculture and Food Science. Axelos M.A., Van De Voorde M., editors. John Wiley & Sons; 2017. Nanoscience: relevance for agriculture and the food sector; pp. 347–362.

Stampoulis, D., Sinha, S. K., & White, J. C. (2009). Assay-dependent phytotoxicity of nanoparticles to plants. Environmental science & technology, 43(24), 9473–9479. https://doi.org/10.1021/es901695c

Talarico, D., Arduini, F., Amine, A., Cacciotti, I., Moscone, D., &Palleschi, G. (2016). Screen-printed electrode modified with carbon black and chitosan: A novel platform for acetylcholinesterase biosensor development. Analytical and Bioanalytical Chemistry, 408(26), 7299–7309.

Wang, L., Hu, C., & Shao, L. (2017). The antimicrobial activity of nanoparticles: present situation and prospects for the future. International journal of nanomedicine, 12, 1227–1249. https://doi.org/10.2147/IJN.S121956

Wang, W. N., Tarafdar, J. C., &Biswas, P. (2013). Nanoparticle synthesis and delivery by an aerosol route for watermelon plant foliar uptake. Journal of Nanoparticle Research, 15(1), 1417.

Wu, M. Y. (2013). Effects of incorporation of nano-carbon into slow-released fertilizer on rice yield and nitrogen loss in surface water of paddy soil. In Intelligent system design and engineering applications (ISDEA), (2013) third international conference on (pp. 676-681). IEEE, January.

Xu, J., Fan, Q. J., Yin, Z. Q., Li, X. T., Du, Y. H., Jia, R. Y., *et al.*, (2010). The preparation of neem oil microemulsion (Azadirachtaindica) and the comparison of acaricidal time between neem oil microemulsion and other formulations in vitro. Veterinary Parasitology, 169(3–4), 399–403.

Xu, J., Luo, X., Wang, Y.,&Feng, Y. (2018). Evaluation of zinc oxide nanoparticles on lettuce (Lactuca sativa L.) growth and soil bacterial community. Environmental Science and Pollution Research, 25(6), 6026–6035.

Yang, R., Zhou, Z., Sun, G., Gao, Y., Xu, J., Strappe, P., *et al.*, (2015). Synthesis of homogeneous protein-stabilized rutinnanodispersions by reversible assembly of soybean (Glycine max) seed ferritin. RSC Advances, 5(40), 31533–31540.

Yin, Y. H., Guo, Q. M., Han, Y., Wang, L. J., & Wan, S. Q. (2012). Preparation, characterization and nematicidal activity of lansiumamide B nano-capsules. Journal of Integrative Agriculture, 11(7), 1151–1158.

Zhang, T., Lv, C., Chen, L., Bai, G., Zhao, G., &Xu, C. (2014). Encapsulation of anthocyanin molecules within a ferritin nanocage increases their stability and cell uptake efficiency. Food Research International, 62, 183–192.

Zhang, W. X. (2003). Nanoscale iron particles for environmental remediation: an overview. Journal of Nanoparticle Research, 5, 323-332.

Zheng, Z., Li, X., Dai, Z., Liu, S., & Tang, Z. (2011). Detection of mixed organophosphorus pesticides in real samples using quantum dots/bi-enzyme assembly multilayers. Journal of Materials Chemistry, 21(42), 16955–16962.

17

Nano Biopesticides: A Promising Alternative to Chemical Pesticides

Tripta Jain*, Deepali Chittora, Pushpa Gehlot, Jyoti Yadav and Poonam Meena

Microbial Research Laboratory, Department of Botany, University College of Science Mohanlal Sukhadia University, Udaipur-313001, Rajasthan, India

Abstract

Over the years synthetic pesticides like herbicides, insecticides nematicides etc. have been used to improve crop yield. Indiscriminate application of pesticides ultimately leads to their excess discharge into water bodies during rainfall and causes death of fish and other aquatic life. Also consumption of seafood/fishes may cause biomagnification of chemicals in human body and leads to several lethal diseases like cancer, diabetes, neurodegenerative disorders like Parkinson, Alzheimer, and amyotrophic lateral sclerosis *(ALS),* birth defects *and reproductive disorders. Likewise synthetic pesticides also damage soil texture, soil microbes and soil fertility. Their residues enter into plants and animal systems and further move into animals or humans body through food chain. The deleterious impact of synthetic pesticides have necessitated the search for ecofriendly alternatives such as use of organic pesticides (biopesticides) which are economical, environment friendly and sustainable. Biopesticides are originated from biological sources like microbes (e.g. metabolites), plants (e.g. plant exudates/ essential oil or extracts from bark, root, and leaves) and nanoparticles obtained through green synthesis (e.g., silver and gold nanoparticles). The nanobiopesticides are very effective against the insect pests due to their small size (1-100 nm), large surface area, high stability, cost-effectiveness, fewer toxicity, sustainability and easy field application. Nanobiopesticides have higher pesticidal activity and targeted or controlled release with top-notch biocompatibility and biodegradability. In this chapter, we focused on types of biopesticides especially nanobiopesticides, their synthesis and advantages over synthetic pesticides.*

Keywords: *Biopesticide, Chemical pesticide, Nanobiopesticides, Nano Technology, Sustainable agriculture.*

* *Correspondence author e-mail: tripta26jain@yahoo.com*

1. Introduction

Pesticides play a crucial role in modern agriculture by safeguarding crops against destructive pests, thereby ensuring food security and economic stability. However, the indiscriminate use of conventional synthetic pesticides has raised significant concerns regarding their adverse effects on human health, the environment and non-target organisms (Ankit *et al.*, 2020). In response to these challenges, alternative pest management strategies have gained prominence, including biopesticides and nanobiopesticides (Lade *et al.*, 2019). This chapter aims to elucidate the different types of pesticides, namely synthetic pesticides, biopesticides, and nanobiopesticides, exploring their respective advantages and disadvantages in agricultural applications. Synthetic pesticides, also known as chemical pesticides are chemical compounds synthesized specifically to eliminate or control pests. These pesticides come in various forms including insecticides, herbicides, fungicides, and rodenticides (Abubakar *et al.*, 2020). The primary advantage of synthetic pesticides lies in their rapid and potent action, effectively eradicating pests and minimizing crop damage. However, their indiscriminate use has led to several drawbacks (Begum *et al.*, 2017). Synthetic pesticides often persist in the environment, contaminating soil, water, and ai, and posing risks to human health and non-target organisms. Additionally, pests can develop resistance to synthetic pesticides over time, necessitating increased application rates or the use of more toxic compounds, further exacerbating environmental and health concerns (Jindal *et al.*, 2013). Alternatively biopesticides are derived from natural sources, such as plants, microorganisms or their metabolites and exploit biological mechanisms to control pests. Examples of biopesticides include microbial pesticides (e.g., bacteria, fungi, and viruses), plant-based pesticides (e.g., botanical extracts, pheromones) and biochemical pesticides (e.g., insect growth regulators, enzymes) (Kumar *et al.*, 2021). Biopesticides offer several advantages including reduced environmental impact, lower toxicity to humans and non-target organisms and compatibility with organic farming practices (Khursheed *et al.*, 2022). Moreover, they often exhibit selective modes of action, targeting specific pests while preserving beneficial insects and pollinators. However, biopesticides may have limitations regarding efficacy, stability and persistence, require frequent applications and integrated pest management strategies for optimal results (Ndolo *et al.*, 2019). Nanobiopesticides represent a novel approach that combines nanotechnology with biological agents to enhance pest control efficacy while minimizing environmental and health risks (Lade *et al.*, 2019). These formulations typically consist of nanoparticles loaded with bioactive compounds, such as microbial agents, plant extracts or pheromones designed to target pests at the nanoscale level. Nanobiopesticides offer several advantages such as targeted delivery, enhanced efficacy through improved penetration and bioavailability, reduced environmental persistence and lower toxicity compared to synthetic pesticides (Pan *et al.*, 2023). Moreover, they can mitigate issues of pesticide resistance by leveraging multiple modes of action and enable precision agriculture practices. However, challenges related to nanoparticle toxicity, environmental fate, regulatory frameworks, and scalability of production remain to be addressed (Rodrigues *et al.*, 2017).

In summary while synthetic pesticides have long dominated agricultural pest management, their adverse effects have spurred the development of alternative strategies such as biopesticides and nanobiopesticides. Each type of pesticide presents distinct advantages and disadvantages, highlighting the need for a diversified and integrated approach to sustainable pest management in agriculture.

2. Synthetic/Chemical Pesticides

Pesticides are the chemical compounds applied in agriculture sector for controlling pests and prevent plant associated diseases. Pesticides have different active components like organophosphates, chlorinated hydrocarbons, carbamates, and carbamide derivatives (Cuenca *et al.*, 2020; Yousefi *et al.*, 2021). These active compounds have important applications in veterinary medicine, agriculture, and healthcare sectors. There are five types of pesticides on the basis of their structure such as fungicides, herbicides, insecticides, bactericides and rodenticides as shown in Fig. 1.

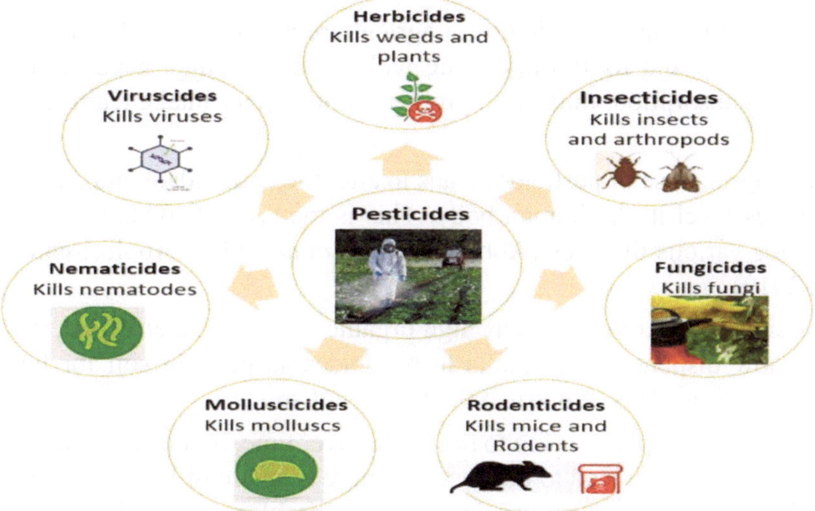

Fig. 1. Types of pesticides based on target pest

The ultimate aim of pesticide production is to manage pests and weeds from cultivated areas. Farmers use pesticides for improving the quality of productive crops (EPA, 2021). But, various farmers and researchers are looking for the harmful effects of chemical on growing crops. In the present era, researchers are found that excess chemicals in agriculture leads to environment pollution and harmful for public health too.

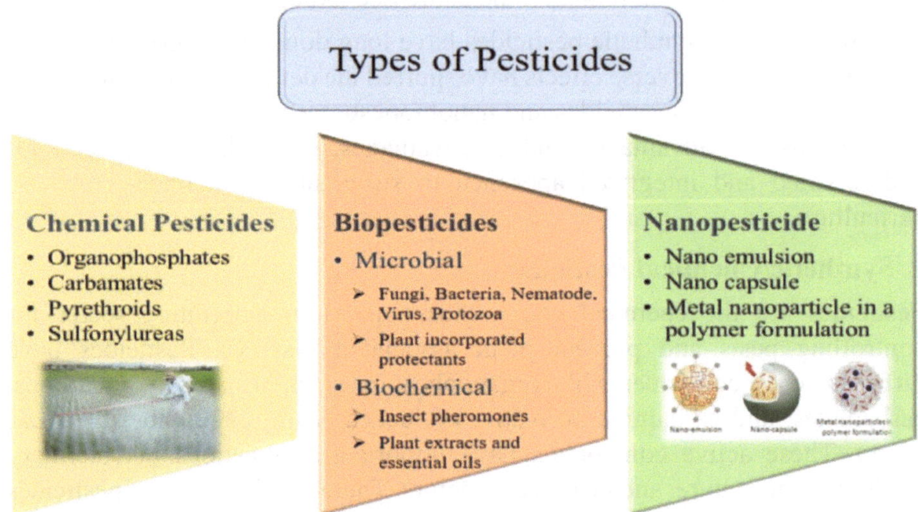

Fig. 2. Types of Pesticides based on origin.

2.1. Problems with Pesticides

- The use of chemical pesticides in agriculture are proven to be harmful for both soil and its fertility along with creating human health issues like skin absorption, inhalation problems and dietary intake (Jain *et al.*, 2023).
- Inappropriate use of pesticides leads to alter the antioxidant and oxidant enzymes level in the human body. This is the type of oxidative stress initiates various diseases (Abbasi-Jorjandi *et al.*, 2020; Abolhassani *et al.*, 2019).
- Pesticide formulations such as high organic solvent concentration, dust drift, low dispensability results in pesticides suspend in soil for long period of time. This will affect the fertility of soil.
- Pests can develop resistance to pesticides over time, rendering them less effective. This can lead to the need for more frequent or higher doses of pesticides, which in turn can intensify the problems of environmental pollution and hazardous impact on human health (Khan *et al.*, 2023)

Among all drawbacks of pesticides, most important is associated with the release of pesticides into the air, only 1% remaining on surfaces. This disorganization leads to severe environment pollution. These pesticides are also detrimental for public health (Saberi *et al.*, 2020). World Health Organization (WHO) also supports restricted use of pesticides. Therefore many pathologists focused their research work towards developing biological substitutes to synthetic chemicals for controlling pest and disease. Biopesticides, nanobiopesticides and nanopesticides are alternative methods for controlling pest growth (Ayilara *et al.*, 2023) (Fig. 2). Biopesticides have low toxic effects, target specific, effective in small amounts and decompose more quickly than conventional pesticides (McGuire *et al.*, 2020). These pesticides are divided into two categories: biopesticides and nano-pesticides.

3. Biological Pesticides/ Biopesticide

Biopesticides are the best alternative of chemical pesticides for controlling plant pests growth. Biopesticides can be employed instead of conventional chemical pesticides because they are environmental friendly, target specific and does not affect non target organisms. Because of these properties, they can be effectively used for management of pests. (Damalas *et al.*, 2018; Pathak *et al.*, 2018; Gonçalves, 2021). Following table describe the various drawbacks of conventional chemical pesticides over biopesticides.

Table 1. Differences between Chemical Pesticides and Biological Pesticides

Chemical Pesticides	**Biopesticides**
Synthesized from chemicals	Synthesized from living systems
They have negatively impact on environment, adversely affect organisms, and are not environmental friendly.	They are environment friendly and are target-specific.
Costly chemicals	Relatively cheaper
Microorganisms and insects become resistant towards chemical pesticides	Pests do not develop resistance
High market value	Not preferred in the market
Contaminate natural resources	Does not contaminate natural resources
Chemicals are bioaccumulated	Not bioaccumulated

Over the past 70 years plant extracts and essential oils have been used as synthetic insecticides for insect pest management. Well-known examples are neem and lemongrass oil i.e., commonly used in vegetable market. Neem oil and fungi, such as *Beauveria bassiana*, have been effectively employed to combat pests that feed on vegetables (Li *et al.*, 2024). *Eichhornia crassipes* is capable of preventing *Spodoptera litura*—a nuisance *lepidopteran*—from growing. Similar to this, duckweed and water lettuce suppress *Chenopodium album* (Nazeer *et al.*, 2023). These examples show how related plants, or weeds, have extremely strong inhibitory effects against pathogens due to their allelopathic compounds (Damalas & Koutroubas, 2018). Hence, plant extracts/ essential oils can be used in place of traditional pesticides. In sustainable agricultural system, microalgae-based biopesticides can be used for boosting crop productivity. The antibacterial properties of microalgae derived biologically active molecules make them potentially useful as biopesticides (Gomiero, 2018; Ranglová *et al.*, 2021). They can increase plant growth and also protect agricultural crops (Ranglova *et al.*, 2021). Biopesticide activity of cyanobacteria like *Nostoc piscinale* and two single-celled green algae *Chlamydopodium fusiforme* and *Chlorella vulgaris* against certain pathogens has been reported. In the process of growing spices, some significant microalgae have been used for their advantageous biopesticide properties. (Kumar *et al.*, 2013; Chittora *et al.*, 2020). Several aquatic plants, including hydrilla (*Hydrilla verticillate*), water lettuce (*Pistia stratiotes*), muskgrass (*Chara* spp.), duckweed (*Lemna minor*), water hyacinth (*Eichhornia crassipes*), and filamentous algae (*Lyngbya wollei*) are used as biopesticides.

Allelopathic substances or secondary metobolites produced by these plants have the ability to inhibit the growth, germination, survival, and reproduction of pathogenic organisms. (Costa *et al.*, 2019; Shirgapure and Ghosh, 2020).

Table 2. List of biological pesticides, their source and Affected Pest Type

Biological Agent	Source	Organism	Affected pest type	Reference
Microbe and microbial metabolites	**Bacteria**	*Bacillus thuringiensis var kurstaki*	Caterpillars, fungi (*Botrytis*)	Beris and Korkas, 2021
		B. thuringiensis var tenebrionis	Elm Leaf Beetle, Alfalfa weevil	Saberi *et al.*, 2023
		Bacillus subtilis	*Botrytis* spp.	Bolivar-Anillo *et al.*, 2021
	Fungi	*Beauveria bassiana*	Whitefly	Javed *et al.*, 2019
		Coniothyrium minitans	*Sclerotinia* **spp.**	Gams *et al.*, 2004
		Trichoderma harzianum	*S sclerotiorum.*	Dolatabadi *et al.*, 2011
		Chondrostereum purpureum	Cut stumps of hardwood trees and shrubs	Bailey, 2014
		Paecilomyces lilacinus	Plant-parasitic nematodes in soil	Moreno-Gavíra *et al.*, 2020
	Virus	*Cydia pomonella* granulovirus	Codling moth	Kadoi´c Balasko *et al.*, 2020
	Microalgae	*Anabaena laxa* and *Calothrix elenkinii*	Increase in fungicidal activity	Ranglová *et al.*, 2021
Plant extract/ essential oils	*Neem (Azadirachta indica)*	Azadirachtin	Aphids, scale, thrips, whitefly, leafhoppers, weevils	Chaudhary *et al.*, 2019
	Reynoutria sachalinensis (giant knotweed) extrac*t*	–	Powdery mildew, downy mildew, Botrytis, late blight, citrus canker	Marrone, 2017
	Plant essential oils	–	Ragwort, many arthropods	Isman, 2020

Nanobiopesticide	Silver nanobiopesticide	None	*Alternaria alternata*, *A. solani*	Narware *et al.*, 2019
	Sargassum muticum **derived NPs**	None	*Ariadne merione*, a Lepidopteran pest	Narware *et al.*, 2019; Rodrigues *et al.*, 2019
	Caulerpa scalpelliformis and *Mesocyclops longisetus*-derived NPs	None	*Culex quinquefasciatus*	Narware *et al.* 2019

Source: Kumar *et al.*, 2021

4. Nanotechnology in Agriculture

Nanoparticles (NPs) are 1-100nm size particles i.e., available in various form such as monomeric to polymeric form with definite scale of size. Nanotechnology has been continuously used in agriculture sector for the production of biopesticides. NPs of Ag, Cu, SiO^2, and ZnO have significant properties of pest control and also reduce water, soil, and environmental pollution. They have many applications in textile industries, geosensing machinery, plant protection, manage soil fertility, control weeds, improve soil nutrients, biofuel and biomass production, biological composites and agrochemical industries to create novel products with an extensive range of applications (Chittora *et al.*, 2021). Metal based nanoparticles have advantages as they are stable, slow kinetics so as to scale up for bulk preparations at room temperature and the generation of environment friendly products for biopesticide formulations. Other important applications of nanotechnology in agriculture sector is the development of fertilizers assist in foliar application or to develop species-specific pest repellents. For example, zinc nanoparticles (Zn-NPs) obtained from biodegradable substrate is ecofriendly and used to improve soil fertility. Likewise, fungicides, herbicides and chemicals can be combined with nanoparticles to serve as magic bullets that target particular plant parts for pest control. Nano-encapsulation is another important application for effective control of pests, preventing the accumulation of residues in the soil and its run off to water bodies. Additionally, it protects the active ingredient in the pesticide from deterioration, increases efficacy and significantly reduces pesticide input by at least 10–15 times compared to traditional formulations (Yadav *et al.*, 2020). The clay nanotube (halloysite) is another advanced way to makes better contact with plants i.e., cost effective, slow released and long term availability. It also reduces pesticide use by 70%–80%, thereby making it an economically viable alternative (Yadav *et al.*, 2020).

4.1 Nanopesticides

Pesticides are created using nanometer scale materials to find applications in the agricultural field are known as nanopesticides. They are made with hybrid substrates such as functionalized nanocarriers for enzyme-mediated triggers or encased in a matrix to respond towards external stimuli. These nanoparticles are anticipated to analyse pesticide activities because of their unique form and characteristics. Numerous components including silica, lipids, polymers, copolymers, ceramic, metal, carbon and others, make up the novel nanocarrier composition. (Ishikawa, 2013; Chen, 2018). These nanopesticides have following important properties such as:

- Nano-pesticides can improve the efficacy of pest control due to their increased surface area, which allows for better adhesion to target pests and surfaces. This enhanced efficacy can result in lower application rates and reduced pesticide usage (Dangi and Verma, 2021).
- Nanotechnology enables the precise targeting of pests while minimizing exposure to non-target organisms. Nano-pesticides can be engineered to release their active ingredients selectively, reducing off-target effects and minimizing environmental contamination (Wang *et al.*, 2022).
- Nano-pesticides can provide better protection against pests, diseases, and environmental stressors, leading to improved crop yields and quality. Their ability to penetrate plant tissues more efficiently can result in prolonged and sustained pest control (Jogaiah, *et al.*, 2020)
- Nano-pesticides can help to mitigate the development of pest resistance by offering novel modes of action and reducing the reliance on conventional pesticides. Rotating nano-pesticides with different modes of action can slow down the development of resistance in pest populations (Asif *et al.*, 2021).

It's important to note that while nano-pesticides offer promising benefits, they also raise concerns regarding potential environmental and health risks associated with nanomaterials. Therefore, careful evaluation and regulation are essential to ensure the responsible development and use of nano-pesticides.

Table 3. Chemical and physical carrier agents in nanopesticides formulation.

Covalent Bond	**Carrier System**	**Formulation**	**Pesticides**
Chemical system			
co-monomers	Hybrid materials	(CNT-g-PCA)	Zineb Mancozeb
multifunctional system	Peptide-polymer	Trypsin-PEG	Modulating oostatic factor
Ionic bond	Hallow sphere	Calcium-alginate	Cypermethrin
Electrostatic complex	Polyelectrolyte complex	Clay-gelatine pGPMA-dsRNA	MCPA dsRNA
Cluster	Metallic nanoparticles	Cu—TM Cu^0, Ag^0	Thiophanate methyl
Physical System			
Encapsulation	Coprecipitation Polycondensation Vesicle	Polyelectrolytic interaction. Cation vesicular surfactants	Trichlorfon Acetochlor Benzoylurea-paraquat DNA, RNA Copper

Emulsion	Mixed micelles Pickering emulsion Nanoemulsion Liquid crystal Liposome	mPEG13–b–PLGA5–3 Alginate-Ca^{++} Water-in-oil Monoolein 18-99 PC-chitosan	Pyrethrin γ-cyclodextrin Citronella Phytantriol α-cypermethrin
Matrix system	Hybrid materials	mPEG-PLGA	Metolachlor
Porous system	Grafted-NP. Sol-gel composite	4-ethylortho-Silicate ATP-biochar colloidal silica	Benzoylurea- Fe_2O_3 Glyphosate
Foams	Polymeric emulsion	Poly(alkylene-oxide] alkanol	Glyphosate acid Acetochlor
Osmotic pumps	Polymeric coating	Cellulose ester/ PEG/ Inorganic salt	Diazinon

Source: Jitendra *et al.*, 2021

4.2. Nano-Biopesticides

Nanobiopesticides are biologically derived chemical complexes with nanoparticles, which are intended to be used as pesticides (fig 3). Nanobiopesticides have broad range of reaction efficiency i.e., improved by using polymers, metal oxides and active particles combined with micelles etc. Reactive nanobiopesticides must have intrinsic properties to solubilize the weakly soluble active ingredients and target-oriented gradual release of active components (Not degraded prematurely) (Kuhlbusch *et al.*, 2018).

Nanobiopesticides are available in low quantity i.e., not affect their efficiency. They are target specific and its outcome may be prominent using information of the life cycle and the behavior of the target pathogen or pest. The formulation of nanobiopesticides has significantly affect the properties and fate of its active components. Variation in biological component of nanobiopesticides is affecting their efficiency. The size, shape, nature and surface properties of the nanoparticle are the characteristic criteria for selection nanobiopesticides (Kuhlbusch *et al.*, 2018). Nanobiopesticides include nanoparticles of either pesticidal active element or manually engineered structures with useful pesticidal properties. The concept of nano-biopesticides is well known. These small, biologically active particles can either destroy or repel bacteria, limiting their growth (Hazafa *et al.*, 2021). Nanoencapsulation, nanocontainers, and nanocages are the important property of nanoparticles to bind with biological components. This will enhance the stability and efficiency of nanobiopesticides against pest control.

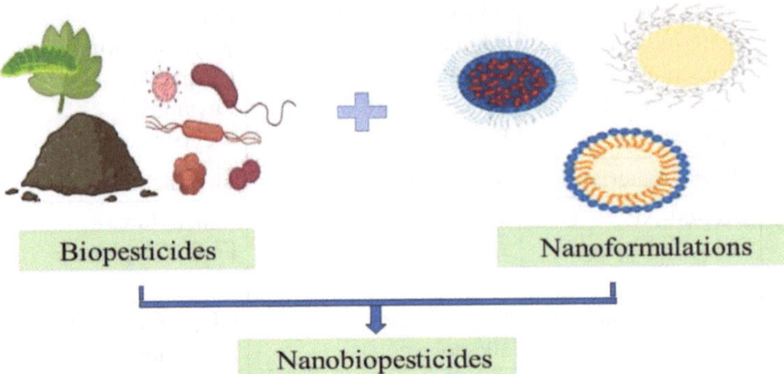

Fig. 3. Nano biopesticides

Because of their innate antibacterial properties, the metallic components of nanobiopesticides (NPs) primarily mitigate the damage that phytopathogens produce. Nanobiopesticides contain metals such as titanium, nickel, gold, silver, and zinc.

The increased solubility and target-oriented delivery of nanobiopesticides over biopesticides results in their greater advantages. Bacteria, fungi and plant extracts are used for the synthesis of nanobiopesticides. For example silver nanobiopesticides (AgNPs) can be synthesized using marine organisms such as *Sargassum muticum*, *Mesocyclops longisetus*, and *Caulerpa scalpelliformis* (Narware *et al.*, 2019). Microorganisms are more advantageous to use in the manufacture of nanoparticles (NPs) than plants because they can withstand higher metal concentrations and their management is much easier, as compared to the plants (Narware *et al.*, 2019).

Bioherbicides are the next important component of nanobioherbicides. For example the ability of Photorhabdus luminescence's metabolites to regulate parasitic and entomopathogenic nematodes (Kremer, 2019).

Similarly, nanofungicides have also been developed to control various pathogenic fungi include *Bipolaris sorokiniana, Fusarium sp.* and *Alternaria alternata.* They are readily soluble, very economical, eco-friendly and safe (Kremer, 2019). All of the above nanobiopesticides are the examples of green synthesized biological nanoparticles that can be safely used for agriculture applications for improving the yield and quality of crops.

5. Conclusion

In the face of mounting concerns over the detrimental effects of chemical pesticides on human health, biodiversity and the environment, the emergence of nanobiopesticides stands as a beacon of hope for sustainable pest management practices. Through the fusion of nanotechnology with biological agents, these innovative formulations offer a multifaceted solution that addresses the limitations of traditional pesticides while capitalizing on the unique properties of nanomaterials. One of the most compelling advantages of nanobiopesticides lies in their ability to deliver bioactive compounds with unprecedented precision and efficiency. By encapsulating biological agents within nanoscale carriers, such as nanoparticles or nanoemulsions, nanobiopesticides can target pests more selectively, minimizing off-target effects and reducing the quantity of active ingredients required for effective pest control. This targeted delivery not only enhances the efficacy of pest management strategies but also minimizes the risk of environmental contamination and non-target organism exposure. Furthermore, nanobiopesticides exhibit enhanced stability and longevity compared to their conventional counterparts, prolonging their shelf life and ensuring consistent performance under varying environmental conditions.

The biodegradable nature of many nanomaterials used in nanobiopesticide formulations holds promise for reducing the environmental persistence of pesticide residues and mitigating long-term ecological impacts. However, despite their immense potential, the widespread adoption of nanobiopesticides still faces several challenges that must be addressed. These include concerns regarding the safety of nanomaterials, potential unintended consequences on non-target organisms, regulatory hurdles and cost-effectiveness. Overcoming these obstacles will require interdisciplinary collaboration among scientists, regulators, industry stakeholders and policymakers to ensure the responsible development, deployment and regulation of nanobiopesticides.

References

Abbasi-Jorjandi, M., Asadikaram, G., Abolhassani, M., Fallah, H., Abdollahdokht, D., Salimi, F. (2020). Pesticide exposure and related health problems among family members of farmworkers in southeast Iran. A casecontrol study. Environmental Pollution. 267: 115-424.

Abolhassani, M., Asadikaram, G., Paydar, P., Fallah, H., Aghaee-Afshar, M, Moazed V., (2019). Organochlorine and organophosphorous pesticides may induce colorectal cancer; a case-control study. Ecotoxicology and Environmental Safety. 178:168–77.

Abubakar, Y., Tijjani, H., Egbuna, C., Adetunji, C. O., Kala, S., Kryeziu, T. L.,and Patrick-Iwuanyanwu, K. C. (2020). Pesticides, history, and classification. In Natural remedies for pest, disease and weed control (pp. 29-42). Academic Press.

Ankit, Saha, L., Kishor, V., & Bauddh, K. (2020). Impacts of synthetic pesticides on soil health and non-targeted flora and fauna. Ecological and practical applications for sustainable agriculture,65-88.

Asif, M., Islam, S., Malik, M. A., Dar, Z. M., Masood, A., Shafi, S. & Sidique, S. (2021). Nano Pesticides Application in Agriculture and their Impact on Environment. In Agricultural Development in Asia-Potential Use of Nano-Materials and Nano-Technology. IntechOpen.

Ayilara, M. S., Adeleke, B. S., Akinola, S. A., Fayose, C. A., Omole, R. K., Uthman, Q. O., & Babalola, O. O. (2023). Biopesticides as a promising alternative to synthetic pesticides: A case for microbial pesticides, phytopesticides, and nanobiopesticides. Frontiers in Microbiology, 14, 1040901.

Begum, A., Alam, S. N., & Jalal Uddin, M. (2017). Management of pesticides: Purposes, uses, and concerns. Pesticide Residue in Foods: Sources, Management, and Control, 53-86.

Beris, E., & Korkas, E. (2021). Additive and synergistic interactions of entomopathogenic fungi with Bacillus thuringiensis for the control of the European grapevine moth Lobesia botrana (Denis and Schiffermüller) (Lepidoptera: Tortricidae). Egyptian Journal of Biological Pest Control, 31(1), 109.

Bolivar-Anillo, H. J., González-Rodríguez, V. E., Cantoral, J. M., García-Sánchez, D., Collado, I. G., & Garrido, C. (2021). Endophytic bacteria Bacillus subtilis, isolated from Zea mays, as potential biocontrol agent against Botrytis cinerea. Biology, 10(6), 492.

Chaudhary, S.; Kanwar, R.K.; Sehgal, A.; Cahill, D.M.; Barrow, C.J.; Sehgal, R.; Kanwar, J.R. (2017) Progress on Azadirachta indica Based Biopesticides in Replacing Synthetic Toxic Pesticides. Frontiers in Plant Science, 8, 610.

Chittora, D., Meena, B.R., Jain, T., and Sharma, K. (2021). Chapter 14- Go Green with Nanotechnology, book Microbial Synthesis of Nanomaterials, Nova Science Publishers, Inc., Editor Sudhir S. Shende, Vishnu D. Rajput, Andrey V. Gorovtsov, Tatiana M. Minkina, Svetlana N. Sushkova, ISBN: 978-1-53619-896-6.

Chittora, D., Meena, M., Barupal, T., Swapnil, P., Sharma, K. (2020). Cyanobacteria as a source of biofertilizers for sustainable agriculture. Biochemistry and Biophysics Reports. 22(100737): 1-10, 2405-5808.

Costa, J.A.V.; Freitas, B.C.B.; Cruz, C.G.; Silveira, J.; Morais, M.G. (2019). Potential of microalgae as biopesticides to contribute to sustainable agriculture and environmental development. Journal of Environmental Science and Health Part B, 54, 366–375.

Cuenca JB, Tirado N, Vikstrom M, Lindh CH, Stenius U, Leander K, *et al.* (2020) Pesticide exposure among Bolivian farmers: associations between worker protection and exposure biomarkers. Journal of exposure science & environmental epidemiology. 30(4):730–42.

Damalas, C.A.; Koutroubas, S.D. (2018). Current Status and Recent Developments in Biopesticide Use. Agriculture, 8(1): 13.

Dangi, K., & Verma, A. K. (2021). Efficient & eco-friendly smart nano-pesticides: Emerging prospects for agriculture. Materials Today: Proceedings, 45, 3819-3824.

Dolatabadi, K.H., Goltapeh, E.M., Varma, A., Rohani, N. (2011), In vitro evaluation of arbuscular mycorrhizal-like fungi and Trichoderma species against soil borne pathogens. Journal of Agricultural Technology, 7, 73–84.

EPA. Ingredients Used in Pesticide Products: Pesticides. What Are Biopesticides? https://www.epa.gov/ingredients-used-pesticide-products/what-are-biopesticides

Eski, A., Demir, D., Sezen, K., Demirbag, Z.A.,(2017). New biopesticide from a local Bacillus thuringiensis var. tenebrionis (Xd3) against alder leaf beetle (Coleoptera: Chrysomelidae). World Journal of Microbial Biotechnology. 33, 95.

Gams, W. Diederich, P. Poldmaa, K. (2004). Fungicolous Fungi. In Biodiversity of Fungi; Mueller, G.M., Bills, G.F., Foster, M.S., Eds.; Academic Press: Cambridge, MA, USA, pp. 343–392.

Gomiero, T. (2018). Food quality assessment in organic vs. conventional agricultural produce: Findings and issues. Applied Soil Ecology, 123, 714–728.

Gonçalves, A.L. (2021). The Use of Microalgae and Cyanobacteria in the Improvement of Agricultural Practices: A Review on Their Biofertilising, Biostimulating and Biopesticide Roles. Applied Science, 11, 871.

Guerra, F., Sepúlveda, S. (2021). Saponin Production from Quillaja Genus Species. An Insight into Its Applications and Biology. Scientia Agricola. 78, 305.

Gurr, G.M. Thwaite, W.G. Nicol, H.I. (1999). Field evaluation of the effects of the insect growth regulator (tebufenozide) on entomophagous arthropods and pests of apples. Australian Journal of Entomology. 38, 135–140.

Hazafa, A., Murad, M., Masood, M. U., Bilal, S., Khan, M. N., Farooq, Q., & Naeem, M. (2021). Nano-biopesticides as an emerging technology for pest management. In Insecticides-Impact and Benefits of Its Use for Humanity. IntechOpen.

Ishikawa, S. (2013). Integrated disease management of strawberry anthracnose and development of a new biopesticide. Journal of General Plant Pathology. 79, 441–443.

Isman, M.B. (2020). Bioinsecticides based on plant essential oils: A short overview. Zeitschrift für Naturforschung C. 75, 179–182.

Jain, T., Gehlot, P., Yadav, J., & Chittora, D. (2023). Molecular basis of biotic and abiotic stress management attributes of plant growth promoting rhizobacteria. Journal of Postharvest Technology, 11(3), 29-55.

Jake, C.R., Serge M.G. (2021). Fisheries, food security, climate change, and biodiversity: characteristics of the sector and perspectives on emerging issues. ICES Journal of Marine Science, 68(61):1343-353,

Javed, K., Javed, H., Mukhtar, T., & Qiu, D. (2019). Efficacy of Beauveria bassiana and Verticillium lecanii for the management of whitefly and aphid. Pakistan Journal of Agricultural Sciences, 56(3).

Jindal, V. I. K. A. S., Dhaliwal, G. S., & Koul, O. P. E. N. D. E. R. (2013). Pest management in 21st century: roadmap for future. Biopesticides International, 9(1), 22.

Jitendra, K., Ayyagari, R., Dharmendra, M. and Vachaspati, M. (2021). An overview of Some Biopesticides and their Importance in plant protection for commercial acceptance, Plants, 10, 1185.

Jogaiah, S., Singh, H. B., Fraceto, L. F., & De Lima, R. (Eds.). (2020). Advances in nano-fertilizers and nano-pesticides in agriculture: A smart delivery system for crop improvement. Woodhead Publishing.

Kadoi´c, B.M., Bažok, R., Mikac, K.M., Lemic, D., Paja˘c, Ž.I. (2020). Pest Management Challenges and Control Practices in Codling Moth: A Review. Insects, 11, 38.

Khan, T., Khan, H. A. A., Haider, M. S., Anwar, W., & Akhter, A. (2023). Selection for resistance to pirimiphos-methyl, permethrin and spinosad in a field strain of Sitophilus oryzae: Resistance risk assessment, cross-resistance potential and synergism of insecticides. Environmental Science and Pollution Research, 30(11), 29921-29928.

Khursheed, A., Rather, M. A., Jain, V., Rasool, S., Nazir, R., Malik, N. A., & Majid, S. A. (2022). Plant based natural products as potential ecofriendly and safer biopesticides: A comprehensive overview of their advantages over conventional pesticides, limitations and regulatory aspects. Microbial Pathogenesis, 173, 105854.

Kirk, W.W., Schafer, R.S. (2015). Efficacy of new active ingredient formulations and new biopesticides for managing Fusarium root rot disease of gladiolus hybrids. Acta Hortic. 1105, 55–60.

Kremer, R.J. (2019). Bioherbicides and nanotechnology: Current status and future trends. In Nano-Biopesticides Today and Future Perspectives; Koul, O., Ed.; Academic Press: Cambridge, MA, USApp. 353–366.

Kuhlbusch, T.A.J., Wijnhoven, S.W.P., Haase, A. (2018). Nanomaterial exposures for worker, consumer and the general public. NanoImpact, 10, 11–25.

Kumar, J., Ramlal, A., Mallick, D., & Mishra, V. (2021). An overview of some biopesticides and their importance in plant protection for commercial acceptance. Plants, 10(6), 1185.

Kumar, M. Prasanna, R. Bidyarani, N.,Babu, S., Mishra, B.K., Kumar, A.., Adak, A., Jauhari, S., Yadav, K. Singh, R. *et al.* (2013). Evaluating the plant growth promoting ability of thermotolerant bacteria and cyanobacteria and their interactions with seed spice crops. Scientia Horticulturae.164, 94–101.

Lade, B.D., Gogle, D.P., Lade, D.B., Moon, G.M., Nandeshwar, S.B., & Kumbhare, S. D. (2019). Nanobiopesticide formulations: Application strategies today and future perspectives. In Nano-biopesticides today and future perspectives (pp. 179-206). Academic Press.

Lehr, P. (2014). Global Markets for Biopesticides; Report Code CHM029E; BCC Research: Wellesley, MA, USA,

Li, Y., Mbata, G. N., Simmons, A. M., Shapiro-Ilan, D. I., & Wu, S. (2024). Management of Bemisia tabaci on vegetable crops using entomopathogens. Crop Protection, 106638.

Marrone, P. (2002). An effective biofungicide with novel modes of action. Pestic. Outlook, 13, 193–194.

Mensah, R.; Moore, C.; Watts, N.; Deseo, M.A.; Glennie, P.; Pitt, A. (2014). Discovery and development of a new semiochemical biopesticide for cotton pest management: Assessment of extract effects on the cotton pest Helicoverpa spp. Entomologia experimentalis et applicata. 152, 1–15.

Moreno-Gavíra, A., Huertas, V., Diánez, F., Sánchez-Montesinos, B., Santos, M. (2020). Paecilomyces and Its Importance in the Biological Control of Agricultural Pests and Diseases. Plants, 9, 1746.

Narware, J.; Yadav, R.N.; Keswani, C.; Singh, S.P.; Singh, H.B. Silver nanoparticle-based biopesticides for phytopathogens: Scope and potential in agriculture. In Nano-Biopesticides Today and Future Perspectives; Koul, O., Ed.; Academic Press: Cambridge, MA, USA, 2019; pp. 303–314.

Nazeer, A., Ahmad, F., Verma, N., & Ahmad, S. (2023). Pesticides and Crop Protection. In Targeted Delivery of Nanopesticides and Nanofertilizers in Sustainable Agricultural Farming (pp. 99-113). Cham: Springer Nature Switzerland.

Ndolo, D., Njuguna, E., Adetunji, C. O., Harbor, C., Rowe, A., Den Breeyen, A., & Hospet, R. (2019). Research and development of biopesticides: Challenges and prospects. Outlooks on Pest Management, 30(6), 267-276.

Pathak, J. Maurya, P.K. Singh, S.P. Häder, D.P. Sinha, R.P. (2018). Cyanobacterial farming for environment friendly sustainable agriculture practices: Innovations and perspectives. Frontiers in Environmental Science, 6, 7.

Ranglová, K. Lakatos, G.E.; Câmara Manoel, J.A.; Grivalský, T.; Suárez, Estrella, F.; Acién Fernández, F.G. Molnár, Z. Ördög, V. Masojídek, J.(2021) Growth, biostimulant and biopesticide activity of the MACC-1 Chlorella strain cultivated outdoors in inorganic medium and wastewater. Algal Research. 53, 102136.

Rodrigues, D.; Costa-Pinto, A.R.; Sousa, S.; Vasconcelos, M.W.; Pintado, M.M.; Pereira, L.; Rocha-Santos, T.A.P.; Costa, J.P.D.; Silva, A.M.S.; Duarte, A.C.; (2019). Sargassum muticum and Osmundea pinnatifida Enzymatic Extracts: Chemical, Structural, and Cytotoxic Characterization. Marine Drugs, 17, 209.

Rodrigues, S. M., Demokritou, P., Dokoozlian, N., Hendren, C. O., Karn, B., Mauter, M. S. & Lowry, G. V. (2017). Nanotechnology for sustainable food production: promising opportunities and scientific challenges. Environmental Science: Nano, 4(4), 767-781.

Saberi, F., Marzban, R., Ardjmand, M., Pajoum Shariati, F., & Tavakoli, O. (2023). Optimization of the Culture Medium, Fermentation Process, and Effectiveness of a Biopesticide from an Iranian Bacillus thuringiensis var. tenebrionis (BN2). Journal of Agricultural Science and Technology, 25(2), 469-484.

Saberi, F., Marzban, R., Ardjmand, M., Pajoum, S.F., Tavakoli, O. (2020). Optimization of culture media to enhance the ability of local Bacillus thuringiensis var. tenebrionis. Journal of the Saudi Society of Agricultural Sciences. 19, 468–475.

Shirgapure, K. H., & Ghosh, P. (2020). Allelopathy a tool for sustainable weed management. Archives of Current Research International, 20(3), 17-25.

Tripathi, S., Srivastava, P., Devi, R.S., Bhadouria, R. (2020). Influence of synthetic fertilizers and pesticides on soil health and soil microbiology. In Agrochemicals Detection, Treatment and Remediation: Pesticides and Chemical Fertilisers; Prasad, M.N.V., Ed.; Butterworth-Heinemann: Oxford, UK, pp. 25–54.

Wang, D., Saleh, N. B., Byro, A., Zepp, R., Sahle-Demessie, E., Luxton, T. P., ... & Su, C. (2022). Nano-enabled pesticides for sustainable agriculture and global food security. Nature nanotechnology, 17(4), 347-360.

Yadav, R.K., Singh, N.B., Singh, A., Yadav, V., Bano, C., Khare, S., Niharika. (2020). Expanding the horizons of nanotechnology in agriculture: Recent advances, challenges and future perspectives. Vegetos, 33, 203–221.

Yousefi, F., Asadikaram, G., Karamouzian, S., Abolhassani, M., Moazed, V., Nematollahi. (2021). M.H. MGMT methylation alterations in brain cancer following organochlorine pesticides exposure. Environmental Health Engineering and Management Journal, 8(1): 47–53.

18

Nanotechnology: Research, Patenting Trends and Challenges

***Namita Ashish Singh*[1]*, Jaya*[1]*, Shakshi*[1]*, Nafees Ahmad*[1] *Chhavi Bhalothia*[2] *and Girima Nagda*[2*]**

[1]*Department of Microbiology, Mohanlal Sukhadia University, Udaipur-313001 Rajasthan, India*

[2]*Department of Zoology, Mohanlal Sukhadia University, Udaipur-313001, Rajasthan India*

Abstract

Nanotechnology epitomizes one of the most dynamic and transformative fields in modern science and technology. This multidisciplinary domain bridges chemistry, physics, biology, along with engineering, enabling the creation of materials and devices with extraordinary properties and applications. The book chapter reflects the research and patenting trends in nanotechnology. USA leading the global share for most patents granted followed by South Korea, Japan, China and Germany for the top five spots contributing more than 76% of all the granted patents by EPO & USPTO related to nanotechnology. Nanotechnological products in relation to sub-industrial areas of medicine, food, agriculture, cosmetics, environment, and renewable energies are also highlighted in chapter. This capter also highlights the commercialization of nanotechnology products and the major challenges facing the implementation and dissemination of nanotechnology. Innovations such as smart nanoparticles for targeted therapy, nanoscale biosensors for early disease detection, and nanorobots for precise surgical interventions are likely to transform healthcare. Nanotechnology research is not strongly linked to viable commercial production which could be improved by enhancing intellectual property rights protection, regulations as well as ethical guideline. Continued research and innovation in the area of nanotechnology, coupled with strategic patenting and effective management of emerging challenges, will drive the sustainable and ethical growth of this transformative field.

Keywords: *Nanotechnology, Patent, Commericalization, patenting issues*

* *Corresponding author e-mail: girima.nagda@mlsu.ac.in*

1. Introduction

Nanotechnology, rooted in the principles of science and engineering, facilitates the design and production of products at atomic, molecular, and nanoparticle (NP) levels. Defined by the US National Nanotechnology Institute as research and technology within the 1-100 nanometre range, it unlocks unique properties in materials and structures due to their diminutive size. Over time, its scope expanded, encompassing larger products engineered for specific traits. Its applications, spanning diverse sectors like medicine, food, and cosmetics, offer notable advantages such as enhanced bioavailability and packaging efficiency. Since Richard Feynman's seminal discourse in 1959, nanotechnology has catalyzed transformative changes in manufacturing, leading to substantial economic growth (Anjum *et al.*, 2021).

The interest in nanotechnology burgeoned following Feynman's visionary lecture, setting the stage for manipulating matter at the atomic scale. Its essence lies in creating and employing materials, devices, and systems at nanometer scales, driving innovations across various domains. Nanotechnology's current applications, including microchips and computer hard drives, underscore its potential for groundbreaking advancements with profound societal and economic impacts. As products and services emerge, policymakers increasingly emphasize the need for comprehensive policy research to explore the technology's implications fully, as highlighted by the US Congressional Research Service (Malik *et al.*, 2023).

The evolution of nanotechnology into a ubiquitous technology has seen the emergence of four generations of nanomaterials, facilitating applications across industries globally. While developed regions like Europe, China, and America lead in investments, developing nations encounter hurdles due to economic constraints. Nevertheless, both developed and developing scientific communities recognize nanotechnology's potential as the next technological frontier, necessitating further industrial investment and advancement (Mazumder *et al.*, 2014). Nanobiotechnology, an amalgamation of molecular biology and nanotechnology, focuses on nanoscale structures or functional materials, driving advancements in medical imaging, diagnostics, and therapies. NPs, including metal, lipid, ceramic, and polymeric variants, hold promise for targeted drug delivery and cancer treatment. Despite its transformative potential, nanotechnology's full spectrum of applications in healthcare remains untapped, presenting avenues for future exploration (Malik *et al.*, 2023).

In food and agriculture domains, nanotechnology is crucial for improving crops ensuring food quality, and enhancing human health. Nanomaterials showcase unique characteristics, finding utility in medicine, agro-food industries, and addressing environmental issues. However, concerns regarding safety and potential adverse effects on human and animal health underscore the need for comprehensive risk assessment strategies(Nile *et al.*, 2020).

Nanostructure types: Nanotechnology encompasses various common nanostructures, including NP's, carbon nanotubes, dendrimers, nanoprobes quantum dots, nano-diamonds, and nanowires, each with unique properties and applications. NP's, ranging from 1 to 100 nm, are classified into metal lipid, ceramic, polymeric, and semi-conductor types, each with distinct characteristics.

- Metal NP's, derived from metal precursors, exhibit unique optoelectrical properties. NP's made of ceramics exhibit diverse characteristics, including amorphous, polycrystalline, hollow, and solid states. They function as potent catalysts in dye photodegradation and contribute to advancements in imaging technology. Semiconductor NP's combine features of both metallic and non-metallic NP's, making them valuable in applications such as photo-optics and electronic devices.
- Carbon nanotubes (CNTs) are structures which are cylindrical in shape formed from graphite sheets, featuring closed terminal sections sealed by fullerene caps. Renowned for their remarkable mechanical resilience, effective absorption of magnetic radiation, and catalytic capabilities. CNTs are used in modern healthcare systems for targeted drug and protein delivery, as they can easily cross cell membranes which are partially permeable. Dendrimers, naturally biodegradable nanopolymers, have a 3D globular shape and are ideal for gene and drug delivery due to their small size, globular structure, and ability to penetrate cell membranes.
- Nano-diamonds (NDs) consist of individual diamond crystal nanostructures, each measuring around 4-5 nm in size, boasting high thermal conductivity and biocompatibility. Featuring a modifiable surface and expansive surface area, they serve as valuable imaging probes in diagnostics. Quantum dots (QDs), synthetic nanostructures, emit varied coloured light contingent upon their dimensions. Graphene and carbon QDs stand out for their safety and stability, particularly in applications within the healthcare sector.
- Nanofilms represent polymeric membranes with expansive surface areas and thicknesses measuring only a few nanometers. They find application in the construction of multilayered biofilms across various fields. Liposomes, spherical vesicles comprising one or more lipid bilayers, serve as highly effective drug delivery systems, owing to their resemblance to cellular membranes and their capacity to encapsulate diverse substances(Anjum *et al.*, 2021).

This chapter critically reviews the trajectory of nanotechnology research, patenting trends, and associated challenges to provide insights for navigating the complexities of this dynamic field. Through a thorough analysis of the major advancements, noteworthy facts, and overall dynamics in this burgeoning sector, major insights can be gained. These will cast light on the main factors that are accelerating nanotechnology forward, as well as on novel possibilities and chances for growth.

2. Research and Patenting Trends in Nanotechnology

Since the dawn of nanotechnology, an entirely novel era of scientific discovery and research covering almost every field of human endeavour has begun. From being only a theoretical idea, it has developed into a vivacious, multidisciplinary field with cutting-edge applications that are a part of our everyday lives. The unparalleled implications of manipulating matter at the nanoscale are both ubiquitous and indispensable varying from the downsizing of electronics to targeted administration of drugs and sustainable energy solutions. This emerging field has triggered an ecstasy of patent activity and a vigorous pursuit for research as countries, businesses, and academic institutions compete to claim a technological frontier full of groundbreaking potential.

Researchers, scientists, and inventors who are persistent in expanding the boundaries of science and technology have been leading the way in the current advancement in nanotechnology. Through their innovative efforts, novel nanomaterials, nanodevices, and nano-systems that surpass conventional boundaries have been made feasible. Every invention and every patent awarded reflects a significant step forward in humanity's pursuit of influencing and conquering nature's fundamental components. Innovations in medical, technology, energy, manufacturing, and other fields have been driven by the cascading impact of breakthroughs and gradual advances toward progress.

2.1. Research & publications

A fundamental aspect of advancement in nanotechnology is our capacity to understand and evaluate the patterns guiding scientific endeavours in this field. Publications on nanotechnology and its relevant domains have been published at an exponential rate during the last few decades. The evaluation of articles published in this field have shown astonishing increase in publications which were consistent with the report of Zhou *et al.* who analysed a bibliometric study from 1991-2005. Another recent study carried out by Chakraborty & Rathi in 2018 focusing on the research areas & evolving trends in nanotechnology. Their work identifies the multi-disciplinary nature of nanotechnology research including its comprehensive implementations & impact among different sectors such as nanomedicines, nanobiotechnology, nanomaterials, etc. The keen interest in the inspection of massive potential and increased funding in nanotechnology led to this exponential upsurge in publications in nanotechnology.

Nanotechnology applications extents to a vast range of industries & fields spanning from health science to energy and beyond. Varied studies in the past have been carried out to shed light on different applications of nanotechnology in different fields like Bhaskar *et al.* (2014) explored the role of medicinal nanotechnology exhibiting its uses in diagnostics, drug delivery, therapeutics, etc. and role of nanomaterials in imaging, biosensing, targeted drug delivery were highlighted in enhancing healthcare consequences. Thakkar *et al.* (2018) gave the perceptions of nanotechnological uses in agriculture emphasising its potential to transform crop production & food security, enhancing delivery of nutrients, soil remediation, pest management via the use of nanomaterials.

Nanotechnology importantly affects extensive fields including biomedicine, energy, electronics, agriculture, food, defence, cosmetics, optics, etc. but its mammoth growth is correlated to its widespread applications in energy, biomedical research, and electronics (Nagda *et al.*, 2023; Bhalothia and Nagda 2022; Pokrajac *et al.*, 2021; Nayfeh *et al.*, 2018). Similarly, Table-1 focuses on different important areas of nanotechnology, their crucial achievements, and their application & impact areas.

Table-1: Prominent Areas of Nanotechnology Research & Advancements

S. No.	Research Field	Overview	Key Milestones	Applications & Implications
1.	Nanobiotechnology	Blend of biology & nanotech for biomedical implementations	Tissue engineering, nano-biosensors, nanotoxicology, nano-biodevices, nanomedicine, nanoparticle drug delivery	Biomedical devices, targeted diagnostics & therapeutics, regenerative medicine
2.	Nano-Agri/Food Tech	Applications of nanotechnology for increased yields and sustainability in food production, processing, safety, and agriculture.	These include edible nanocoatings, nanoagrochemicals/fertilizers, nanobiosensors for pollutant detection, and nanoencapsulation of nutrients.	Nano-enabled food, precision agriculture, smart food packaging
3.	Nanodiagnostics	Tools and platforms based on nanotechnology for molecular diagnostics, biomarker profiling, bioimaging, and disease monitoring in vivo.	Nanoarray chips for multiplexed biomarker detection, magnetic nanoparticle-based bioseparation and nanosensors, and quantum dot probes for fluorescence imaging.	Image-guided surgery, companion diagnostics, theranostics, early disease detection
4.	Nanobiorobotics	Combining biological systems with nanorobots and nanomachines for biomedical uses such as surgery and targeted therapy.	Nanorobots for in-vivo imaging and biosensing, DNA nanowalkers for molecular transport, and bacteria/virus-based nanorobots.	Precision nanomedicine, targeted drug delivery, bio-hybrid robots
5.	Energy Nanotechnology	Nanotech based solutions for energy technology	Nanostructures solar cells, thermoelectrics, supercapacitors, batteries, hydrogen storage, perovskites	Renewable energy, conservation, energy storage
6.	Nano-Bioelectronics	Biological systems and nanoelectronic devices are interfaced to monitor, control, and mend biological processes.	Nanosensors embedded in cyborg cells, bioelectronic implants, nanoelectronic neural interfaces, and biomolecular field effect transistors.	Biological computing, Electroceuticals,prosthetics, neuroscience

7.	Nanobiosensors	devices with great sensitivity and selectivity for detecting and quantifying biological substances, processes, or diseases using nanomaterials.	The biosensors that use fluorescent nanoparticles, nanowire/nanotube arrays, localized surface plasmon resonance, and CRISPR-based SARS-CoV-2 nanosensors are all being developed.	Environmental monitoring, food safety, biodefence, disease diagnostics
8.	Nano-photonics/ optics	Interaction of nanomaterials with light	Nano-photonics, chiral nano-optics, metamaterials, quantum dots, metasurfaces, plasmonics	Optical sensing, communications, displays, imaging, photovoltaics
9.	Nanodrug Delivery	Drug solubility, stability, circulation time, and targeted distribution to particular cells/tissues are all enhanced by nanoscale carriers.	Liposomes, dendrimers, stimuli-responsive nanocarriers that react to pH, temperature, and enzymes, and ligand-targeted nanocarriers are examples of polymeric nanoparticles.	Vaccine development, gene delivery, anti-microbial therapies, targeted cancer therapy

Nanomaterials exhibits exceptional chemical and physical properties mainly because of their nanoscale size & large surface area. The optical properties of nanomaterials are affected by the size which imparts different colours among visible range of absorption (Khan *et al.*, 2019). Even the size is also responsible for the biological & physiological properties of nanomaterials and together with shape & structure of nanomaterials the toughness, reactivity, and their other attributes are affected (Baig *et al.*, 2021; Jeevanandam *et al.*, 2018).

Hundreds of different nanomaterials have been developed in the last decade via enhanced research & patent trends in nanotechnology, focusing on implying the published data on nanomaterials in converting them in patented technology which finally gets translated in products to be used worldwide. The data on number of publications for different nanomaterial, number of patents granted to those nanomaterials and number of products translated from these nanomaterials (StatNano) and it also indicates that trend in patents granted to different nanomaterials is consistent and nearly analogous to the trend in publications of different nanomaterials (Table 2).

Table 2: Data of articles published, patents granted and products developed for different nanomaterials

S.No.	Nanomaterial	Publications	Patents	Products
1.	Nanotube	272297	46063	6
2.	Nanocomposite	266801	18607	1
3.	Nanocrystalline materials	204469	12326	8
4.	Graphene	189223	24563	332
5.	Nanoporous materials	140386	10274	66
6.	Nanoplate (Nanosheet)	136997	5810	13
7.	Nanowire	128867	18703	21
8.	Quantum dot	119491	17542	-
9.	Graphene Oxide	101852	4854	50
10.	Nanofibers	82446	15481	119
11.	Fullerene	53745	6976	41
12.	Dendrimer	23458	5910	1
13.	Aerogel	21387	5847	66
14.	Nanocolloid	14415	2254	118
15.	Porous Silicon	10856	3161	66
16.	Nanofilm	9290	640	11
17.	Nanocapsule	7574	1934	35
18.	Nanoliposome	1518	136	219

Nanotechnology related publications indexed in the Web of Science (WOS) have increased more than three times over the last fifteen years around the globe.

China is leading the graph for most nanotechnology publications followed by USA, India, South Korea and Germany for the top five spots contributing more than 70% of all the published articles related to nanotechnology in the world as depicted in Figure 1 (StatNano).

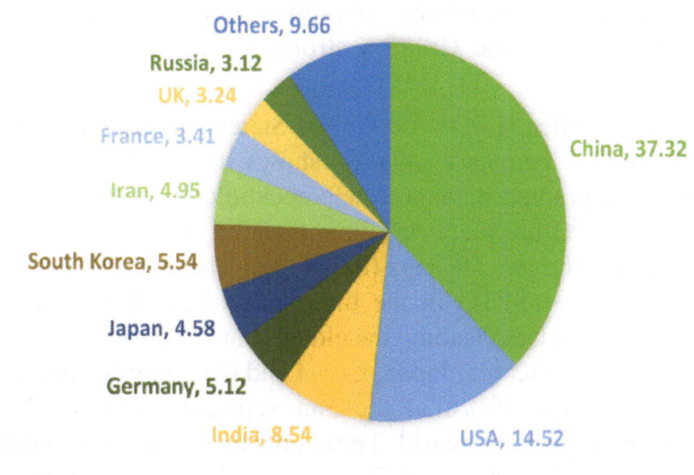

Fig. 1: Percentage Global Share of Different Countries in the Nanotechnology Related Articles Published from 2010-2023

2.2. Patents in Nanotechnology

Alongside the increase in research yield, patenting in nanotechnology has arisen as an important characteristic of scientific novelty and commercialization. The multidisciplinary traits of nanotechnology introduces new technical and legal concerns for the evaluation, classification, and analysis of patent systems worldwide. In order to make sure that their ideas are properly safeguarded, inventors and researchers must abide by the laws currently in effect in both their home country and foreign nations (Bleeker *et al.*, 2004). The wide-ranging study of worldwide patent databases by Sabadash *et al.*, 2012 displayed an important upsurge in patent applications mainly in fields like nanomaterials, nanomedicines, and nanoelectronics reflecting the course of research advancements in the field of nanotechnology and shedding light on trends in nanotechnology patent filings.

The global nanotechnology patenting strategies highlighting the strategic importance of patent portfolio management were investigated by Katz &

Martin (2009) for establishing market position & nurturing innovation. They emphasized how patents safeguard intellectual property, provide businesses a competitive edge, and draw funding for research and development in nanotechnology.

A comparative study of China's and the USA's patenting trends in nanotechnology by Li and Wang (2008) brought spotlight on the changing dynamics of innovation and intellectual property rights worldwide. Their analysis suggested that government investment, collaborative research, and growth schemes had made China a stronger contender in the global nanotechnology setting.

The European Patent Office (EPO) and United States Patent and Trademark Office (USPTO) granted patents evaluation shows that USA is the leading country in number of granted patents in nanotechnology field. The number of granted patents by EPO has increased more than 3.5 times and USPTO has increased by 3 times in the last 15 years (StatNano). The data of of combined granted patents by EPO+USPTO in the time span of 2010-2023 has been depicted in figure 2, where USA leading the global share graph for most patents granted followed by South Korea, Japan, China and Germany for the top five spots contributing more than 76% of all the granted patents by EPO & USPTO related to nanotechnology in the world. These conclusions are comparable to the studies of Chen *et al.* (2008) and Dang *et al.* (2010) who investigated the period among 1976-2006 and 1991-2008 respectively. India despite being ranked third and contributing around 8.54% to the global share in number of nanotechnological publications, contributes only 0.54% to the global share of granted patents (StatNano).

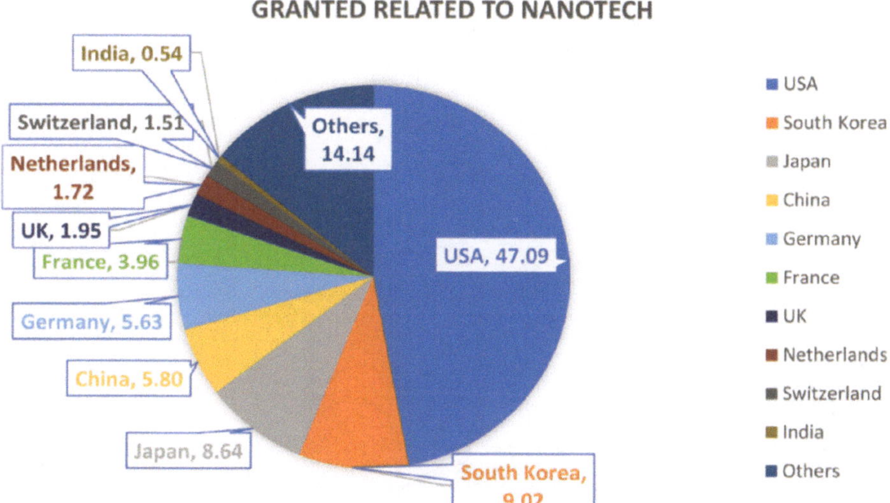

Fig. 2: Percentage Global Share of Different Countries in the Number of Patents Granted related to nanotechnology from 2010-2023

3. Nanotechnology Based Commercially Available Products in the Global Market

Nanotechnological products information which are presently employed in extensive array of industrial applications are visualized in Figure-3 which includes 11,171 products developed by 3,910 companies around 68 countries in the world in more than 15 different fields of industries (Nanotechnology Products Database NPD by StatNano). Most number of products are developed in the sub-industrial field of electronics having 1962 products followed by medicine, construction, and cosmetics sector with 1318, 1170, & 1011 products.

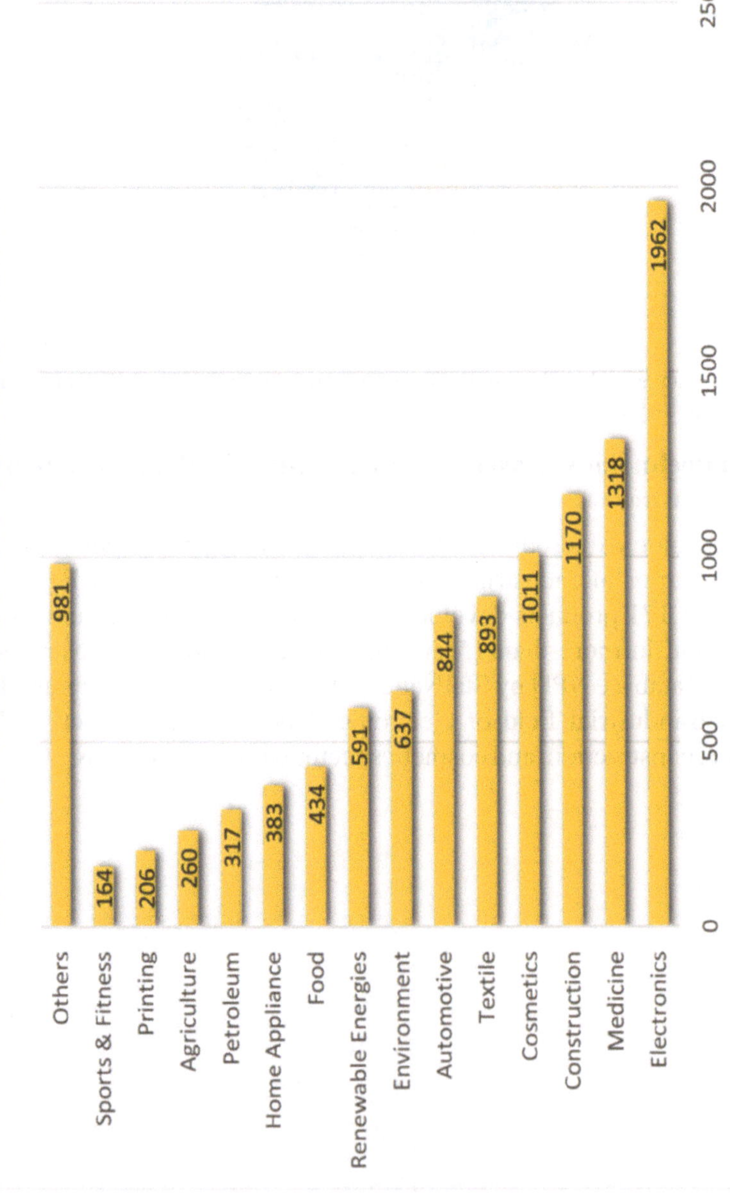

Fig. 3: Bar Graph Depicting Number of Nanotechnological Products Utilized in Different Industrial Fields

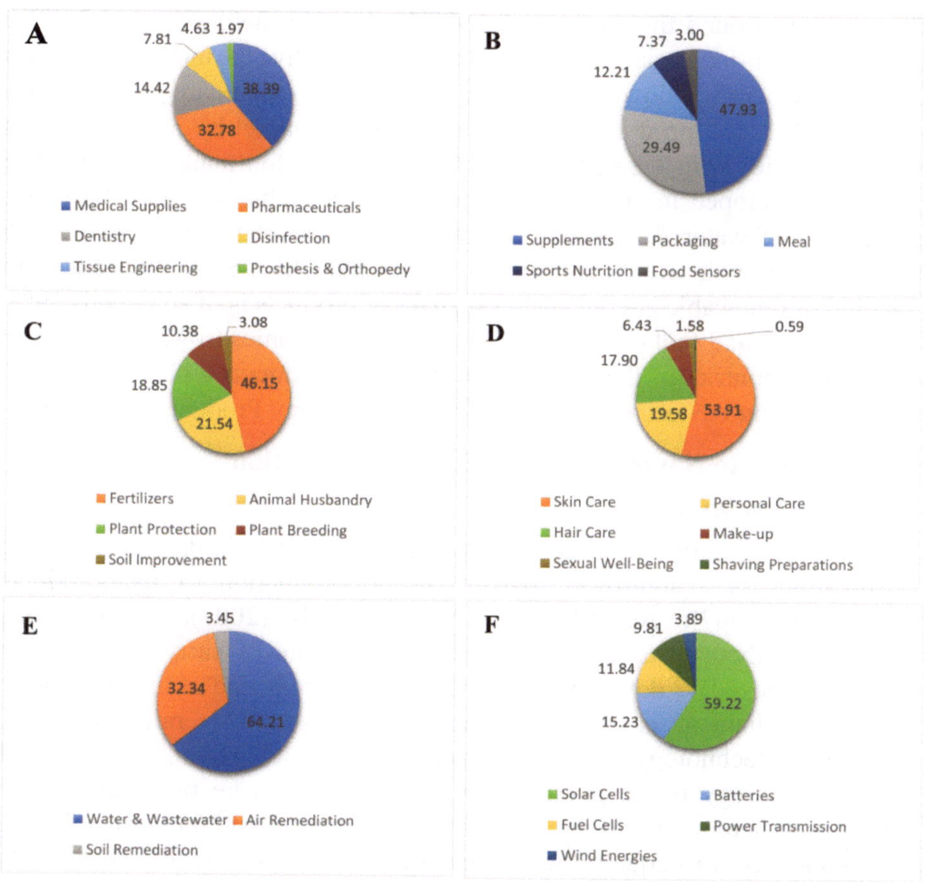

Fig. 4: Different Pie Chart Depicting Nanotechnology Products Implemented in **A)** Medicine; **B)** Food; **C)** Agriculture; **D)** Cosmetics; **E)** Environment; and **F)** Renewable Energies (Stat nano up to 2023)

Nanotechnological products in relation to sub-industrial areas of medicine, food, agriculture, cosmetics, environment, and renewable energies are outlined in Figure-4 (A-F). In the field of medicine, largest share is of medical supplies 38% & pharmaceuticals 33% (Figure-4A). Companies (n-547) located in 48 diverse countries of the world developed these products with USA, Germany, and China being the top manufacturer. The sub-industrial field in food has the largest share of supplements 48% and packaging materials 29% (Figure-4B). Companies (n-204) of 32 countries are held responsible for the introduction of food nanotech products in the market with USA leading them while Iran, Australia, and India being the next players.

The products in agriculture area, having the largest share of fertilizers 46% and animal husbandry products 22% being developed by 96 companies in

29 countries around the world with India leading the field (Figure-5C). The products of sub-industrial sector of cosmetics which have the most share of skin care products 54% followed by personal care 20% and hair care products 18% (Figure-4D. These products are introduced in the market by 340 companies of 33 countries with USA, Brazil, & UK being the top manufacturers. The products developed in environment sector of nanotechnology has the major share 64% of water & wastewater products (Figure-4E) introduced by 316 companies in 36 different countries with USA and China being the leaders. In the field of renewable energies, largest share of products is of solar cells 59%, batteries 15% and fuel cells 12% (Figure-4F). Companies (n-422) situated in 43 various countries of the world developed these products with USA, China, and India being the top manufacturer (Nanotechnology Products Database).

4. Challenges to Nanotechnology Commercialization

4.1. Infrastructural issue

Lack of an institution or organization for creating standards and awarding certifications is a significant barrier to the production of nanoproducts (Shahverdi, 2007). According to Moulayinasab (2007), organizations that present themselves as authorities are incapable of carrying out these responsibilities. The deficiency of financial resources and a suitable public support mechanism is another infrastructure issue. Generally speaking, mastering a technology or area of expertise does not guarantee successful sales. Technology needs public support before it can be presented to the market, yet there might not be a suitable method for presenting or displaying the technology. Inadequate human resources can also lead to issues, aside from a lack of funding. Because nanotechnology is still relatively new, there aren't enough skilled workers to produce these kinds of goods, and hiring new workers takes money and time (Shahverdi, 2007). Furthermore, having a respected brand is essential for a company to successfully commercialize a new product. Prominent brands foster a favorable perception of the product, which might benefit the brand owner (Shokrieh, 2007). However, because producing technical information is costly and because enterprises don't rely on colleges and research facilities, the majority of businesses choose to import the technical knowledge they need from other nations (Rakhsha, 2007).

4.2. Legal issues

It takes a unique legal framework to support the development of new technologies. Among the things that motivate researchers to innovate and develop new goods and procedures are patents and copy rights (Moulayinasab, 2007). Copyrights and patents are two of the few defenses against corporate infringement that small businesses and startups have against huge organizations.

Patents are once again the primary means of intellectual property protection as businesses expand and their capacity to maintain trade secrets declines. Therefore, it's crucial to remember that, barring the applicant's opting out, a patent application becomes published eighteen months after filing, at which point no foreign patent application may be made for the invention. Numerous federal and state legal statutes safeguard intellectual property rights. If the property is not protected, it enters the public domain and can be utilized without a permission by anybody. A prudent management approach would be to methodically assemble a portfolio of distinct intellectual property rights in order to safeguard the many facets of the business's technological and commercial interests. Intellectual property rights safeguard a company's business interests throughout the whole design, production, and other process.

Bastani *et al.* (2002) For those who grant patents, the intrinsically multidisciplinary nature of nanotechnology poses formidable obstacles. In actuality, examiners who possess the most relevant competence are tasked with reviewing applications related to inventions. Since most patent applications pertaining to nanotechnology cover a variety of scientific and engineering domains, it is improbable that a single examiner will possess the necessary knowledge to properly determine whether an application qualifies for patent protection. This increases the possibility of misjudging the uniqueness or ingenuity of an invention and ignoring pertinent prior art. Additionally, it raises the possibility of awarding shoddy patents that might not hold up in court (Milojevic *et al.*, 2012).

4.3. Cultural issues

A sociocultural problem is the absence of communication within organizations. This idea is referred to as "lack of constructive interaction between researcher and investor" in the context of technology commercialization. Generally speaking, researchers and investors don't g*et al.*ong because of their lack of comprehension and expertise in market and business concerns. Conversely, investors, who assume the significant risk of funding novel technologies and concepts, request further privileges for researchers. This miscommunication can seriously impede the commercialization process (Shahverdi, 2007). Consumer ignorance and unfavorable attitudes towards nanotechnology represent another societal problem. Owing to a lack of knowledge about the potential benefits of nanotechnology, some customers are skeptical of nano-products and occasionally have second thoughts when they hear the word. But if customer are given enough information on how nanotechnology is used in new items, they will just purchase it for its seeming advantages.

4.4. Economic issues

Financial and credit systems are usually reluctant to invest on new technologies, due to the high risk involved and lack of an appropriate system for measurement of future capabilities of these technologies (Salehi vaziri *et al.*, 2003). Currently, there are many investors who will to take risk; however, many of them find it more cost-effective to invest their money on low tech products or even non-productive businesses.The main financial challenges associated with developing and utilizing nanotechnologies are their high production costs; specifically, the process of development, implementation, and commercialization necessitates significant financial and time investments; lack of knowledge regarding the effects of nanotechnologies on living cells (the toxicity of nanomaterials and their biocompatibility); and consequently, society's reluctance to accept novel technologies in the absence of guaranteed safety regulations; relative lack of venture capital; private corporations' lack of readiness for large financial investments into the nanoindustry due to its high level of risk relative to other high-tech productions; and undeveloped infrastructure, which is evident in the inadequate coordination and inefficiency of nano industry entities - from invention to use of nanotechnologies (Pavlov *et al.*, 2014).

4.5. Communication issues

The public's ignorance and lack of interest in nanotechnology is recognized as a concern since it may make information difficult for the public to understand, which in turn makes it challenging to pique the public's interest in the information presented. It is arguable that the general public lacks awareness, is unaccustomed to the technology, holds misconceptions, and finds it difficult to grasp key ideas (Duncan *et al.*, 2011). making it challenging to engage the public in deliberative processes or to reach them. Furthermore, views towards nanotechnology are known to be influenced by values, beliefs, and emotions (Bostrom and Lofstedt 2011; Cormick and Hunter 2014). Since attitudes are created by values, feelings, hopes, and expectations rather than by factual knowledge, it is challenging to predict how information will be perceived. This is why the involvement of values and emotions is characterized as a problem theme.

4.6. Consent issues

Creating secure and useful nano-medical technology requires extensive research and study. Clinical researchers have certain particular difficulties while doing nano-medical research. First, because research into nanomedical devices and technology is still in its early stages and is still innovative, it may be more challenging to effectively get participants' informed consent for the proposed nano-medical device and nano-chemical therapeutic trials. Because many features of how nanotechnology behaves in the human body are unknown or unpredictable, medical practitioners may find it difficult to completely comprehend the implications of testing using nano-medical devices (Resnik and Tinkle, 2007; Roco *et al.*, 2011). Furthermore, clinical research participants may underestimate the possible dangers of medical medications containing nano-particles (Resnik and Tinkle, 2007). This poses challenges for getting participants' fully informed consent. Because the technology related dangers and consequences are still being investigated, those who volunteer to participate in nano-medical research will not have enough context to give informed consent (Atalla *et al.*, 2016). While it is unknown how to appropriately explain these hazards to participants, special attention must be made when talking with them so that they fully appreciate the potential risks (Mazzuckelli *et al.*, 2007; Atalla *et al.*, 2016).

4.7. Healthcare sector issues

Since research studies on nano-medicines and nano-devices are still in their early stages of development, one option to speed up this process is to encourage researchers to focus on creating new techniques to get around related obstacles. The steady advancement of approaches based on nanotechnology has raised hopes for the effective treatment of debilitating and perhaps fatal illnesses in the near future. Priority must be given to closing the gaps left by insufficient efficacy and preclinical safety studies so that we can quickly and fully utilize the enormous promise of nanotechnology, which is still untapped. Although nanotechnology offers solutions for a wide range of issues, this does not negate the existence of any difficulties or restrictions. Toxicology is one of the main barriers to the use of nano-based products for healthcare services in living systems. Numerous nano-materials have caused allergic reactions and other undesirable side effects that may be detrimental to the health. Because it depends on so many different aspects, including morphology, size, dose, surface area, route, and duration of administration, toxicity is a very complex notion in and of itself (Kanwar *et al.*, 2012). This draws our focus to the necessity of standardizing or customizing the application of nonmaterial's, which is another issue. There is still more work to be done in the areas of experiment reliability and reproducibility involving NPs. It's also challenging to regulate these minuscule entities' behavior in delicate settings.

Jha *et al.* (2014), an effective *in vivo* monitoring system can significantly improve biomedical processes, including the diagnosis and treatment of a variety of critical disorders. Mechanisms that enable us to fully comprehend what happens to NPs after they are employed are also required. These include how long they remain in the body, what factors affect how long they take to degrade, how to extend and shorten their stay, what effects they have both short- and long- term, the precise behavior of the body towards these foreign objects, their properties and mechanisms of action, and how to standardize these particles so that experiments can be repeated. Prior to implementing nanotechnologies in the healthcare industry, these need be addressed. In addition to this, there are a lot of other questions whose answers need to be thoroughly researched and tested. To prevent any unanticipated events, we also need to determine the possible risks connected to these nonmaterial's. To obtain the safest and most effective therapy regimen, the numerous nano-medicines and nano formulations targeting different diseases need to be carefully created. We finish with the hope that advances in nanotechnology will lead to the development of more promising treatments for a range of serious illnesses, as well as give researchers useful instruments to clear various obstacles in the healthcare industry.

According to Jamkhande *et al.* (2019), some other drawbacks include their high price, the existence of contaminants, their effects on the environment, etc. These risks could have extremely fatal consequences if they are not handled with caution. We still need to create cutting-edge instruments and tools that can swiftly characterize synthetic nonmaterial's, independent of already-existing analytical methods, even with the advanced equipment and tools that have been produced recently. In order to ensure high yields, stability, and purity as well as compliance with the published security criteria, we also urgently need to establish standardized techniques for the synthesis of these nonmaterial's.

4.8. Challenges in building supply chain

One of the reasons was that the nanotechnology was new to both suppliers and customers, making it challenging to establish a supply chain for it. They went on to say that it might be hard to locate the correct suppliers in either scenario since the components of the new technology might not truly exist or they might need to be modified in order to be utilized as components in the current product. They also emphasized how difficult it could be to choose the best distribution channel. They clarified that distribution is made more difficult by the tiny market size in terms of client accessibility.

However, given that the new technology may have multiple application areas that current supply networks are unable to handle, we discovered that the most challenging task is identifying and developing the appropriate supply chain (Natsheh *et al.*, 2015).

5. Conclusion and Future Prospects

Research in nanotechnology spans a vast array of sectors, including medicine, electronics, energy, and environmental science, driving significant advancements and innovations. Concurrently, the rapid advancement of nanotechnology research has been accompanied by a significant increase in patenting activity as organizations and researchers seek to protect their intellectual property and capitalize on novel nanotechnological inventions. There has been an exponential growth in the number of patent applications related to nanotechnology over the past two decades. This surge reflects the extensive research and development efforts across various sectors, driven by the promise of substantial economic and societal benefits. Further, the patents are held by a broad array of entities, including universities, research institutions, startups, and large multinational corporations. This diversity indicates the collaborative and interdisciplinary nature of nanotechnology, as well as the broad interest in securing intellectual property rights in this field.

However, the rapid evolution of nanotechnology presents unique challenges in patenting, including issues related to patent scope, classification of patents, the harmonization of patent laws across jurisdictions. These challenges necessitate continuous adaptation of regulatory frameworks to foster innovation while ensuring fair competition and the ethical use of nanotechnology. Nanotechnology stands at the forefront of scientific and technological innovation, offering transformative potential across numerous fields. The rapid expansion of nanotechnology research has led to a surge in patent filings, highlighting the commercial importance and competitive nature of this field. However, the challenges associated with patenting nanotechnological innovations ranging from defining patent scope to harmonizing international laws require careful consideration and ongoing adaptation of regulatory frameworks. Addressing these challenges is essential to ensure that nanotechnology continues to drive progress while promoting ethical and sustainable development.

As we look forward, the future of nanotechnology is expected to be characterized by significant research breakthroughs, evolving patenting landscapes, and the emergence of new challenges. The future of nanotechnology research is poised to bring about revolutionary changes in several key areas. In the field of health care and medicine, the development of advanced nanomaterials for drug delivery, regenerative medicine, and personalized treatments is anticipated to make significant strides. Innovations such as smart nanoparticles for targeted therapy, nanoscale biosensors for early disease detection, and nanorobots for precise surgical interventions are likely to transform healthcare.

Continued miniaturization and enhançement of electronic components will drive the evolution of more powerful and efficient devices. Quantum computing, enabled by nanoscale quantum dots and transistors, promises to revolutionize data processing and problem-solving capabilities. Nanotechnology will play a critical role in creating more efficient renewable energy sources and storage systems. Advances in nanomaterials for solar cells, batteries, and supercapacitors will help meet the growing global energy demands sustainably.

As nanotechnology research progresses, the landscape of patenting is expected to evolve in several notable ways. The number of patents related to nanotechnology is likely to continue its upward trajectory as more innovations reach commercial viability. Companies and research institutions will seek to protect their intellectual property and secure competitive advantages. Emerging economies and startups are expected to become increasingly active. This will diversify the pool of patent holders and foster global collaboration leading to interdisciplinary patents. As we move forward, the collaboration between scientists, policymakers, industry leaders, and the public will be key to realizing the full potential of nanotechnology for the benefit of society.

References

Al Natsheh, A., Gbadegeshin, S. A., Rimpiläinen, A., Imamovic-Tokalic, I., & Zambrano, A. (2015). Identifying the challenges in commercializing high technology: A case study of quantum key distribution technology. Technology Innovation Management Review, 5(1).

Anjum, S., Ishaque, S., Fatima, H., Farooq, W., Hano, C., Abbasi, B. H., & Anjum, I. (2021). Emerging applications of nanotechnology in healthcare systems: Grand challenges and perspectives. Pharmaceuticals, 14(8), 707.

Atalla, K., Chaudhary, A., Eshaghian□Wilner, M. M., Gupta, A., Mehta, R., Nayak, A., & Trivedi, P. (2016). Ethical, privacy, and intellectual property issues in nanomedicine. Wireless Computing in Medicine: From Nano to Cloud with Ethical and Legal Implications, 567-600.

Bastani, B., & Fernandez, D. (2002). Intellectual property rights in nanotechnology. Thin Solid Films, 420, 472-477.

Bhalothia, C., & Nagda, G. (2022). Nanotherapeutics a promising approach for treatment of Parkinson's disease. Materials Today: Proceedings, 69, A1–A13.

Bostrom, A., & Löfstedt, R. E. (2010). Nanotechnology risk communication past and prologue. Risk Analysis: An International Journal, 30(11), 1645-1662.

Cormick, C., & Hunter, S. (2014). Valuing values: better public engagement on nanotechnology demands a better understanding of the diversity of publics. NanoEthics, 8, 57-71.

Duncan, T. V. (2011). The communication challenges presented by nanofoods. Nature Nanotechnology, 6 (11), 683-688.

https://product.statnano.com/

https://statnano.com/indicators/nano

https://statnano.com/nanomaterials

https://statnano.com/report/s102

https://statnano.com/report/s103

https://statnano.com/report/s29

Interview with Dr.Shokrieh; Manager of Iranian Composite Institute, Development Center at University of Science and Industries, Tehran, 2007.

Interview with Mr. Moulayinasab; Manager of Green Nanotech Company, Tehran, 2007.

Interview with Mr. Rakhsha; Manager of Amid Co, Tehran, 2007.

Jamkhande, P. G., Ghule, N. W., Bamer, A. H., & Kalaskar, M. G. (2019). Metal nanoparticles synthesis: An overview on methods of preparation, advantages and disadvantages, and applications. Journal of Drug Delivery Science and Technology, 53, 101174.

Jha, R. K., Jha, P. K., Chaudhury, K., Rana, S. V., & Guha, S. K. (2014). An emerging interface between life science and nanotechnology: present status and prospects of reproductive healthcare aided by nano-biotechnology. Nanotechnology Reviews, 5(1), 22762.

Kanwar, J. R., Mahidhara, G., & Kanwar, R. K. (2012). Nanotechnological based system for cancer. Nanomedicne and Cancer Therapies, 2, 1-18.

Malik, S., Muhammad, K., & Waheed, Y. (2023a). Nanotechnology: A revolution in modern industry. Molecules, 28(2), 661.

Mazumder, S., Sarkar, D., & Puri, I. K. (2014). Nanotechnology commercialization: prospects in India. J Mater Sci Nanotechnol, 2(2), 201.

Mazzuckelli, L. F., Methner, M. M., Birch, M. E., Evans, D. E., Ku, B. K., Crouch, K., & Hoover, M. D. (2007). Identification and characterization of potential sources of worker exposure to carbon nanofibers during polymer composite laboratory operations. Journal of Occupational and Environmental Hygiene, 4(12), D125-D130.

Milojevic, S. (2012). Multidisciplinary cognitive content of nanoscience and nanotechnology. Journal of Nanopartical Research 14, 685.

Nagda, G., Bhalothia, C., Patel, P., & Dadheech, K. (2023). Role of gold, silver, and chitosan-based nanoparticles in healthcare. BIOINFOLET-A Quarterly Journal of Life Sciences, 20(3b), 497–504.

Naseri, R., & Davoodi, R. (2011). Commercialization of nanotechnology in developing countries. In 3rd International Conference on Information and Financial Engineering.

Nile, S. H., Baskar, V., Selvaraj, D., Nile, A., Xiao, J., & Kai, G. (2020). Nanotechnologies in food science: applications, recent trends, and future perspectives. Nano-Micro Letters, 12, 1–34.Pavlov, A. J., & Batova, V. N. (2014). Ensuring economic security of the innovative development of nanotechnology in the Russian Federation and foreign countries. Life Science Journal, 11(6), 322.

Resnik, D. B., & Tinkle, S. S. (2007). Ethics in nanomedicine 9.

Roco, M. C., Hersam, M. C., Mirkin, C. A., Nel, A., Grainger, D., Alvarez, P. J., & Wiesner, M. (2011). Nanotechnology environmental, health, and safety issues. Nanotechnology Research Directions for Societal Needs in 2020: Retrospective and Outlook, 159-220.

Salehi Vaziri *et al.*, "Recommendations for Nanotechnology Management in Iran", 2nd Conference on Science &Technology, Strategic Research Center, 2003.

Index